Pharmaceuticals and Care Products in the Environment

About the Cover

End-use (e.g., consumer) sources–origins and environmental fate of pharmaceuticals and personal care products (PPCPs) are depicted in this illustration (a complete version is available at: http://www.epa. gov/nerlesd1/chemistry/pharma/index.htm). All chemicals ingested or applied externally (for both humans, domestic animals such as confined animal feeding operations, and certain agricultural crops) have the potential to be continually excreted or washed into sewage systems and from there discharged to the aquatic or terrestrial environments (or transported from land during wet-weather events). Input to the environ-ment is a function of the efficiency of human–animal absorption–metabolism and the efficiency of any sewage treatment technologies employed. Purposeful disposal of PPCPs to sewage and landfills of unwanted PPCPs (both legal and illicit) is another major source.

The occurrence of PPCPs in the environment highlights the importance of dispersed, diffuse, non-point Adischarges@ of anthro-pogenic chemicals to the environment. Their occurrence in the environ-ment illustrates the immediate, intimate, and inseparable connection of the actions and activities of the individual with the environment, high-lighting the significance of the individual in directly contributing to the combined load of chemicals in environment.

Each source for PPCPs can impact various physical environ-mental Acompartments@. For the environmental chemist, this means that analytical tools are required to detect and quantify individual PPCPs (at very low, trace concentrations) in various environmental Amatri-ces@ (e.g., water, sediment, suspended solids, sludges, tissues, etc.). The immediate environmental dispo-sition for PPCPs is primarily to the terrestrial or aquatic environments. Because of the polar, less-volatile nature of most PPCPs (in contrast to the conventional industrial pollutants), the ultimate sink tends to be the aquatic environment. Some of the major environmental compartments in which PPCPs can occur include surface waters (streams, rivers, lakes, and estuaries receiving treated sewage), ground waters, tissues of aquatic organisms, and to a much lesser degree, drinking water. The ramifications for both human and ecological health are unknown.

ACS SYMPOSIUM SERIES **791**

Pharmaceuticals and Personal Care Products in the Environment

Scientific and Regulatory Issues

Christian G. Daughton, EDITOR
U.S. Environmental Protection Agency

Tammy L. Jones-Lepp, EDITOR
U.S. Environmental Protection Agency

American Chemical Society, Washington, DC

Library of Congress Cataloging-in-Publication Data

Pharmaceuticals and personal care products in the environment : scientific and regulatory issues / Christian G. Daughton, editor, Tammu L. Jones-Lepp, editor.

p. cm.—(ACS symposium series ; 791)

Includes bibliographical references and index.

ISBN 0–8412–3739–5

1. Drugs—Environmental. 2. Toilet preparations—Environmental. 3. Drugs—Toxicology. 4. Toilet preparations—Toxicology.

I. Daughton, Christian G. (Christian Gaaei) II. Jones-Lepp, Tammy L. III. Series.

TD196.D78 .P42   2001
363.738—dc21   2001022568

The paper used in this publication meets the minimum requirements of American National Standard for Information Sciences—Permanence of Paper for Printed Library Materials, ANSI Z39.48–1984.

Foreword

The ACS Symposium Series was first published in 1974 to provide a mechanism for publishing symposia quickly in book form. The purpose of the series is to publish timely, comprehensive books developed from ACS sponsored symposia based on current scientific research. Occasionally, books are developed from symposia sponsored by other organizations when the topic is of keen interest to the chemistry audience.

Before agreeing to publish a book, the proposed table of contents is reviewed for appropriate and comprehensive coverage and for interest to the audience. Some papers may be excluded to better focus the book; others may be added to provide comprehensiveness. When appropriate, overview or introductory chapters are added. Drafts of chapters are peer-reviewed prior to final acceptance or rejection, and manuscripts are prepared in camera-ready format.

As a rule, only original research papers and original review papers are included in the volumes. Verbatim reproductions of previously published papers are not accepted.

ACS Books Department

Contents

Personal Care Products

Waste Treatment

Ecotoxicological Issues

Risk Assessment

Epilog: Linkage between Social Sciences and Environmental Monitoring

Preface

Inescapable, intimate, and immediate connections between our personal activities and the environment in which we live are often overlooked in our daily lives. This is especially true with regard to the use and disposal of consumer chemicals. A significant aspect of our global society that illustrates the potential impact of our lives on the environment is the widespread and escalating use of pharmaceuticals and personal care products — simply referred to as *PPCPs*. Many of these chemicals are specifically designed to elicit potent pharmacological or toxicological effects. In distinct contrast to nearly all agro–industrial chemicals, which are often used on large, relatively confined scales, the end use for PPCPs is highly dispersed and centered around the activities and actions of the individual. PPCPs enjoy worldwide usage and attendant discharge or inadvertent release to the environment. Their introduction to the environment has no geographic boundaries or climatic-use limitations as do many other synthetic chemicals; they are discharged to the environment wherever people live or visit, regardless of the time of year.

It is difficult for the individual to perceive their small-scale activities as having any measurable impact on the larger environment; personal actions, activities, and behaviors are often deemed minuscule or inconsequential in the larger scheme. Yet it is the combined actions and activities of individuals that indeed can significantly impact the environment in a myriad of ways. The personal, individual ingestion–application of chemicals from this very large, diverse group of biologically active substances (and their metabolic-transformation products) leads to their direct and indirect worldwide discharge to the environment through sewage treatment systems (indirectly from excreta and directly via disposal or washing) and from other sources such as terrestrial runoff–leaching from excreta of medicated domestic animals (including pets) and landfills. A factor making it difficult for the individual to perceive the interconnectedness of their lives and the health of the environment is the immediacy of any adverse effects, because most actions have delayed consequences, obscuring the ultimate causes. The direct impact of the individual consumer in adding to the potential pollutant load on the environment is a concept that has not been well appreciated.

Chemical pollution represents one of only a few major classes of stressors that can effect ecological change and impair human health. Up to now (during the past 30 years), environmental science has dwelled almost exclusively on what are most conveniently referred to as the "conventional" pollutants—those that compose various regulatory lists, primarily agrochemicals and large-volume industrial chemicals—some of the most notorious of which are the persistent organic pollutants ("POPs", such as DDT, PCBs, chlorinated dioxins). In terms of the vast number of chemicals in commercial use, however, these conventional pollutants comprise only a small percentage of the total numbers of chemicals that could potentially be found in a hypothetical, extensive environmental survey. Do these conventional pollutants represent the major chemical pollution concern, or are we missing major classes of other chemicals?

For society to gain the ability to prioritize risks, and thereby direct a rational allocation of its limited, economic resources to pollution prevention–mitigation, an expanded understanding of the universe of exposure risks is needed. In any process for evaluating additional chemicals for consideration in relative risk, emphasis needs to be given to the likelihood of environmental presence coupled with the potential to elicit biological effects at low exposure levels—not just in humans, but also in wildlife.

While a select few, heavily used pharmaceutically active chemicals have been known for nearly 30 years to occur (and sometimes persist) in the environment, only in the last decade has a more concerted effort been made to search for any of the myriad of other PPCPs in the environment. Until very recently, most of the environmental work on PPCPs had been done in Europe. European municipalities in general are more congested than in the United States—the higher confluence of sewage treatment outfalls coupled with smaller receiving-river capacities and lower per-capita water use has heightened awareness of this issue in Europe. The literature on occurrence, exposure, effects, transport–fate, engineering mitigation, and pollution prevention is very uneven in its development and is spread over a large and sometimes difficult-to-access literature. Although nearly all the literature is in the form of journals and associated compendia, some of which are reviews, no book or U.S. symposium had ever been dedicated to delineating the overall scope of this potential environmental pollution issue. This current book is partly based on the first-ever all-day symposium in North America devoted to this topic: *Pharmaceuticals in the Environment,* American Chemical Society 219th National Meeting, San Francisco, California, March 27, 2000 (part of the larger symposium entitled "Issues in the Analysis of Environmental

Endocrine Disruptors"; available via: http://www.acs.org/meetings/sanfran2000/ or www.acs.org/meetings/). While a number of PPCPs (representing several therapeutic and consumer classes) have already been identified (especially antimicrobials and hormonal agents) in surface, ground, and sometimes drinking waters, sparse data exists for many of the most widely prescribed drugs, and little data exists for personal care products (other than for musk fragrances, sunscreen agents, and antiseptics). Recent data on occurrence of PPCPs in the environment has only laid a foundation on which to further explore this topic, the initial objective of which is to determine if any concern is warranted.

The scope of knowledge required for understanding the environmental aspects of potential chemical pollution from PPCPs can be compartmentalized into discrete categories of the risk assessment "paradigm", which are subjects of great interest to a very broad spectrum of disciplines. The audience for this book includes analytical and environmental chemists, toxicologists, hydrologists, medical scientists, sanitary engineers, risk assessors, and policy makers. Technical areas of the risk paradigm include: chemical analysis (methods for identification and quantitation), identification of source–occurrence, environmental fate, exposure, effects, risk assessment, mitigation, pollution prevention, regulation, and research planning (inter-organization coordination). In addition to professional chemists, anyone with a keen interest in the environment, as well as educators and graduate students, should be able to use some of the materials presented here to gain new insights and reach new perspectives in their fields. With regard to the published environmental PPCP research, the coverage among the risk paradigm categories is very uneven and within some categories is partial–non-existent (such as for aquatic effects). Although a number of measures could be instituted in the near term to greatly reduce the burden of PPCPs in the environment (ranging from "green drug" design, to better controlling dispensing–use–disposal, to implementing new waste treatment technologies) or to screen for risks, few major actions seem warranted at this time without a better overall understanding of the issue.

The possible significance of PPCPs in the environment triggers a host of questions. This book does not pretend to answer these questions, but rather to promote a focused discussion, lend new perspectives, and foster further research. Some of these questions include: Are the currently regulated lists of "priority pollutants" the only chemicals that deserve our attention regarding environmental fate and effects? What are the occurrence, prevalence, and trends of PPCPs in aquatic environments receiving untreated sewage (overflows–spills and "straight-piping") and liquid effluent from treated sewage? Is there any significance to PPCPs leaching into groundwater from solid waste disposal, septic fields, or sewage sludge land application? Is there a concern regarding possible residual PPCPs in treated wastewater intended for groundwater recharge? If PPCPs occur in the environment, are their concentrations sufficient to trigger toxicity concerns for non-target organisms [those species for which these compounds were never intended to expose (PPCPs were designed for human use and consumption), unlike agrochemicals that have designed effects on "pest" species]? Do we know enough about the potential effects of PPCPs on non-target species, many of which may not possess the same suite of receptors as do humans? Even if the concentration of an individual PPCP is sufficiently low, is there perhaps need for concern if other PPCPs sharing the same mode of action can combine to reach a threshold level? Does the aquatic environment present a special exposure case since its biota are exposed continually (through multiple generations) to any pollutant that happens to be present? Are there cellular mechanisms in aquatic biota that confer protection from continual exposure—protection that can be compromised, thereby promoting disease? While acute toxicity, carcinogenesis, and mammalian endocrine disruption are potential concerns for a portion of environmental pollutants, should more attention be paid to other, less overt toxicological endpoints, such as immuno-disruption and subtle neurobehavioral change? Should the direct disposal of excess or expired PPCPs to the domestic sewage system and the over-prescribing of various drugs continue? Can humans receive significant exposures to multiple PPCPs by consuming contaminated drinking water or shell–fin-fish? In short, are there unanticipated consequences of PPCPs in the environment—where's the evidence—do sufficient data exist to decide whether PPCPs in the environment warrant a more in-depth look?

These are only some of the questions that this book attempts not necessarily to answer, but rather to highlight, in order to stimulate further discussion and research. A major objective of this book is to foster further research so that the scientific community and the public can reach informed decisions. Future research should focus on establishing the sufficient and sound science (especially with regard to extent, magnitude, trends, and diversity of occurrence, as well as exposure, susceptibility–effects, and pollution prevention) so that science-based decisions can be made regarding human and ecological health.

Perhaps one of the more revealing aspects of this book is its attempt to wrest some of the perhaps long overwrought attention that has been devoted to a small subset of conventional pollutants (e.g., POPs) and in posing a bigger question as to whether true environmental risk assessments can be performed without a more encompassing understanding of the universe of chemical stressors to which wildlife and humans are exposed.

Some interesting and potentially important issues relevant to environmental chemistry and ecotoxicology, but which are not frequently discussed, include the notions that chemical stability is not a prerequisite for environmental "persistence", that cumulative subtle toxicological effects could yield major outcomes, and that toxic outcomes could result from non-toxic chemical "potentiators".

Coupled with ecotoxicology's traditional preoccupation with conventional pollutants has been an emphasis on easily measured acute–overt toxicity endpoints. PPCPs serve to illustrate the potential importance of other, more subtle biotic effects (for example, see Chapter 15 by Fong on the effect of antidepressants on aquatic life) and less-known cellular processes (for example, see the review in Chapter 14 by Epel and Smital of multi-drug transport and the consequences of its inhibition in aquatic biota), which are two aspects of environmental toxicology that have received little attention, and the latter of which has some profound implications for aquatic health. The question can be asked as to whether subtle effects can accumulate so slowly that major outward change is not detectable until their cumulative level finally cascades to irreversible change. Perhaps another underappreciated issue is that of combined, cumulative exposures to multiple pollutants (perhaps each at a sub-toxicity concentration) but whose combined doses prove significant. Also to consider is the significance of stressors that do not follow the monotonic sigmoid dose-response curve (e.g., U- and inverted-U-shaped curves). While these concerns are not necessarily unique to PPCPs, their biochemically active nature serves to illustrate these possibilities more so than other pollutants.

Chapter 20 breaks new ground in applying the current body of knowledge (on the occurrence of therapeutic drugs in the environment) by opening a window onto what has long been an aspect of society that has proved difficult to accurately assess (i.e., the magnitude and extent of the use of illicit drugs). The use of sewage-influent monitoring is proposed as a non-intrusive means of calculating illicit drug usage on a community-wide scale. A variety of public and governmental organizations will have to decide whether (and how) they want to implement this newly proposed tool, which is merely the next step in the natural extension of current scientific knowledge.

Acknowledgments

The editors sincerely thank the efforts of the authors in completing their high-quality contributions in a timely manner, and in collaborating with each other to minimize overlap. We thank the numerous reviewers who donated their time and insights to peer review. We also thank the managers from the U.S. EPA's National Exposure Research Laboratory (especially Dr. John G. Lyon, Dr. Rick Linthurst, and J. Gareth Pearson) without whose active support and encouragement this effort simply would not have been possible, and we thank Marilyn Janunas for her help with some of the details involved in preparing this book.

Christian G. Daughton
Office of Research and Development
U.S. Environmental Protection Agency
Las Vegas, NV 89119

Tammy L. Jones-Lepp
Office of Research and Development
U.S. Environmental Protection Agency
Las Vegas, NV 89119

Overview

Chapter 1

Pharmaceuticals and Personal Care Products in the Environment: Overarching Issues and Overview

Christian G. Daughton

Chief, Environmental Chemistry Branch, ESD/NERL, Office of Research
and Development, Environmental Protection Agency, 944 East Harmon Avenue,
Las Vegas, NV 89119 (Telephone: 702-798-2207; Fax: 702-798-2142;
email: daughton.christian@epa.gov)

Summary: While the point-source emissions of pollutants from manufacturing waste streams have long been monitored and subject to controls, the environmental impact of the *public's* (i.e., the *individual's*) activities regarding the use of chemicals is more difficult to assess. Of particular question is the widespread release to sewage and surface/ground waters of pharmaceuticals and personal care products after their ingestion, external application, or disposal. Certain pharmaceutically active compounds (e.g., caffeine, nicotine, and aspirin) have been known for over 20 years to enter the environment, especially in populated geographic locales, by a variety of routes — primarily via treated and untreated sewage effluent. A larger picture, however, has emerged only more recently, where it is evident that numerous personal care products (such as fragrances and sunscreens) and drugs from a wide spectrum of therapeutic classes can occur in the environment and drinking water (albeit at very low concentrations), especially in natural waters receiving sewage. As of 1999, nearly all ecological monitoring studies for pharmaceuticals and personal care products (informally referred to as "PPCPs" [1]) had been performed in Europe.

The full extent, magnitude, and ramifications of their presence in the aquatic environment are largely unknown. Whether PPCPs in the environment pose a risk to humans or wildlife is not known. Aquatic exposures are noteworthy in that they can be continuous and can result solely from the external presence of a stressor. A major issue yet to be addressed by ecotoxicological science is the impact on aquatic biota of stressors eliciting effects sufficiently subtle as to go unnoticed in real time – but whose cumulative impacts eventually yield recognizable outcomes but having no obvious cause. How might the cumulative impacts of unrecognizable subtle effects compare with the overt, acute impacts of known toxicants? Another important issue is the potential impact of inhibitors/inducers of multi-drug transport (efflux) systems (as well as the better-known cytochrome P450 monooxygenase isoforms) in aquatic biota.

"Emerging" Risks and Science Planning: One of the primary goals of the U.S. EPA's Office of Research and Development is to identify and foster investigation of previously "hidden" or potential environmental issues/concerns before they become critical ecological or human health problems — pollution prevention being preferable to remediation/restoration (to minimize public cost and to minimize human and ecological exposure). A major route to achieving this end is to highlight potential environmental issues (an objective of this book), thereby fostering further research, and to compile and integrate the resulting data so that the scientific community and the public can reach informed decisions – ensuring that science provides the foundation for any eventual discussion/decisions (if any) to regulate or not regulate. When stakeholders are included in open discussions, diverse viewpoints are assimilated during early development/planning stages rather than having these viewpoints in conflict at later implementation stages. By being proactive, the process of ruling-in/ruling-out (with the use of fast-failure analysis) allows society to minimize confrontation and inefficiency, and thereby promote cooperation and conservation of economic resources. A more extensive discussion of the topic of "emerging" issues (as well as coverage of the general topic of PPCPs in the environment) is available at a U.S. EPA web site devoted to environmental aspects of PPCPs: http://www.epa.gov/nerlesd1/chemistry/pharma/index.htm.

Drivers of Ecological Change: Ecological change is effected by human activities primarily via three routes — habitat disruption/fragmentation, alteration of community structure (e.g., introduction of alien/nuisance species), and chemical pollution. The scopes of the first two are highly delineated compared with that of chemical pollution. Previously unidentified or underappreciated aspects of chemical pollution involve chemical classes not before recognized as pollutants — the chemicals that compose any of the long-established regulatory "lists" of pollutants are not all-inclusive of those having toxicological significance.

During the last three decades, the impact of chemical pollution has focused almost exclusively on conventional "priority pollutants" (it is important to recognize that the current lists of non-atmospheric priority pollutants were primarily established in the 1970s largely out of expediency — that is, they could be measured with off-the-shelf analytical technology; these priority pollutants were NOT necessarily selected solely on the basis of risk but rather because they could also be measured at sufficiently low detection limits), especially on those collectively referred to as "persistent, bioaccumulative, and toxic" (PBT) pollutants or persistent organic pollutants (POPs). The "dirty dozen" is a ubiquitous, notorious subset of these (e.g., see UNEP at: http://www.chem.unep.ch/pops).

The Larger Picture and Risk Assessment: This diverse "historical" group of persistent chemicals (comprising mainly agricultural and industrial chemicals and synthesis by-products, heavily represented by highly halogenated organics), however, may only be one piece of a larger puzzle. This bigger picture, if it does exist, has been largely unattainable with respect to risk assessments. Many other chemical classes (those that can be loosely referred to as "unregulated bioactive pollutants" or "unassessed" pollutants) must also be considered to gain a better look at the larger puzzle. Pharmaceuticals and personal care products (PPCPs) can be viewed simply as an example of one set of the universe of environmental pollutants that have received little attention with respect to potential impact on either ecological or human health. Of particular concern are effects on non-target organisms (e.g., aquatic biota) — effects caused by unintended/inadvertent sources (e.g., human exposure to PPCPs via drinking water bearing low levels of PPCPs that have survived upgrade treatment systems), and effects brought about by non-target receptors (binding at receptors for which a drug was not designed; while there are currently hundreds of target receptors around which drugs are designed, these possibly represent only but a small fraction of those that might exist, most of which represent the untold numbers of non-target receptors). The added risk of individual chemicals cannot be considered outside the context of all chemicals to which organisms are simultaneously or sequentially exposed.

Limitations of Toxicity Assessment: The primary endpoints of interest to risk assessment (and toxicology) have traditionally been human morbidity and mortality (viz. acute toxicity and carcinogenesis). Comparatively little attention has been paid to the universe of other endpoints through which toxicants can express their action in non-mammalians (esp. aquatic organisms) — some of which might be very subtle but nonetheless could lead to unanticipated, insidious outcomes. Endpoints such as neurobehavioral, immunological, and endocrine homeostasis alterations could lead to previously unrecognized outcomes and to those that are not recognized using current assessment criteria. Subtle endpoints could also be effected by extremely low toxicant concentrations (toxicology's domain has traditionally focused on high doses – e.g., see 2), and responses are not necessarily dictated by linear dose-response extrapolations. For example, effects mediated via hormone-like compounds do not necessarily follow the monotonic sigmoid dose-response curve; alternate, unexpected responses, following U- and inverted-U-shaped curves, are also possible (*3*).

Any effects imparted to non-target organisms could differ between and among each class of therapeutics being that their receptors are different — for example, among antimicrobials, endocrine modulators, selective serotonin re-uptake inhibitors (SSRIs), and antineoplastics. This fact, coupled with a large spectrum of species (both aquatic and terrestrial) that could be exposed, means that a very large array of toxicity screening procedures could be needed — the prospects for a single apical assay are accordingly low. Accounting for wild-type drug-metabolism/transport polymorphisms further complicates any screening approach. Exposure to multiple classes (and to multiple members within individual classes) must also be considered. Additive, synergistic, and cumulative time-course effects cannot be fathomed unless our understanding of the aggregate "exposure universe" is expanded.

It is particularly important to keep in mind that any concern for ecological or human health risks posed by newly considered chemical classes must have a scientific basis not just in environmental occurrence and exposure, but ultimately, in measurable effects having significant outcomes. We must also be diligent to weigh any known untoward effects due to environmental release of PPCPs against their established benefits as consumer products.

Summary of Concern — Pharmaceuticals and Personal Care Products in the Environment

Certain pharmaceutically active compounds (caffeine [e.g., *4*], nicotine, and aspirin, among others) have been known for over 20 years to enter the environment via sewage. Only more recently, however, has a larger picture emerged — where it is evident that numerous drugs and personal care products from a wide spectrum of therapeutic and consumer-use classes (many having potent biochemical activity) can be inadvertently released to the environment (albeit yielding very low concentrations). The routes of release include both direct (disposal and wastage from external application) and indirect (excretion, washing, and swimming) — primarily via treated and untreated sewage effluent — and also by terrestrial runoff (for example from confined animal feeding operations – CAFOs – and excreta from medicated pets) and by wind-borne drift of agriculturally applied antimicrobials to crops. Municipal/domestic sewage, more so than hospital sewage, is the major source for most drugs classes (but not all) and quantities. Only limited subsets of the large spectrum of commercial PPCPs have been documented in the environment so far. Many therapeutic classes have not yet been surveyed for. The scope of the issue is ill-defined — the numbers and types of PPCPs occurring in surface and ground waters across a large number of watersheds and municipalities are only partly known. Few generalizations can be made. It is reasonable to surmise, however, that the occurrence of PPCPs in waters is not a new phenomenon — it has only become more widely evident in the last decade because continually improving chemical analysis methodologies have lowered the limits of detection for a wide array of xenobiotics in environmental matrices and have greatly expanded the ability to detect polar compounds.

The fact that PPCPs can be introduced on a continual basis to the aquatic environment via treated and untreated sewage essentially imparts a quality of "persistence" to compounds that otherwise may not possess any inherent environmental stability — simply because their removal/transformation (by biodegradation, hydrolysis, photolysis, etc.) is continually countered by their replenishment, establishing a pseudo-steady-state in a manner analogous to a bacterial chemostat.

The full extent, magnitude, and ramifications of the presence of PPCPs in the aquatic environment are largely unknown. The toxicological significance of PPCPs in the environment with regard to either humans or terrestrial wildlife is poorly understood for most PPCPs. Lack of comprehensive exposure data (which for the aquatic environment, in contrast to the terrestrial environment, can sometimes be inferred simply from occurrence/concentration data) is a critical limitation to the advancement of risk assessment.

Although there are no monitoring requirements in the U.S. (and most others countries) for PPCPs in water, a preponderance of ecological monitoring studies in Europe has laid the foundation for this field. The problem has been most well defined in Europe because of the higher confluence of sewage outfalls in municipal areas, lower per-capita water usage, and smaller stream flows, all resulting in higher discharge concentrations and lower dilution by receiving waters. In North America (esp. Canada and parts of rural U.S.), however, there may be a high incidence of untreated sewage discharge, and therefore potential for greater variety of chemical types and higher concentrations.

While the misuse (which often includes overuse) and subsequent direct and indirect release (e.g., from CAFOs) of antimicrobials (synthetic and natural antibiotics) and natural/synthetic estrogenic steroids to the environment has generated nearly all the controversy to date regarding pharmaceuticals as potential pollutants (e.g., use in livestock [5]), a plethora of other drug classes, bioactive metabolites and transformation products, and personal care products have yet to be examined. The environmental fate of drugs in the excreta of medicated pets (and in carcasses from drug-euthanized animals, such as with barbiturates) is another area of question. Just as with humans, numerous drug classes are administered to pets – both those approved by the U.S. FDA (see: FDA's "Green Book" of Approved Animal Drugs: http://dil.vetmed.vt.edu/ or http://www.fda.gov/cvm/fda/greenbook/elecgbook.html) and those administered off-label by veterinarians and pet owners: including antimicrobials/parasitics, antihistamines, non-steroidal anti-inflammatories, "behavior modifiers"/antidepressants (an example is the recently approved use of clomipramine, a tricyclic antidepressant, for treating pet anxiety),

CNS agents, gastric agents, heart drugs, hormones, antineoplastics; "off-label" refers to drug use directed at conditions for which the FDA has not deemed effective and therefore for which no information appears in the manufacturer's package insert. Terrestrial run-off of excreta could serve as a highly dispersed route of drug/metabolite introduction to both the terrestrial and aquatic environments.

The concerns regarding antimicrobials (esp. promotion of pathogenic resistance, e.g., see *6*), hormones (esp. reproductive/developmental effects), and musks (ubiquitous persistence) in the environment are better established than for other PPCPs. Antimicrobials (esp. low concentrations) impose selective pressure for resistance (unabated growth), or more commonly, tolerance (temporary growth stoppage, but continued viability) among potentially pathogenic microorganisms; significantly (as discussed later), promotion of antimicrobial resistance is at least partly caused by the proliferation or over-expression of cellular "multidrug resistance" systems (efflux pump-mediated drug resistance), which serve to minimize the intracellular concentrations of toxicants. Moreover, the acquired resistance or tolerance can be permanent— it is frequently genetically conserved, persisting in the absence of continued selective pressure by the antimicrobial. Antimicrobials also have the potential to alter microbial species diversity, leading to altered successional consequences.

In comparison, the ecological concerns that we might have (if any) regarding other classes of PPCPs are ill-defined. There are, however, known mammalian effects for certain drug classes that do point to the need for further investigation. For example, various antiepileptic drugs (e.g., phenytoin, valproate, and carbamazepine, the latter of which is frequently identified in sewage effluents) are becoming more recognized as human neuroteratogens, triggering extensive apoptosis in the developing brain (during critical developmental "exposure windows"), and consequently leading to general neurodegeneration (e.g., see *7,8* and references cited therein; *9*); such data prompt the question of whether these compounds affect non-target species.

To better serve the public, effort needs to be applied to performing the appropriate and sufficient science (which includes establishing occurrence, exposure, susceptibility/effects) so that sound decisions can be made regarding human and ecological health— decisions based on "sound" science (*10*) — rather than on "policy" or arbitrary assumptions that result in measures that are overly or under-protective of ecological or human health and that thereby waste economic resources or, conversely, jeopardize health; note that establishment of "occurrence" of a PPCP includes not just its structural identification, but also its concentration, frequency, and geographic extent and distribution for a given environmental matrix.

Clarification & "Disclaimer": As a result of various science planning activities (within and outside government), confusion often develops with regard to the relationship between PPCPs and "endocrine disrupting compounds". Only a small subset of PPCPs are known or suspected of being *direct-acting* endocrine disrupting compounds (EDCs)[†] (primarily synthetic steroids and other synthetic hormones, acting as hormone or anti-hormone modulating mimics — agonists or antagonists, respectively). While many xenobiotics can have a wide range of *ultimate, indirect* effects on the endocrine system, few have direct effects (i.e., serve as immediate endocrine agonists/antagonists at the hormone-receptor level). As an example, the inhibition or induction (such as by triazine herbicides) of P450 aromatase can effect changes in androgen/estrogen ratios (*11*); this effect is not at the receptor level. It is important to note that PPCPs and direct-acting EDCs are NOT synonymous, and the toxicological concerns are usually totally different.

[†]a.k.a: environmental estrogens, endocrine-disruptors, endocrine-modulators, estrogenic mimics, ecoestrogens, environmental hormones, xenoestrogens, hormone-related toxicants, hormonally active agents (phytoestrogens being a subset)

Furthermore, the endocrine system (and its interconnected signaling pathways) is extraordinarily complex and cannot be easily distilled to a simple issue of "disruption" or "modulation". While "disruptors" can act directly at the hormone-receptor level, they can also act indirectly via a plethora of alternative routes (e.g., nervous system, immune system, specific cellular transporter systems), most of which are not always considered in the scope of many of the current definitions of EDCs. Endocrine disruption, in general, is narrowly viewed as a reproductive/developmental issue. An excellent overview of EDCs can be found at the "Environmental Estrogens and other Hormones" web site (Bioenvironmental Research at Tulane and Xavier Universities): http://www.tmc.tulane.edu/ECME/eehome. Whether EDCs represent a meaningful way to classify toxicants with respect to environmental risk — whether they play a significant toxicological role in environmental exposure (especially for humans) — continues to be being actively debated (e.g., see *12*).

Another important point is that the focus of this overview is on the use and disposal of PPCPs as originating from the activities and actions of individuals and hospitals — not those from the PPCP manufacturing sector, whose waste streams are more confined and controllable. Any overview of a potential environmental pollution issue cannot encompass all of the aspects of environmental chemistry, toxicology, and other disciplines. With that perspective, this book does not pretend to be anything other than a catalyst and guide for promoting further exploration of the literature by the reader.

Current State of Knowledge Regarding PPCPs in the Environment

PPCPs as Environmental Pollutants? PPCPs are a diverse group of chemicals that have received comparatively little attention as potential environmental pollutants. PPCPs comprise all drugs (whether available by prescription or "over-the-counter", including the new genre of "biologics" —

proteinaceous drugs), diagnostic agents (e.g., X-ray contrast media), "nutraceuticals" (bioactive food supplements such as huperzine A), and other consumer chemicals, such as fragrances (e.g., musks), sun-screen agents (e.g., methylbenzylidene camphor), and skin anti-aging preparations (e.g., retinoids). While already used in vast quantities, the consumption and usage of PPCPs is only expected to increase, driven partly by new, previously unforeseen applications of existing drugs — examples include "chemopreventatives" (to reduce the chances of disease or slow its onset; e.g., tamoxifen for breast cancer; aspirin for colon cancer) — and the growing practices of off-label prescribing and promotion of "physician samples" (the latter's combined U.S. 1999 quantities amounted to over $7B in equivalent retail sales) (13).

It is difficult to gain good estimates of total nationwide drug usage in the U.S. (ethical prescription and over-the-counter [OTC], both human and veterinary) and much more difficult to obtain regional usages. According to PhRMa's "Industry Profile 2000" report (http://www.phrma.org/publications/industry/profile00/ PhRMA_Tables.pdf), sales in the U.S. of ethical pharmaceuticals (by U.S. and foreign companies) during a 12-month period in 1999/2000 were roughly one-half of worldwide (ca. $100B of $200B total), veterinary sales being roughly 1-2% of total, and OTC sales ca. 2% of total. In 1997, the U.S. per capita expenditures on drugs were the fourth highest in the world ($319), with Belgium, Japan, and France being the top three, ranging from $321 to $351, respectively. Pharmaceutical sales in 1997 for the U.S. were 1.4% of gross domestic product. These figures reflect very large usage rates.

Many PPCPs are extremely bioactive compounds and are unwittingly introduced to the environment as complex mixtures via a number of routes — especially sewage effluent (both treated and raw). Drugs differ from agrochemicals in that they often have multiple functional groups (including ionizable groups and more frequent and extensive fluorination) and often lower effective doses (sub mg/kg), complicating fate/transport modeling and lending an extra dimension to the analytical techniques required for monitoring; drug structure also spans the spectrum from very simple low-molecular weight to large, complex molecules. In contrast to the conventional PBTs, most drugs are usually neither bioaccumulative nor volatile. Personal care products, however, such as the musk fragrances and sun screen agents, tend to be more lipophilic.

To further pursue the issues related to PBTs, a good place to start is with the integrated set of tools developed under the EPA's Office of Pollution Prevention and Toxic Substances' PBT program — the "PBT Profiler" (http://www.epa.gov/ opptintr/pbt/toolbox.htm), which estimates environmental persistence, bioconcentration potential, and aquatic toxicity from chemical structure; tools such as this could perhaps be used to target potential PPCPs for monitoring. Indeed, such a modeling approach was used by Blok et al. (14) in an extensive "screening" evaluation for PBTs, where it was found that numerous, previously unevaluated candidates ("non-assessed" chemicals) are found to be pharmaceuticals.

The Published Literature: The world's published literature on the topic of PPCPs in the environment is disparate (spanning numerous unrelated disciplines), disjointed, and uneven in its coverage. It is also more difficult to locate and to examine in a holistic, comprehensive manner than for "conventional" pollutants. Several compendia and reviews appearing since 1999 (*1, 15-18*) serve as good places to start and have also served for much of the perspective for this overview. Also worth noting are the pre-1999 review articles cited in Daughton and Ternes (*1*). Finally, two brief articles are available that summarize some of the salient points from the seminal ACS meeting that served as the foundation for this book (*19,20*).

Being that most environmental scientists are not as familiar with many of the aspects of PPCPs as they are with POPs, a good place to access numerous databases and other information in addition to the *Merck Index* (*21*) or the *Physicians Desk Reference* (*22*), can be found in various world wide web databases provided through links at "Coreynahman.com Pharmaceutical News and Information" (http://www.coreynahman.com/index.html); for example, comprehensive compilations of physician materials for individual drugs in current use can be found at "RxList - The Internet Drug Index" (http://www.rxlist.com), "Scholz HealthCare" (http://www.ditonline.com/monograph), or "Gold Standard Multimedia Clinical Pharmacology 2000" (http://www.gsm.com). Information on new and anticipated drugs can be found at "Lexi-Comp's Clinical Reference Library" (http://www.lexi.com/new_drugs.htm). These and additional links are provided at the U.S. EPA's PPCPs web site: http://www.epa.gov/nerlesd1/chemistry/ppcp/relevant.htm.

Inter-Connectedness of Humans and the Environment: Perhaps more so than with any other class of pollutants, the occurrence of PPCPs in the environment highlights the intimate, inseparable, and immediate connection between the actions, activities, and behaviors of individual citizens and the environment in which they live. PPCPs, in contrast to other types of pollutants, owe their immediate origins in the environment directly to their worldwide, universal, frequent, highly dispersed, and individually small but cumulative usage by multitudes of individuals — as opposed to the larger, highly delineated industrial manufacturing/usage of most high-volume synthetic chemicals. PPCPs enjoy true worldwide usage and concomitant potential for discharge to the environment. Their introduction to the environment has no geographic boundaries or climatic-use limitations as do many other synthetic chemicals. They are discharged to the environment wherever people (having access to medication or medical care) live or visit, regardless of season.

Manufactured and used in large quantities worldwide (from several to hundreds of kilograms, to thousands of tons – for each individual substance), PPCPs comprise a diverse array of pollutants — usage rates of many are on par with agrochemicals. Escalating introduction to the marketplace of new pharmaceuticals; expanding usage of existing drugs (antidepressants, stimulants, and other physchotropic drugs to modify child behavior; off-label prescribing for both humans and pets; conversion of established drugs to OTC status as a result of the growing "self-care" movement being only three examples); new uses for "retired" drugs (e.g., thalidomide for inflammation); and the growing practice of procuring prescription drugs from other countries, are each adding perhaps exponentially to the already large array and amounts of PPCP chemical classes, each with distinct and expanding modes of biochemical action (many of which are poorly understood, especially in wildlife, and some needing only very low concentrations to impart effects). It is also interesting to note that some chemicals serve double duty as both existing/experimental drugs and as pest-control agents (e.g., 4-aminopyridine: an avicide and experimental multiple sclerosis drug; warfarin: a rat poison and anticoagulant; triclosan: general biocide and gingivitis agent used in toothpaste; azacholesterols: antilipidemic drugs and avian/rodent reproductive inhibitors [e.g., Ornitrol]; certain antibiotics: used for orchard pathogens; acetaminophen: an analgesic and useful for control of brown tree snakes); the potential significance of these alternative uses as sources for environmental release has never been explored. Usage of certain drugs also continuously increases as a result of abusing prescription drugs for non-medical, "recreational" purposes and because of unneeded dispensing ("imprudent use" and unfounded patient demands or physician expectations). Another source of highly bioactive chemicals, but whose impact on the environment is totally unknown, is the clandestine manufacturing and use of illegal drugs — which constitutes a potentially large but highly dispersed source of totally different pollutants through surreptitious manufacture and direct disposal; see Streetdrug.org (available: http://www.mninter.net/%7epublish/index2.htm) and National Institute of Drug Abuse (NIDA) (available: http://www.nida.nih.gov/NIDAHome1.html) for comprehensive information regarding "recreational"/street drugs. Also note the concluding chapter in this book (*23*), which focuses specifically on illicit drugs.

Sewage and domestic wastes are the primary sources of PPCPs in the environment (posing concerns for drinking water supplies?): These bioactive compounds are continually introduced to the environment (primarily via surface and ground waters) from human and animal use largely through sewage treatment works systems (STWs), failed septic fields (e.g., see: http://www.starnews.com/news/articles/0717_SEPTIC.html/), leaking underground sewage conveyance systems, and wet-weather runoff — either directly by bathing/washing/swimming (via discharge of externally applied PPCPs, such as fragrances or sun-screen agents, or those excreted in sweat) or indirectly by excretion in the feces or urine of unmetabolized parent compounds. Bioactive metabolites (including reconvertible conjugates) are also excreted. Excretion of parent drug and metabolites is partly a function of the age/health, timing of dose (diurnal aspects of absorption/transport/metabolism; gestational/developmental stage), and constitution of the gut (microbial contribution to endogenous metabolism can be significant). It should be noted, however, that pharmacokinetics is not the only predictor for the fraction of a drug that is excreted unmetabolized. Actual drug formulation has the potential to prevent full dissolution of a drug in the gut, leading to minimized uptake and therefore significantly increased excretion of the parent drug; for example, the use of certain excipients, such as stearic acid derivatives in tablets, is reported to sometimes greatly reduce the dissolution of a drug, resulting in excretion of undissolved pills (*24*). Disposal via municipal refuse serves as another route of introduction to the environment (e.g., via leaching to groundwater). Other routes to the environment include storm overflow events, residential "straight piping", and disposition of the massive quantities of drugs contributed to humanitarian assistance projects (e.g., *25*).

While aspirin and caffeine have long been known to occur in sewage, only since the 1980s have other PPCPs been identified in surface and ground waters — even certain drinking waters. As a recent example, low ng/L levels of clofibric acid, diazepam, and tylosin have been identified in Italian drinking water (*26*). Portions of the free excreted drugs and derivatives can escape degradation in municipal sewage treatment facilities. Removal efficiency in an STW or drinking water upgrade facility is a joint function of the drug's structure (including its stereochemistry, which can lead to chiral enrichment by preferential bioalterations) and the treatment technology employed (e.g., ozonation, chlorination, irradiation, carbon sorption). Treatment effectiveness would be expected to vary greatly across drug classes (as a function of chemical structure). For waste streams, one might expect especially poor removals for certain antimicrobials and antineoplastics (because of acute toxicity to various microbial species) and for highly sterically hindered compounds (e.g., iodinated contrast media); effectiveness could also be a function of the waste stream's origin (e.g., hospital waste versus domestic waste, especially with respect to higher concentrations and more acutely toxic drugs, such as genotoxicants used primarily at hospitals). Any metabolically conjugated derivatives released to the environment can eventually be converted back to the free parent drug; therefore, while not detectable as the parent compound, conjugates can serve as surreptitious reservoirs of additional toxicant, serving to regenerate the parent. General, overall STW effectiveness can fluctuate due to time of day (both composition and volume of sewage influent, the latter of which is largely a function of diurnal population activity, precipitation/runoff input, and industrial contributions) and season (treatment efficiency influenced by temperature and nutrient loads/physicochemical conditions, and dilution of effluent as a function of receiving water volume/flow). The intricacies involved with enhancing treatment effectiveness are complex, as exemplified by oxidative treatment of drugs in water by ozonation (*27*).

The low concentrations of individual PPCPs (possibly below the catabolic enzyme affinities of sewage microbiota), coupled with metabolic "novelty" to microorganisms (possibly an issue with newly introduced drugs), could lead to incomplete removal from STWs; removal efficiencies from STWs can span the entire spectrum — from complete to ineffective. Introduction of many PPCPs to individual STWs is in the multi-g or -kg/day range, depending on the population served. While lipophilic PPCPs could partition to the solids (sludge), resulting in subsequent disposal in various ways on land (e.g., use of sewage sludge as agricultural fertilizer), little data is yet available; worth noting, however, is that living plants have the potential to bioaccumulate drugs (irrespective of polarity), giving the potential for subsequent transfer to animals and humans (*28*). While veterinary pharmaceuticals also are excreted (or washed from skin/hair), their route to the environment tends to be directly via wet-weather runoff (e.g., from CAFOs and medicated pets) from land to surface waters or ground waters, as well as directly via losses from aquaculture; the introduction of a wide array of antibiotics via aquaculture has been well documented, and these compounds show considerable persistence and a large spectrum of non-target effects (*29*), including phytotoxicty at ppb levels (*30*). Sometimes, the uptake of aquaculture medication is greater by nearby wild fin fish and shellfish than by the targeted cultured fish because the latter's diseased states minimizes uptake and absorption (*31*).

Whether PPCPs survive in natural waters sufficiently long down-gradient to be taken up in untreated drinking water, or whether they survive drinking water treatment, creating the potential for long-term exposure of humans, has received even less investigation than has environmental occurrence. Certain drugs/metabolites, however, have been documented in potable waters in Europe (see references cited in Daughton and Ternes [*1*]; *26*). The extremely low concentrations (parts per trillion, ng/L), orders of magnitude below therapeutic threshold levels, might be expected to have minuscule (but still unknown) health consequences for humans, even for those who continually consume these waters over the course of decades; the primary concern, if any, would focus on those with heightened drug responses or the health-impaired (e.g., fetuses, infants, and children, or aged or diseased individuals). The estrogenicity of STW effluents in the UK (as measured by vitellogenin production in male fish – primarily a function of natural and synthetic steroid estrogens) has been shown to persist for several kilometers downstream of effluent discharges (*32*), maximizing the potential for drinking water uptake in rural areas that do not use upgrading facilities or that practice minimal treatment. It is important to keep in mind, however, that most drinking water is subjected to yet further treatment (upgrading), which lessens even further the chances of PPCP contamination. The issue of potable water contamination is more pertinent to situations where drinking water is not upgraded — for example, when withdrawn directly from contaminated surface or ground waters (e.g., private wells); recharge of groundwater using treated sewage with residual PPCPs may pose a particular concern deserving further attention. The National Research Council (*33*) recently pointed to pharmaceuticals as one of several major chemical classes pollutants that had not received sufficient attention as potential water pollutants and recommended that they be considered among the larger universe of previously unevaluated pollutants for future versions of the Environmental Protection Agency's Drinking Water Contaminant Candidate List (CCL; see: http://www.epa.gov/OGWDW/ccl/cclfs.html).

In addition to the sources of drugs in the environment discussed here, one should reasonably expect to see a number of other routes of introduction of drugs to the environment with the advent of new technologies for discovery, manufacturing, and administration. For example, transgenic production of "functional food" drugs/vaccines (but mainly proteinaceous) by genetically altered plants (a.k.a. "molecular farming", "biopharming") could pose additional concerns for the unintended or uncontrollable release of drugs to the environment (e.g., see: 2nd International Molecular Farming Conference 1999 (29 August - 1 September 1999, London, Ontario, Canada; http://res2.agr.ca/initiatives/mf99/main_en.html).

Ubiquitous, persistent, and bioaccumulative? Very few PPCPs (primarily certain personal care products) and their metabolites have been found to display all the qualitites of conventional POPs. Many display one or two of these qualities, but few (such as musks) occur widely, persist, and bioaccumulate. Some (such as clofibric acid) occur at concentrations in surface waters on par with that of the widely recognized and ubiquitous organochlorine POPs (e.g., DDT, PCBs). Concentrations in natural surface waters (including oceans) generally range from ppb (μg/L) to ppt (ng/L). Some PPCPs are extremely persistent and introduced to the environment in very high quantities (e.g., see chapters in this book on musks and on polyiodinated X-ray contrast media). Others act as if they were indeed long-lived persistent pollutants simply because their source (sewage effluent) continually replenishes any removal caused by way of natural environmental processes (e.g., microbial degradation, photolysis, particulate sorption). Continual replenishment effectively sustains perpetual, multi-generational life-long exposures for aquatic organisms. Chiral enrichment (by selective biodegradation of chiral isomers) is another consideration. While most PPCPs display degradation in surface waters (preventing their accumulation but not full-time presence), their fate in ground waters (where the likelihood of persistence is maximized) is less understood (see chapter in this book by Drewes and Shore). As such, there is an imperative that we examine those possible environmental contamination situations that can be reversed only with difficulty. Groundwater recharge using treated wastewaters is one area that may need to be specifically scrutinized — to ensure that contamination of these critical resources is minimized; extreme, unanticipated persistence in groundwater has been noted in some situations (e.g., pentobarbital; *34*). Of particular noteworthiness is the almost complete absence of monitoring data on the presence (or absence) of PPCPs (other than musks) in fin- and shell-fish. While the polarity of most drugs precludes any concern about their bioconcentration in lipid, occurrence as adducts in various tissues cannot be excluded. Understanding of metabolic fates in aquatic biota is basically unknown.

A myriad of chemical classes, ranging from endocrine disruptors, antimicrobials, antidepressants ... to lipid regulators and synthetic musk fragrances: Excluding the antimicrobials and steroids (which include many members), over 50 individual PPCPs or metabolites (from more than 10 broad classes of therapeutic agents or personal care products) had been identified as of 1999 in environmental samples (mainly surface and ground waters) (see: Table I, summarized from Daughton and Ternes [*1*]); concentrations generally range from the low ppt- to ppb-levels. It is important to note that these only comprise a subset of those in wide use — members of most classes have never been searched for in the environment (see: Table II).

Significantly, many of these compounds have no published aquatic toxicity data; some might have the potential for significant effects (e.g., antiepileptics, antineoplastics). Conversely, some PPCPs (such as the SSRI antidepressants, calcium-channel blockers, and efflux pump inhibitors) that do have published aquatic effects data (albeit limited) have yet to be surveyed in environmental samples (and therefore are not listed in Table I). Still others have great potential for profound aquatic effects but have neither the aquatic toxicological database nor any occurrence data (e.g., psychoactive agents and street drugs).

Majority of PPCP classes have no environmental survey data: The number of PPCP classes (and individual members) that have been surveyed in various environmental samples (Table I) serve as a foundation that needs to be built upon. There are many other classes of drugs that have yet to be subjected to environmental surveys (see examples in Table II), and many of the most widely prescribed members of the classes in Table I have not been reported from environmental surveys. Note that while the literature is silent regarding these classes, it cannot be concluded whether this is because of an absence of data or because of a failure to report "data of absence"; it is important to note that investigators need to value reporting negative results. Many of these drugs are among the most widely prescribed in the U.S. (see: http://www.rxlist.com/top200.htm — data compiled from *American Druggist*). Also note that the numerous ingredients in personal care products and the many bioactive compounds in "nutraceuticals" (e.g., the active ingredient[s] in St. John's wort) are not even considered here. Attention is only just beginning to be focused on the plethora of ingredients in personal care products. Phthalate esters is a case in point (*35*); mercury (from dental amalgam and toiletries) is another (*36*). The list

Table I. Representative classes (and members) of PPCPs reported in STWs and environmental samples (see more detailed table in Daughton and Ternes [1]).

therapeutic class	example generic name	example Brand name
analgesics/non-steroidal anti-inflammatories (NSAIDs)	acetaminophen (analgesic) diclofenac ibuprofen ketoprofen naproxen	Tylenol Voltaren Advil Oruvail Naprosyn
antimicrobials	e.g., sulfonamides, fluoroquinolones	many
antiepileptics	carbamazepine	Tegretal
antihypertensives (betablockers, beta-adrenergic receptor inhibitors)	bisoprolol metoprolol	Concor Lopressor
antineoplastics	cyclophosphamide ifosfamide	Cycloblastin Holoxan
antiseptics	triclosan	Igrasan DP 300
contraceptives	β-estradiol 17α-ethinyl estradiol	Diogyn Oradiol
β_2-sympathomimetics (bronchodilators)	albuterol	Ventolin
lipid regulators (anti-lipidemics; cholesterol-reducing agents; and their bioactive metabolites)	clofibrate (active metabolite: clofibric acid) gemfibrozil	Atromid-S Lopoid
musks (synthetic)	nitromusks polycyclic musks reduced metabolites of nitromusks	musk xylene Celestolide substituted amino nitrobenzenes
anti-anxiety/hypnotic agents	diazepam	Valium
sun screen agents	methybenzylidene camphor avobenzene octyl methoxycinnamate	Eusolex 6300 Parsol A Parsol MOX
X-ray contrast agents	diatrizoate	Hypaque

of drugs newly identified in environmental samples continually expands — new ones are continually added to the literature. For example, since Table II was compiled, several drugs on that list (plus others) have now been identified. For example, Zuccato et al. (*26*) identified in appreciable amounts up to hundreds of ng/L in Italian rivers of atenolol (antianginal, antihypertensive, beta blocker), ranitidine (histamine H2-receptor antagonist; e.g., Zantac), and furosemide (sulfonamide-type, loop diuretic). Snyder et al. (*37*, see chapter in this book) identified in lake water at the tens-to-hundreds of ng/L range a number or previously unreported drugs, including phenytoin, phenobarbital, primidone, hydrocodone, codeine, and others; some of these have also been reported by Möhle and Metzger (see chapter in this book). There is no reason to think that the incidence of discovery will not continue to grow; it is also important to note that the spectrum of PPCPs identified in one region or country can differ from those in another as a result of prescribing and usage patterns.

Table II. Representative distinct classes of drugs for which concerted environmental surveys have not been performed
(bolded names are among the top 200 most prescribed in U.S.: http://www.rxlist.com/top200a.htm) [cardiovascular and antihypertensive drugs include alpha blockers, angiotensin converting enzyme (ACE) inhibitors, anti-arrhythmics, beta blockers, calcium-channel blockers, centrally acting agents, diuretics, nitrates, and peripheral vasodilators]

therapeutic class	example generic name (many drugs cross over into multiple classes)	example Brand name
adrenergic receptor inhibitors (anti-BPH agents)	**terazozin, doxazosin,** finasteride	Hytrin, Cardura, Proscar/Propecia
amyotrophic lateral sclerosis	riluzole	Rilutek
analgesics (non-NSAIDs and narcotics)	**tramadol, propoxyphene, oxycodone, hydrocodone**	Darvon, Ultram, Tylox
anorexiants (diet drugs)	fenfluramine, orlistat	Pondimin, Xenical
antiarrhythmics	disopyramide, flecainide, amiodarone, sotalol	Norpace
anticoagulants	**warfarin**	Coumadin
antidepressants	esp. SSRIs (**sertraline, paroxetine, fluoxetine**, fluvoxamine), tricyclics (desipramine), MAOIs (phenelzine)	Zoloft, Paxil, Prozac, Luvox, Wellbutrin (**bupropion**), Serzone (**nefazadone**), Effexor (**venlafaxine**)
antidiabetic agents	insulin sensitizers, antihyperglycemic (e.g., sulfonyluereas)	Rezulin (**troglitazone**), Glucophage (**metformin**), Glucotrol (**glipizide**), Diaβeta (**glyburide**)

Continued on next page.

Table II. *Continued*

therapeutic class	example generic name (many drugs cross over into multiple classes)	example Brand name
antihistamines (H-1 blockers)	**fexofenadine, loratadine, cetirizine,** terfenadine	Allegra, Claritin, Zyrtec, Seldane
histamine (H-2) blockers	**famotidine, ranitidine, nizatidine**	Pepcid, Zantac, Axid
decongestants	ephedrines	
anti-infectives	many special disease classes (amebicides, anti-fungals, -malarials, -tuberculosis, -leprosy, -viral) and chemical classes	Diflucan (**fluconazole**)
antimetabolites	methotrexate	Rheumatrex
antipsychotics, CNS agents	**alprazolam, zolpidem, clonazepam, risperidone, temazepam** thioridazine, trifluoperazine	Xanax, Ambien, Klonopin, Risperdal, Restoril
calcium-channel blockers	**diltiazem, nifedipine, amlodipine, verapamil**	Cardizem, Procardia, Norvasc
digitalis analogs	**digoxin,** digitoxin	Lanoxin
diuretics	thiazide (hydrochlorothiazide, chlorthalidone); loop (furosemide, bumetanide); potassium-sparing (spironolactone, triamterene)	Lasix (**furosemide**) Dyazide (**hydrochlorothiazide, triamterene**)
dopamine agonists	anti-Parkinsonian agents (e.g., pramipexole, ropinirole)	Mirapex, Requip
expectorants	**guaifenesin**	Entex
gastrointestinal agents (ulcer drugs)	**omeprazole, lansoprazole, cimetidine**	Prilosec, Prevacid, Tagamet
HIV drugs	protease inhibitors, anti-retrovirals (nucleoside analogs/reverse transcriptase inhibitors)	Crixivan (indinavir), Retrovir (zidovudine)

hormonally active agents
   androgens
   anti-acne agents
   adrenocortico steroids
   inhalable steroids
   estrogen antagonists
	
	fluoxymesterone
	isotretinoin, tretinoin
	prednisone, triamcinolone
	fluticasone
	tamoxifen
		
		Accutane, Retin-A
		
		Flovent
		Nolvadex

muscle relaxants	cyclobenzaprine	Flexeril
osteoporosis agents (biphosphonates)	**alendronate** sodium	Fosamax
prostaglandin agonists	**latanoprost**	Xalatan
psychostimulants (amphetaminelike)	methylphenidate, dextroamphetamine	Ritalin
retinoids	**tretinoin**	Retin-A; Vesanoid

Table II. *Continued*

therapeutic class	example generic name (many drugs cross over into multiple classes)	example Brand name
sexual function agents	**sildenafil** citrate	Viagra
vasodilators (esp. angiotensin converting enzyme [ACE] inhibitors)	**lisinopril, enalapril, quinapril, benazepril** **losartan, fosinopril**, ramipril	Zestril, Vasotec, Accupril, Lotensin Cozaar, Monopril
street drugs (illicit, illegal, recreational)	many: e.g., see listing at: "Streetdrug.org" (http://www.mninter.net/%7epublish/index2.htm); National Institute of Drug Abuse (NIDA): http://www.nida.nih.gov/ NIDAHome1.html	
newly approved, upcoming, and investigational drugs	ongoing; see listing at: "LexiComp.org" (http://www.lexi.com/new_drugs.htm)	
"chemosensitizers", efflux pump inhibitors (EPIs)	verapamil (and others from diverse classes; e.g., see: http://www.microcide.com/ICAAC99Posters/icaac99_posters.html)	
cytochrome P450 inhibitors/inducers	http://dml.georgetown.edu/depts/pharmacology/clinlist.html	

Environmental survey — Data for a comprehensive approach are largely lacking: Environmental surveys for PPCPs (which analytes to target) can be guided somewhat by ranking the *expected* prevalence in STW effluents of each PPCP and using the ranked results to design a target-analyte approach. For personal care products, this can be accomplished by determining which of the myriad ingredients might be bioactive and then using industry production figures for combined members of each consumer-chemical "class". For medicinals, the objective is more complex. Production/consumption figures for individual locales are largely confidential or not available. Prescriptions filled and amounts consumed are difficult to acquire (for some general direction, refer to *American Druggist*, e.g., RxList Top 200: http://www.rxlist.com/top200.htm); usage figures for regional/local levels are nearly impossible to acquire (and the types of drugs can vary from region to region, country to country – for example, according to the age structure or usage conventions of the local population). Knowledge is needed of the excretion efficiency for the unaltered parent drug and its respective conjugates (or bioactive metabolites). But this must be tempered with the possibility that a drug's formulation (e.g., with stearic acid) may reduce absorption and thereby increase excretion of the parent drug. Drug interactions with other chemicals and the user's physiological/disease condition can also dramatically reduce uptake and thereby enhance excretion (or expulsion, e.g., through vomiting) of the parent drug; for example, chelation, alteration in gastrointestinal mobility, or alteration of gastric pH (e.g., chelation of tetracycline by dairy products or of fluoroquinolones by divalent cations). Data also required for these excreted compounds include their expected relative partitioning between STW solids and aqueous effluent, and their expected biodegradation efficiency within the STW. Data for other potential transformation pathways (e.g., photolysis) would be useful. While knowing the effective dosage for a drug's therapeutic target-effects in humans (and domestic animals) is useful, non-target, non-therapeutic effects (e.g., side effects) must also be considered, and these can occur at much lower doses (below low- to sub-µg/kg level — or ppb-ppt). Knowledge of potential aquatic effects would be most useful for directing final selections.

Aquatic organisms — captive to continual, life-cycle chemical exposures: Any chemical introduced to the aquatic domain can lead to continual exposure for aquatic organisms. Chemicals that are continually infused to the aquatic environment essentially become "persistent" pollutants even if their half-lives are short. Their supplies are continually replenished and this leads to life-long multi-generational exposures for aquatic organisms. Continual exposure can add an extra dimension to the exposure dose required for eliciting an effect (by reducing it even further). Of aquatic life, fish in particular are sufficiently mobile (as a result of natural behavior or avoidance responses) as to avoid continuous exposures to toxicants. This ability can obviate the course of what would ordinarily be captive, continuous exposure to any perpetual situation, leading to reduced, intermittent exposures. But, depending on the mechanism of action for a toxicant, such transient, intermittent exposure can still have comparable – and persistent – effects. For example, the estrogenic response (vitellogenin production from exposure of fathead minnows to estradiol) from intermittent exposure exceeds that which occurs from an equivalent time-integrated dose from a continuous exposure (*38*). Therefore, for at least certain toxicants, intermittent exposure (resulting either from episodic/fluctuating concentrations or from contact avoidance) does not necessarily lessen the degree or duration of the effect.

Effects — from acute to subtle: By their nature, pharmaceuticals are mostly designed to be highly bioactive — many exquisitely so. Their intended biological targets (e.g., receptors) are frequently extremely specific and sensitive; "receptors" essentially serve as switches that upon receiving an incoming "command" then generate an outgoing signal, the value of which is determined by whether the command is blockage by an antagonist or facilitation by an agonist. Furthermore, both the target receptors and intended effects can be very different from those of currently regulated pollutants (e.g., industrial chemicals and agrochemicals). Unintended, unexpected effects (e.g., adverse side effects; idiosyncratic drug reactions) can be caused by previously unrecognized drug-receptor interactions, previously unidentified receptors, previously unrecognized inter-drug interactions (e.g., via efflux-pump blockers/inducers or cytochrome P450 inducers/inhibitors), and by a broad diversity in drug-metabolizing/transport phenotypes (genetic polymorphisms). Many accomplish their therapeutic effects by mechanisms that have yet to be elucidated; and mechanisms of action can change depending on dose (2). Finally, drugs often tend to receive market approval based on their short-term effectiveness against "surrogate endpoints" (e.g., blood pressure), not on the basis of long-term efficacy against the target malady (e.g., heart disease).

While our understanding of the spectrum of these variables in humans is not fully understood, they are poorly characterized in aquatic organisms. Effects on non-target organisms (e.g., wildlife – not laboratory organisms) are almost completely uninvestigated. Furthermore, the important point, which is totally obscure to current knowledge, is the significance that these variables might have at the population level (effects on individuals being important only as far as an ultimate impact on the population is concerned). Extremely little is known about the effects of these substances on non-target organisms, many of which have different metabolic pathways and different potential receptors. Effects on aquatic (non-target) species could range from easily detectable acute symptomology to unnoticeable subtle change. What little that is known serves to show that rather low concentrations at least have the potential to exert substantive effects on aquatic life. A few PPCPs are already known to elicit profound effects on aquatic life at very low concentrations. For example, selective serotonin re-uptake inhibitor antidepressants (SSRIs) such as fluoxetine and fluvoxamine can induce reproductive behavior in certain shellfish at 10^{-10} M (ca. 30 ppt); see the extensive discussion of SSRIs in the chapter in this book by Fong (39). Universal regulatory pathways, such as the control of intracellular Ca^{2+} concentration for activating sperm, could be affected by any of the many calcium channel blockers. For example, pimozide and penfluridol inhibit sperm activation in sea squirts (40); this perhaps is not surprising since infertility in both human sexes has been linked to numerous drugs (ca. over 150) over the years (including members of chemotherapeutics, anti-psychotics, beta-blockers, antimicrobials, etc., and of course, a wide range of hormones). Male infertility, however, is not commonly screened for using lab animal models. Nifedipine (another calcium channel blocker) has been known to impair the ability of sperm to fertilize (reversible effect). Whether any of these compounds occur in the environment (structural stability might be a major limitation) is not known.

While only very sparse data have been generated so far on the subtle ways in which drugs could affect aquatic life, this line of concern is bolstered and illustrated by the substantially more data that is available for pesticides and other "conventional" pollutants. For example, brief exposure of salmon to 1 ppb of certain carbamate or organophosphorus insecticides is known to affect signaling pathways (via olfactory disruption and pheromone detection), leading to alteration in homing behavior (with obvious implications for predation, feeding, mating, and olfactory imprinting) (e.g., 41,42).

Another consideration is that of additive effects. For therapeutic classes comprising multiple members (SSRIs, calcium channel blockers, and anticholinergics being three examples), even if the individual concentrations of each drug were low in the aquatic environment, the combined concentrations of all members sharing a common mode of action could prove significant (i.e., summation of "toxic units"; see *43*). This consideration could be particularly important for those drugs having low therapeutic indices (margins of safety, or therapeutic "windows"). As an example of additive effects, anticholinergics are known for their potential to create an insidious additive "anticholinergic load" problem for elderly patients simply because this broad class of drugs comprises about 20 distinct, otherwise unrelated, therapeutic classes – the combination of multiple drugs as opposed to individual members is problematic (*44*). Analogous issues can be drawn for the aquatic environment.

Finally, the issue of sub-therapeutic exposures is only a part of the overall concern being that continual, life-long exposure to trace levels of any substance is a relatively unexplored domain of human or environmental toxicology. This is one reason that trends data are important. Even where PPCP concentrations are currently low, no data sets exist for revealing long-term trends. For any particular PPCP at a given geographic location, whether its concentration over time is decreasing, remaining constant (steady state), or increasing could dictate whether any action needs to be contemplated to control its introduction to the environment.

Could subtle effects accrete to unnoticed change? A major issue yet to be addressed by ecotoxicological science (and of potentially critical importance for risk assessment) is the impact on non-target species (such as aquatic biota) of stressors eliciting effects (perhaps via low but continual concentrations) sufficiently subtle as to go unnoticed in real time – but whose cumulative impacts eventually yield recognizable outcomes having no obvious cause (*1*). How would the impact of such cumulative, unrecognizable subtle effects compare with the overt, acute impacts of known toxicants? Unequivocal environmental impacts or change, by their obvious nature, are easily detected, and measures to mitigate untoward outcomes can be engineered and implemented. Change that takes a surreptitious tack, however, grows unnoticed and can result in outcomes not recognized as such — seeming to arise from nowhere or perhaps mistaken as a natural result of adaptation or evolution within the interconnectedness of nature. While any immediate biological actions on non-target species (unintentional effects, especially in the aquatic habitat) might be imperceptible, we must ask whether PPCPs nonetheless have the potential to lead to adverse impacts as a result of subtle effects ("silent toxicity") from low, ppb-ppt concentrations (μg-ng/L) that impart latent damage. This problem would be exacerbated if the resultant outward change were rationalized simply as resulting from natural evolutionary change or variance and no remediative messages implemented.

This type of ecological impact could persist indefinitely, with no mitigative measures ever put in place. Antineoplastics, SSRIs, and calcium-channel blockers are three drug classes that are known to have the potential for long-term aquatic effects; neuroteratogenic anti-epileptics have the potential for adverse aquatic effects. Subtle, unnoticed effects could accumulate over time until any additional incremental burden imposed by a new, unrelated stressor could possibly trigger sudden collapse of a particular function or behavior across a population. While PPCP residues by themselves might well be below even the thresholds required for subtle, latent effects, their presence may add to the overall aggregate of other, unrelated toxicants – leading to an unacceptable overall burden. An associated risk is the loss of genetic diversity caused by selection against those organisms with insufficient xenobiotic detoxification pathways.

In a range of articles on behavioral and neuro toxicology (see *45,46,* and references cited therein), a cogent argument is presented that neurobehavioral effects (with humans being the subjects of concern) are particularly apt to display with such subtlety that they can escape detection by conventional quantitative approaches. Many of these effects are manifested as organisms age and mature — through subtle alterations in functions such as capacity to learn, motor-sensory performance, and reproduction, coupled with a gradual age-dependent decline in their requisite compensating mechanisms. This makes detection of these slow, degenerative effects, and their distinction from the "natural aging process", extremely difficult.

Weiss makes the argument that truly "abnormal" behavior has the potential to masquerade as seemingly normal deviation within a natural statistical variation. Change can occur so slowly that it appears to result from natural events – with no reason to presume artificial causation. Connections of cause and ultimate effect are difficult to draw, in part because of the ambiguous and subjective nature of the effects, but especially when they are confounded as aggregations of numerous, unrelated interactions.

Effects that are sufficiently subtle that they are undetectable or unnoticed present a challenge to risk assessment (especially ecological). Deviations in behavior within statistical variations are particularly problematic. Advances are required in development and implementation of new aquatic toxicity tests to better ensure that such effects can be detected. While certain fundamental biochemical pathways (e.g., various developmental signaling pathways) are evolutionarily conserved – being the same across disparate species – toxicokinetic differences are still substantial between humans and animals. Just as animal models are frequently called into question for their relevance to human health, mammalian toxicity data (e.g., for PPCPs) may not necessarily be transferrable to aquatic organisms, especially with respect to behavior. The diversity and importance of animal behavior is extremely complex (the focus of ethology, and can be explored in depth at: http://www.york.biosis.org/zrdocs/zoolinfo/behav.htm). Even the correlation between mammalian toxicity (on the basis of dose) and aquatic toxicity (on the basis of concentration) for conventional tests is non-existent for well-known PBTs (*47*). It should also be emphasized that the priorities for selecting PPCPs for toxicological evaluation can not be based solely on their relative rankings with respect to environmental concentrations because drugs can dramatically vary with respect to the concentrations at which they impart effects — sometimes by orders of magnitude.

Data supporting this subtle-effects scenario (posed by Daughton and Ternes [1]) was recently published (32). Thresholds for estrogenic responses (vitellogenin production in male fish) are shown to be much lower for real-world chronic exposure (i.e., for wild fish) than for short-time study exposures (i.e., for caged fish). Responses are a function of not just the dose, and the timing of dose, but also duration of exposure. Response thresholds (no-effect concentrations) are continually reduced as exposure times increase. Moreover, no overt changes are noted in any fish (e.g., condition factors, survival, growth) regardless of the STW effluent exposure concentration or duration — only the less obvious vitellogenin concentrations were found to be affected. Longer exposures led to exponentially higher vitellogenin concentrations with concomitantly lower response thresholds.

Efflux pumps: A foundation for aquatic health in polluted waters — Efflux pump inhibitors: Health risks to aquatic biota? Over the last decade, the identification and characterization of membrane-based active transport systems that "eject" or "pump" toxicants from inside cells have revealed a number of important roles for these systems. These protein-based ejection systems are often referred to as "efflux pumps" or "multi-drug transporters", the best characterized being the P-glycoprotein-like (Pgp) transporter systems (see discussion in Chapter by Epel and Smital [48]; and in Bard [49]). First spotlighted by Daughton and Ternes (1) for their potential significance with respect to PPCPs in the aquatic realm, it is becoming evident that these drug (or multi-xenobiotic) efflux pumps, which only started to be characterized in the 1980s for mammalian cells (esp. tumors) and in bacteria, may be broadly distributed across the entire biotic spectrum — including aquatic organisms.

These pumps are responsible for actively transporting toxicants (esp. amphiphilic molecules) that gain access to (or are created within) the interior of cells back to the outside of the membrane, thereby preventing the intracellular accumulation of toxicants and bioactive metabolites. "Over expression" of these proteinaceous efflux pumps in tumor cells and pathogenic bacteria is widely recognized as enabling resistance to chemotherapeutic agents and antimicrobials. These systems can also be induced (up-regulated) by one drug, leading to enhanced excretion of other drugs (50). Conversely, any of a diverse array of certain chemicals that are able to inhibit these pumping systems can lead to intracellular penetration by, and accumulation of, toxicants in general, and thereby potentiate adverse effects from extracellular concentrations of toxicants that would otherwise prove benign. Because toxicants cannot be readily removed by an exposed organism whose transport system(s) has been impaired, exposure time to the toxicants is thereby lengthened by intracellular accumulation. Another important metabolic enzyme system that also needs to be considered is that comprising the better-known cytochrome P450 monooxygenase isoforms (see listing at: http://dml.georgetown.edu/depts/pharmacology/clinlist.html); this system is intimately involved with drug metabolism, and numerous drug-drug interactions occur as a result of its inhibition or induction. Inhibition of any of the many cytochrome P450 monooxygenases (or their helper enzymes) can modulate steroid metabolism and lead to population-wide reproductive effects in certain aquatic invertebrates, such as molluscs (51). Numerous drugs and other toxicants are ligands for the ubiquitous P450 enzyme superfamily. This is the route, for example, by which tributyltin causes imposex in molluscs, thereby serving as an indirect endocrine disruptor; P450-mediated steroid metabolism is also used by crustaceans (51).

While efflux pumps are best known for making it more difficult to get therapeutic doses of drugs to target organs (because of reduced drug absorption/uptake, enhanced drug elimination), much less appreciated is the fact that they provide a fundamental level of protection for aquatic biota — organisms ranging from fish and crustaceans (see Chapter by Epel and Smital [48], and discussion in Daughton and Ternes [1]) to ciliates (52), all of which can suffer from continual exposure to any toxicants present. Now recognized for enabling a significant portion of the growing incidence of antimicrobial resistance among bacteria, these systems play a critical role in protecting cells from toxicants, especially in those environments such as the aquatic realm where filter-feeding organisms in particular suffer continual, maximal exposure to toxicants.

Another consideration is the prevalence in the aquatic environment of chemicals that can induce the expression of efflux pump systems. This would add yet another dimension to these considerations by selectively enriching populations for those individuals with inducible efflux systems and thereby placing the entire population at maximum risk should they eventually be exposed to potent inhibitors of these systems. It is also believed that inducement of efflux pump expression is a means by which resistance to antiseptics can be promoted; this mechanism of resistance is also applicable to antimicrobials that supposedly have non-specific mechanisms of actions (e.g., general membrane disruption/denaturation) and that were previously believed to not be amenable to promoting resistance. Efflux pump induction is a recently recognized new avenue of drug-drug interaction in humans, leading to enhanced excretion of unmetabolized drug (50).

A number of substances (primarily pharmaceuticals) are known to inhibit efflux pumps (via a variety of mechanisms), some of the more potent being verapamil, reserpine, and cyclosporin. These inhibitors are also known as efflux pump "reversal agents", "chemosensitizers", "efflux pump blockers", or "efflux pump inhibitors" (EPIs). Given the fundamental role that efflux pumps may play in the health of biota throughout the aquatic food-web (and therefore entire aquatic ecosystems), attention may need to be devoted prior to environmental assessment for screening new chemicals (esp. drugs) for EPI activity. Little is known, however, about which existing xenobiotics can act as "chemosensitizers" in the aquatic environment — or their frequency of occurrence in the environment. Evidence shows that samples from polluted surface waters impair the functioning of multi-xenobiotic transport in aquatic organisms more than the analogous samples from less-polluted waters (see references cited in Daughton and Ternes [1]).

Recent developments show that much more potent, broad-spectrum EPIs (which do not necessarily possess any intrinsic bioactivity) are being designed to resurrect or expand the usefulness of existing drugs (including antimicrobials that are no longer effective) as well as enhance the effectiveness of new drugs. These, currently research-only investigational compounds can show orders of magnitude greater EPI activity than verapamil (see the posters prepared by Microcide Pharmaceuticals, Inc., for the 39th Interscience Conference for Antimicrobial Agents and Chemotherapy: http://www.microcide.com/ICAAC99Posters/ icaac99_posters.html). The potential for release of such potential future EPI drugs may require careful assessment for environmental impact. This is but one example of a new and promising route to disease therapy that may require active balancing of the clear benefits for human health against the high potential for aquatic harm.

Questions also arise as to whether the combined additive/synergistic action of numerous EPIs (whether drugs or other xenobiotics) in the aquatic environment could be one (of a number) of possible causes for sudden, unexplained mass die-offs of various species. Could systems in apparent good health possibly collapse simply by exposure to one or a series of EPIs (which by themselves would not prove toxic) — by potentiating the action of toxicants that were already present? How far-reaching could the action of EPIs in the aquatic realm be? The possibly widespread expression of efflux pumps in aquatic organisms also begs the question as to the relevance of trying to correlate lipid burdens of toxic pollutants with organisms' health (e.g., see43) if these concentrations have little if any bearing on the concentrations that actually effect intracellular exposure. Can fat-accumulated chemicals act as reservoirs of toxicants that gain access to intracellular receptors once efflux pumps are shut down?

Finally, it is worth noting that aquatic toxicity is traditionally measured in the absence of efflux pump inhibitors. This means that EC_{50} values may represent conservative, upper estimations for a given toxicant being that any EPIs that may be present will effectively increase the actual dosage that enters exposed cells — thereby lowering the EC_{50}.

Are the current approaches to risk assessment comprehensive?

Questions can be raised as to whether the approaches to environmental risk assessments and epidemiological studies sufficiently consider the "universe" of toxic substances involved in exposure or whether the focus on conventional "priority pollutants" gives a narrow perspective. By ignoring exposure to chemicals with significant bioactivity (such as pharmaceuticals), can the usefulness of such studies be questioned? What might be the significance of not factoring individual risks from previously unidentified pollutants in discussions of overall risk? This question is addressed in the ecological risk rankings published by the EPA's Science Advisory Board (53). Gross within-class differences with respect to aquatic effects possibly make the approach of assessing eco risk on a class-by-class basis untenable. For example, some SSRIs are extremely potent with regard to shellfish reproductive behavior while others have almost no effect. By not factoring in multi-chemical effects (such as from efflux pump inhibition), potentially large effects could be overlooked.

Why the concern over exposures that are well below therapeutic doses?

It is important to recognize that recommended therapeutic dosages can be higher than the therapeutic dose actually required (a dual consequence of not adjusting doses downward from those set in clinical trials and of not individualizing therapy). Non-target, unintended effects cannot be discounted at sub-therapeutic dosages (unintended effects can occur at lower concentrations than do therapeutic effects, especially during long-term maintenance therapy). Additive/synergistic effects from simultaneous consumption of numerous drugs are essentially unknown (intake of multiple drugs adds to the complexity of additional burden for a patient already taking medications with low therapeutic indexes). Finally, continual, life-long exposure to trace levels is an unexplored domain of toxicology.

Relevant to these points regarding exposure to low environmental concentrations is the very important fact that trends data are lacking. Even where PPCP concentrations are currently low, no data sets exist for revealing long-term trends. For any particular PPCP at a given geographic location, is its concentration over time decreasing, remaining constant (steady state), or increasing?

Continually expanding uses and consumption of individual PPCPs points to concentrations of existing drugs increasing in addition to adding to the burden of discrete drug types in the environment. Will those PPCPs with longer half-lives tend to accumulate? Drug use can be expected to continue to rise due to a confluence of drivers: increased per capita consumption, expanding population, expanding potential markets (partly due to mainstream advertising/marketing), expiration of patents, increased use of physician samples, new target age groups, and new uses for existing drugs. Old therapeutics are being used not just for additional clinical conditions (those for which they were not originally developed) but also for non-disease states — for example, medical manipulation or alteration of personality traits and satisfaction of certain social needs — referred to as "cosmetic pharmacology."

Finally, there are potential, intangible benefits in being proactive versus reactive — the Precautionary Principle – the principle of precautionary action that redistributes the burden of proof because the science required for truly and fully assessing risks lags far behind the requisite supporting science (e.g., see: http://www.biotech-info.net/uncertainty.html).

Future considerations — Drug sewage monitoring as a tool: A significant aspect of our society that illustrates the potential impact of our lives on the environment — as well as the potential impact of our activities on ourselves and others — is the widespread use of illicit drugs. The potential for the occurrence of illicit drugs (and drugs of abuse) in sewage influent presents a unique and powerful opportunity for a new tool to raise the public's consciousness of drug use — down to the local community level. By monitoring sewage influent for illicit drugs (and perhaps by back calculating with pharmacokinetic assumptions), estimates of drug use at the community level could be ascertained, while preserving the anonymity of all individuals. Such data have the potential for not only lending a new perspective to illicit drug use, but also for providing a new tool to enhance society's knowledge and perspective regarding the manufacture, trade, and use of these substances. This idea is expanded upon in the last chapter of this book (23).

Summary of Under-Addressed Issues Regarding Environmental Ecotoxicology

In an examination of the world's disparate and disjointed literature involving pharmaceuticals in the environment, Daughton and Ternes (1) highlight a number of issues not frequently encountered in discussions of environmental toxicology. These and others may deserve further attention and debate.

> ▸ **Multi-Drug Resistance as a Vulnerable Line of Defense for Aquatic Organisms:** Multi-drug transporters (drug efflux pumps) are currently recognized as target receptors of opportunity in human medicine for improving the uptake of drugs by tumor and healthy cells alike. Less appreciated is the importance of these same efflux systems for protecting aquatic biota that can suffer continual exposure to toxicants (especially important for filter feeders). Those pharmaceutical compounds having the ability to inhibit these active transport systems, required for minimizing intracellular exposure of many aquatic organisms ("multi-xenobiotic resistance"), can promote intracellular accumulation of toxicants whose extracellular concentrations would not prove toxic in the absence of the inhibitor. Aquatic biota in homeostasis with their environment could experience rapid toxicological effects upon exposure to an efflux pump inhibitor. Ecotoxicological studies need to factor this phenomenon into aquatic assessments.

▸ **Shared Modes of Action Can Add to Risk:** While the individual concentration of a given drug in the aquatic environment might well be very low (ng-µg/L), the combination of numerous drugs sharing the same purported mode of action (e.g., SSRIs, calcium channel blockers) could be significant (especially true for therapeutic classes comprising many widely used drugs such as NSAIDs, anticholinergics, lipid regulators, antihypertensives, calcium-channel blockers, etc., and for those with low therapeutic indices); this is analogous to the concept of "toxicity equivalency factors" for dioxin and PCP congeners.

▸ **Subtle Effects Could Prove Significant:** Historically, toxicological endpoints of xenobiotic exposure have usually been restricted to acute (and overt), easily measured effects (with facile cause/effect correlations) such as mortality and cancer — little attention has been paid to the universe of other endpoints through which toxicants can express their action; focus has also been on relatively high concentrations of toxicants. PPCPs have the potential to exert very subtle (e.g., neurobehavioral, immunologic) effects that may escape detection but which accumulate sufficiently slowly over time as to eventually result in substantive, outward change that is mistakenly attributed to, or rationalized as resulting from, normal natural processes or ascribed simply as being part of "natural variation". The actual effects are not noticed in real time — only at some future point when they culminate in untoward consequences. Subtle effects/changes may not be observable via short-term snap-shot studies, but rather require more costly long-term, continuous monitoring. This issue presents a major challenge especially to ecological risk assessment.

▸ **Chemical Stability Not Required for "Persistence":** Continuous infusion of a pollutant to the aquatic environment is the sole ingredient necessary for effecting continual, multi-generational life-cycle exposures of sensitive aquatic species. Actual structural persistence (as with DDT and other persistent organic pollutants) is not necessary as long as the pollutant is continually introduced (such as from sewage treatment plants). Current criteria and approaches (as summarized by Rodan et al. [54]) for establishing the importance of pollutants keyed to chemical stability as the prime measure of "persistence" may be overlooking entire classes of significant pollutants.

▸ **Is Risk Assessment "Holistic"?** PPCPs are a very large and diverse suite of potential pollutants. They rival agrochemicals in their usage rates and diversity of chemical classes. A true, "holistic" risk assessment process must take into account all bioactive compounds to which an organism is exposed. Up to now, PPCPs comprise a major group that has been excluded from the risk assessment process. PPCPs have never been subject to any water monitoring program (whether domestic waste or drinking) in the U.S. While true, all-encompassing risk assessments can never be performed, attention to the continually developing picture of the exposure universe could bring us asymptotically closer.

▸ **The Need to Reach Past Our Historic Focus on Conventional Pollutants:** During the last three decades, the study of environmental chemical pollution has maintained a steady fixation on conventional "priority pollutants," including those collectively referred to as "persistent, bioaccumulative, and toxic" pollutants (PBTs) or as "persistent organic pollutants" (POPs); the "dirty dozen" is a ubiquitous, notorious subset of these, comprising halogenated organics, such as DDT and PCBs. Continued focus on this area will add but incrementally to our overall knowledge base of pollutants in the environment. At least a portion of this focus might more profitably be diverted to understanding the scope and significance of additional bioactive, anthropogenic chemical classes that have long escaped attention.

▸ **While Aquatic Effects of PPCPs in the Environment May Pose the Most Obvious Risk, Human Exposure Risk Cannot Be Discounted:** Finally, despite the extremely low concentrations of drugs in domestic potable waters, we must be wary of dismissing the toxicological significance for humans of low (ppt) concentrations of drugs in drinking water (amounting to merely nanograms-per-day intakes via drinking water) on the sole basis that "such concentrations are orders of magnitude lower than the therapeutic dosages." This remains a common assertion (e.g., *26*) in predictions of potential minimal human health effects. A question develops as to whether this view is potentially flawed because (i) non-target, unintended effects can not be discounted at sub-therapeutic dosages, (ii) additive/synergistic effects from the simultaneous consumption of numerous drugs are essentially unknown, (iii) continual, life-long exposure is an unexplored domain of toxicology, and (iv) the trends in environmental concentrations for all drugs are completely unknown.

Conclusions and Recommendations

The future for research on PPCPs in the environment: The poorly characterized ramifications of PPCPs in the environment (occurrence, trends, fate, transport, effects), or their relative importance compared with conventional chemical pollutants such as PBTs, warrants a more precautionary view on their environmental disposition. A portion of the effort that continues to be invested in elucidating the environmental transformation and fate of POPs/PBTs might more profitably be redirected to PPCPs.

A major task of the environmental science community might be to examine each therapeutic or consumer-use class of PPCPs (or those grouped according to purported modes of action) and rule-in or rule-out possible deleterious environmental effects on the basis of those that are known to occur in the environment at significant individual or combined concentrations, or those whose concentrations are trending upward. One area that could be pursued immediately is an exhaustive search of the literature for unintended, unexpected effects on non-target species, especially aquatic. Particular scrutiny should be given to the efflux-pump-inhibitory (and efflux pump-inducing) activity of existing and all new drugs with regard to aquatic organisms. Future designs of STWs should perhaps focus on the most difficult-to-remove, toxicologically significant compounds so as to maximize the removal of all anthropogenic pollutants (even those not currently recognized). This work will involve simultaneous contributions from both exposure and effects scientists working in parallel and in sequence (an iterative process).

Near-term objectives to consider for minimizing introduction of drugs to the environment or to minimize potential environmental effects:

Screening for EPI Potential: A question deserving close scrutiny — "Could unexplained aquatic mass die-offs (e.g., fish kills) result not necessarily from increased burdens of acutely toxic substances already present or from sudden presentation of new toxic substances, but rather from the introduction of ordinarily benign substances (e.g., chemosensitizers, such as EPIs) that potentiate the bioactivity of toxicants present at levels not otherwise toxic?" Serious consideration should be given to developing new aquatic testing procedures specifically for evaluating possible aquatic effects of potent, new-generation EPIs.

Environmental "Friendliness": Environmental proclivity could be factored into PPCP design/marketing. For example, maximize biodegradability/photolability to innocuous end products (e.g., controlling stereochemistry), minimize therapeutic dose ("calibrated dosing"; facilitation of membrane permeability to allow lower doses, e.g., use of efflux pump inhibitors or alteration of structure), single-enantiomers — "green" PPCPs (e.g., see 55).

New Approaches to Pollution Prevention: Approaches to minimizing PPCP discharge to the environment can extend beyond the obvious alternatives such as reducing dosages and practicing proper disposal. An example that serves to illustrate innovative alternatives is that of "drug mining" human excreta, whereby highly toxic drugs (such as genotoxicants) are reclaimed ("harvested") from excreta for recycling purposes (e.g., see approach used by Tru-Kinase, Inc., at: http://www.toilettoilet.com/pharm_recovery.htm).

Imprudent Use: Better inform physicians (and public) as to environmental consequences of over-prescribing medications — minimize misuse/overuse. Engage medical community to develop guidelines. Judicious/justifiable use of antibiotics — minimize "imprudent use" (e.g., identify pathogens prior to prescribing antibiotics; minimize prophylactic use, such as for purported prevention of secondary bacterial infection during the course of viral infections). This might also serve to address a major aspect of the spread of rampant antimicrobial resistance. Minimize use of "physician samples". Minimize misuse of antibiotics by veterinarians, aquaculturalists, and agriculturalists to lessen incidence of resistance in native bacteria. This topic has been thoroughly discussed by many organizations (e.g., see *6*).

Internet Dispensing: Consider the regulation/control of filling prescriptions (and dispensing without a prescription) over the Internet (to minimize unneeded drug use and attendant disposal); this would also lessen the public's exposure to unnecessary (and perhaps risky) medication. The FDA's current view/approach is available at: http://www.fda.gov/oc/buyonline.

Individualization of Therapy: Encourage drug manufacturers to provide the medical community with the necessary information to tailor drug dosages to the individual (esp. long-term maintenance drugs) on the basis of body weight, age, sex, health status, and known individual drug sensitivities — individualization of therapy; this objective will be greatly aided by pharmacogenomics and its use of single nucleotide polymorphisms (SNPs). Identify lowest effective dosages ("calibrated dosing") — effective doses are often lower than generic "recommended doses", which are usually based on higher-dose clinical studies; this would also serve to reduce patient side effects.

Alternative Dosing: Research already underway on alternative modes and routes of drug delivery (e.g., via the lungs) could be expanded. More constant dosing can be achieved via continually advancing the technology involved in formulation, sustained release (e.g., dermal patches), or implant delivery of drugs. This also allows the design of more labile drugs (e.g., those that would ordinarily be degraded by, or poorly transported across, the gut) (e.g., see *56*). Also, the use of alternative excipients could improve the dissolution of medication in the gut, especially in persons of compromised health; for example, stearic acid derivatives can greatly reduce the dissolution of medication, resulting in excretion of undissolved pills (*24*).

Proper Disposal: Inconsistent and potentially environmentally unsound guidelines for non-controlled drug disposal in the U.S. (coupled with deferral by the potentially responsible federal and state authorities) has led to a patchwork of nationwide confusion on the part of all parties involved with the use or disposal of drugs. In a rare survey of the U.S. public's practices on the disposition of unneeded non-controlled drugs (*57*), very few return them to pharmacies, and the vast majority (nearly 90%) dispose of them in municipal garbage or via domestic sewage. In the same survey, some pharmacies were found to have developed standard disposal protocols, but the majority of them (68%), just as with the public, disposed of unneeded drugs via municipal waste or sewage and advised their customers to do the same.

Educate pharmacy industry to provide proper disposal instructions to end-user for unused/expired drugs. Better guidance is needed for disposition of non-controlled substances by disposal companies. Consider implementing Extended Producer Responsibility (EPR) guidelines (e.g., *58*; also: http://www.epa.gov/epaoswer/non-hw/reduce/epr/index.htm). Being that drug disposal is a major problem for humanitarian assistance projects (see guidelines at: http://www.drugdonations.org/eng/index.html), various aspects of drug disposal are discussed by WHO (*25*).

Importance of Individual Action: Educate public on (i) how their individual actions, activities, and behaviors each contributes to the burden of PPCPs in the environment, (ii) how PPCPs can possibly affect aquatic biota (and even impact drinking water), and (iii) the advantages accrued by conscientious and responsible disposal and usage of PPCPs.

Literature Examination of Potential Transformation Products: Potential environmental impact for any given PPCP can be greatly complicated for those parent compounds that yield numerous transformation products (e.g., nitro musk fragrances can yield numerous photoproducts [*59*]). Waste treatment by chlorination or ozonation is known to generate a plethora of products from a single parent compound.

Use of Drugs as Environmental Markers of Sewage: Capitalize on the occurrence of certain, more easily degraded PPCPs to serve as conservative environmental markers/tracers of discharge (early warning) of raw (or insufficiently treated) sewage.

New Paradigm for Maximizing Waste Treatment Efficiency: Given the extreme complexity of ensuring the maximum degradation of myriads of toxicants entering STWs, consideration could be given to establishing "performance-based" treatment guidelines for the design of future water treatment systems.

Suites of significant potential pollutants known to be refractory to treatment could guide design of maximally effective plants.

The implementation of many of these considerations might contribute not just to minimizing the burden of PPCPs in the environment, but might also benefit end-users and patients by way of minimizing their exposure to PPCPs via end-use and the concomitant universe of side-effects and bacterial resistance development. By holding as a major objective the maximization of the usefulness of toxicological studies for risk assessments, such assessments would become more useful to the good of the public and the environment. In the final analysis, we need to ask whether the current defacto, no-cost approach for dealing with the introduction of PPCPs to the environment – namely dilution by receiving waters – is the best, or if one or more proactive approaches should be implemented.

REFERENCES

1.   Daughton, C.G. and Ternes, T.A. "Pharmaceuticals and personal care products in the environment: Agents of subtle change?" *Environ. Health Perspect.* **1999**, *107*(suppl 6), 907-938 [available: http://www.epa.gov/nerl esd1/chemistry/pharma/index.htm].

2.   Conolly, R.B.; Beck, B.D.; Goodman, J.I. "Stimulating research to improve the scientific basis of risk assessment," *Toxicological Sci.* **1999**, *49*, 1-4.

3.   National Research Council, Hormonally Active Agents in the Environment, Washington, DC: National Academy Press, 1999, pp 22-23 [available: http://books.nap.edu/books/0309064198/html].

4.   Sievers, R.E.; Barkley, R.M.; Eiceman, G.A.; Shapiro, R.H.; Walton, H.F.; Kolonko, K.J.; Field, L.R. "Environmental trace analysis of organics in water by glass capillary column chromatography and ancillary techniques," *J. Chromatogr.* **1977**, *142*, 745-754.

5.   Ferber, D. "Superbugs on the hoof?" *Science* **2000** *288*(5467), 792-794.

6.   WHO "World Health Organization Report on Infectious Diseases 2000: Overcoming Microbial Resistance" 2000, WHO/CDS/2000.2 [available: http://www.who.int/multimedia/antibiotic_res/index.html].

7.   Olney, J. "Response to 'Induced damage in the developing brain'," *Science* **2000**, *288*, 977-978.

8.   Olney, J.W.; Farber, N.B.; Wozniak, D.F.; Jevtovic-Todorovic, V.; Ikonomidou, C. "Environmental agents that have the potential to trigger massive apoptotic neurodegeneration in the developing brain," *Environ. Health Perspect.* **2000**, *108*(suppl 3), 383-388.

9.   Samren, E.B.; van Duijn, C.M.; Koch, S.; Hiilesmaa, V.K.; Klepel, H.; Bardy, A.H.; Mannagetta, G.B.; Deichl, A.W.; Gaily, E.; Granstrom, M.L.; Meinardi, H.; Grobbee, D.E.; Hofman, A.; Janz. D.; Lindhout, D. "Maternal

use of antiepileptic drugs and the risk of major congenital malformations: A joint European prospective study of human teratogenesis associated with maternal epilepsy," *Epilepsia* **1997**, *38*(9), 981-990.

10. Society of Environmental Toxicology and Chemistry (SETAC), **1999**, "Sound Science Technical Issue Paper," Pensacola, FL, USA. [available: http://www.setac.org/sstip.html].

11. Sanderson J.T.; Seinen W.; Giesy J.P.; Van den Berg M. "2-Chloro-s-triazine herbicides induce aromatase (CYP19) activity in H295R human adrenocortical carcinoma cells: A novel mechanism for estrogenicity?" *Toxicol. Sci.* **2000**, *54*(1), 121-127.

12. Barlow, S.; Kavlock, R.J.; Moore, J.; Schantz, S.L.; Sheehan, D.M.; Shuey, D.; Lary, J.M. "Teratology Society Public Affairs Committee Position Paper: Developmental toxicity of endocrine disruptors to humans," *Teratology* **1999**, *60*, 365–375.

13. Hensley, S.; Murray, S. "Use of samples in drug industry raises concern," *Wall St. J.* **2000**, 19 July, B1,B4.

14. Blok, J.; Balk, F.; Okkerman, P.C. "Identification of persistent, toxic and bioaccumulating substances (PTBs) – Report to: The Netherlands Ministry of Housing, Spatial Planning and the Environment (VROM)," Haskoning Consulting Engineers and Architects, and BKH Consulting Engineers (Netherlands), 13 July 1999 (G1260.AO/ROO4/FBA/AS), 58 pp.

15. Ayscough, N.J; Fawell, J.; Franklin, G.; Young, W. *Review of Human Pharmaceuticals in the Environment*, R&D Technical Report P390, UK Environment Agency, Rio House, Bristol, June 2000, ISBN 1 85705 411 3; pp. 110 [available at: http://www.wrcplc.com/ea/rdreport.nsf].

16. Hutzinger, O. (Jørgensen S.E. and Halling-Sørensen, Eds.) Drugs in the Environment, special issue of *Chemosphere* (devoted to drugs in the environment) **2000**, *40*(7), pp. 691-793.

17. *Toxicology Letters* (special issue devoted to musks in the environment) **1999**, *111*(1-2), 1-187.

18. Ternes, T.; Wilken, R-D. (Eds.) Drugs and Hormones as Pollutants of the Aquatic Environment: Determination and Ecotoxicological Impacts. **1998** *The Science of the Total Environment 225*(1-2), 176 pp.

19. Raloff, J. "More waters test positive for drugs," *Sci. News* **2000**, 1 April, *157*(14), 212.

20. Zuer, P.S. "Drugs down the drain," *Chem. Eng. News* **2000**, 10 April, *78*(15), 51-53 [available: http://pubs.acs.org/email/cen/html/041000113838.html].

21. The Merck Index, 2000, 12th ed., Merck ; Stationery Office, Whitehouse Station, N.J., 1741 pp. ISBN 0911910123.

22. Physicians' Desk Reference, 2000. 54th ed. Medical Economics Company, Inc., Montvale, NJ, 2000, 3355 p. ISBN: 1563633302 .

23. Daughton, C.G. "Illicit Drugs in Municipal Sewage – Proposed New Non-Intrusive Tool to Heighten Public Awareness of Societal Use of Illicit/Abused

36

Drugs and Their Potential for Ecological Consequences," in Pharmaceuticals and Personal Care Products in the Environment: Scientific and Regulatory Issues; Daughton, C.G. and Jones-Lepp, T., Eds.; American Chemical Society: Washington, DC, 2001 (see concluding chapter in this book).

24. Czap, A. "Supplements facts do not equal all the facts. What the new label does - and doesn't - disclose," *Alternat. Med. Rev.* **1999**, *4*(1), editorial [and personal communication to Daughton, May 2000].

25. WHO "Guidelines for Safe Disposal of Unwanted Pharmaceuticals in and after Emergencies," WHO/HTP/EDM/99.2, March 1999, 36pp [available: http://www.drugdonations.org/eng/index.html/].

26. Zuccato, E.; Calamari, D.; Natangelo, M.; Fanelli, R. "Presence of therapeutic drugs in the environment," *The Lancet* 2000, *355*, 1789-1790.

27. Zwiener, C.; Frimmel, F.H. "Oxidative treatment of pharmaceuticals in water," *Wat. Res.* **2000**, *34*(6), 1881-1885.

28. Migliore, L; Brambilla, G.; Casoria, P.; Civitareale, C.; Cozzolino, S.; Gaudio, L. "Effect of sulphadimethoxine contamination on barley (*Hordeum distichum* L., Poaceae, Liliopsida)," *Agricult. Ecosys. Environ.* **1996**, *60*(2-3), 121-128.

29. Hektoen, H.; Berge, J.A.; Hormazabal, V.; Yndestad, M. "Persistence of antibacterial agents in marine sediments," *Aquaculture* **1995** *133*(3-4), 175-184.

30. Migliore, L.; Cozzolino, S.; Fiori, M. "Phytotoxicity to and uptake of flumequine used in intensive aquaculture on the aquatic weed, *Lythrum salicaria* L.," *Chemosphere* **2000**, *40*(7), 741-750.

31. Kennedy, D.G.; Cannavan, A.; McCracken, R.J "Regulatory problems caused by contamination, a frequently overlooked cause of veterinary drug residues," *J. Chromatogr. A* **2000**, *882*(1-2), 37-52.

32. Rodgers-Gray, T.P; Jobling, S.; Morris, S.; Kelly, C.; Kirby, S.; Janbakhsh, A.; Harries, J.E.; Waldock, M.J.; Sumpter, J.P.; Tyler C.R. "Long-term temporal changes in the estrogenic composition of treated sewage effluent and its biological effects on fish," *Environ. Sci. Technol.* **2000**, *34*(8), 1521-1528.

33. National Research Council, Identifying Future Drinking Water Contaminants, Washington, D.C.: National Academy Press, 1999, 272 pp. [available: http://books.nap.edu/catalog/9595.html].

34. Eckel, W.P.; Ross, B.; Isensee, R.K. "Pentobarbital found in ground water," *Ground Wat.* **1993**, *31*(5), 801-804.

35. Blount, B.C.; Silva, M.J.; Caudill, S.P.; Needham, L.L; Pirkle, J.L; Sampson, E.J.; Lucier, G.W.; Jackson, R.J.; Brock, J.W. "Levels of seven urinary phthalate metabolites in a human reference population," *Environ. Health Perspect.* **2000**, *108*, 979-982.

36. Association of Metropolitan Sewerage Agencies (AMSA) *Evaluation of Domestic Sources of Mercury,* August 2000, AMSA Mercury Workgroup, 31 pp [available: http://www.amsa-cleanwater.org/pubs/mercury/mercury.htm].

37. Snyder, S. Kelly, K.L.; Grange, A.; Sovocool, G.W.; Synder, E.; Giesy. J.P. "Pharmaceuticals and Personal Care Products in the Waters of Lake Mead, Nevada," in Pharmaceuticals and Personal Care Products in the Environment: Scientific and Regulatory Issues; Daughton, C.G. and Jones-Lepp, T., Eds.; American Chemical Society: Washington, DC, 2001 (see chapter in this book).

38. Panter, G.H.; Thompson, R.S.; Sumpter, J.P. "Intermittent exposure of fish to estradiol," *Environ. Sci. Technol.* **2000,** *34*(13), 2756-2760.

39. Fong, P.P. "Antidepressants in Aquatic Organisms: A Wide Range of Effects,"in Pharmaceuticals and Personal Care Products in the Environment: Scientific and Regulatory Issues; Daughton, C.G. and Jones-Lepp, T., Eds.; American Chemical Society: Washington, DC, 2001 (see chapter in this book).

40. Butler D.M.; Allen K.M.; Garrett F.E.; Lauzon L.L.; Lotfizadeh A.; Koch R.A.. "Release of Ca2+ from intracellular stores and entry of extracellular Ca2+ are involved in sea squirt sperm activation," *Develop. Biol.* **1999,** *215*(2), 453-464.

41. Scholz, N.L.; Truelove, N.K.; French, B.L.; Berejikian, B.A.; Quinn, T.P.; Casillas, E.; Collier, T.K. "Diazinon disrupts antipredator and homing behaviors in chinook salmon (*Oncorhynchus tsawytscha*)," *Can. J. Fish. Aquat. Sci.* **2000,** *57*(9), 1911-1918.

42. Waring, C.P.; Moore, A.P. "Sublethal effects of a carbamate pesticide on pheromonal mediated endocrine function in mature male Atlantic salmon (*Salmo salar* L.) parr," *Fish. Physiol. Biochem.* **1997,** *17,* 203-211.

43. Dyer, S.D.; White-Hull, C.E.; Shephard, B.K. "Assessments of chemical mixtures via toxicity reference values overpredict hazard to Ohio fish communities," *Environ. Sci. Technol.* **2000,** *34*, 2518-2524.

44. Mintzer, J. "Anticholinergic side-effects of drugs in elderly people," *J. Royal Soc. Med.* **2000,** *93*, 457-462.

45. Weiss, B. "A risk assessment perspective on the neurobehavioral toxicity of endocrine disruptors," *Toxicol. Indust. Health,* **1998,** *14*(1-2), 341-359.

46. Weiss, B. "Vulnerability of children and the developing brain to neurotoxic hazards," *Environ. Health Perspect.* **2000,** *108*(suppl 3), 373-381.

47. Blok, J.; Balk, F.; van der Zande-Guinée; Okkerman, P.C.; Groshart, C.P "Selection of PTBs phase 3: Detailed evaluation of automatically selected PTB candidate substances – Report to: the Netherlands Ministry of Housing, Spatial Planning and the Environment (VROM)," Haskoning Consulting Engineers and Architects, and BKH Consulting Engineers (Netherlands), 9 July 1999 (G1260.AO/ROO3/FBA/AS), 44 pp.

48. Epel, D.; Smital, T. "Multidrug/multixenobiotic transporters and their significance with respect to environmental levels of pharmaceuticals and personal care products," "in Pharmaceuticals and Personal Care Products in

the Environment: Scientific and Regulatory Issues; Daughton, C.G. and Jones-Lepp, T., Eds.; American Chemical Society: Washington, DC, 2001 (see chapter in this book).

49. Bard, S.M. "Multixenobiotic resistance as a cellular defense mechanism in aquatic organisms," *Aquatic Toxicol.* **2000**, *48*(4), 357-389.

50. Westphal, K. et al. "Induction of P-glycoprotein by rifampin increases intestinal secretion of talinolol in human beings: A new type of drug/drug interaction," Clin. Pharmacol. Therapeut. 2000, 68(4), 345-355.

51. Snyder, M.J. "Cytochrome P450 enzymes in aquatic invertebrates: recent advances and future directions," *Aquat. Toxicol.* **2000** *48*(4), 529-547.

52. Bamdad, M.; Brousseau, P.; Denizeau, F. "Identification of a multidrug resistance-like system in *Tetrahymena pyriformis*: evidence for a new detoxication mechanism in freshwater ciliates," *FEBS Lett.* **1999**, *456*(3), 389-393.

53. U.S. EPA Science Advisory Board "Toward Integrated Environmental Decision-Making," SAB Integrated Risk Project, August 2000, EPA-SAB-EC-00-011[available:http://www.epa.gov/science1/fiscal00.htm].

54. Rodan, B.D; Pennington, D.W.; Eckley, N.; Boethling, R.S. "Screening for persistent organic pollutants: techniques to provide a scientific basis for POPS criteria in international negotiations," *Environ. Sci. Technol.* **1999**, *33*(20), 3482-3488.

55. Lee W.; Li Z.-H.; Vakulenko S.; Mobashery S. "A light-inactivated antibiotic," *J. Medicinal Chem.* **2000**, *43*(1), 128-132.

56. Mort, M. "Multiple modes of drug delivery," *Mod. Drug Discov.* **2000**, *3*(3), 30-32,34 [available: http://pubs.acs.org/hotartcl/mdd/00/apr/mort.html].

57. Kuspis, D.A.; Krenzelok, E.P. "What happens to expired medications? A survey of community medication disposal," *Vet. Human Toxicol.* **1996**, *38*(1), 48-49.

58. Hanisch, C. "Is extended producer responsibility effective?" *Environ. Sci. Technol.* **2000**, *34*(7), 170A-175A.

59. Butte, W.; Schmidt, S.; Schmidt, A. "Photochemical degradation of nitrated musk compounds," *Chemosphere* **1999**, *38*(6), 1287-129.

ACKNOWLEDGMENTS: The author thanks the following people for their valuable time in providing helpful review of both technical and policy aspects of this manuscript: Octavia Conerly (U.S. EPA Office of Water), Michael Firestone (U.S. EPA Office of Pollution, Prevention, and Toxics), Jerry Blancato (U.S. EPA National Exposure Research Laboratory), Thomas Ternes (ESWE, Germany), and ACS's anonymous reviewers, all of whom contributed to improving the quality of the manuscript.

NOTICE: The U.S. Environmental Protection Agency (EPA), through its Office of Research and Development (ORD), funded and performed the research described. This manuscript has been subjected to the EPA's peer and administrative review and has been approved for publication. Mention of trade names or commercial products does not constitute endorsement or recommendation by EPA for use.

Chapter 2

Pharmaceuticals and Metabolites as Contaminants of the Aquatic Environment

Thomas Ternes

ESWE-Institute for Water Research and Water Technology, Soehnleinstrasse 158, D-65201 Wiesbaden, Germany (Telephone: +49 611-7804343; Fax: +49 611-7804375; email: thomas.ternes@ESWE.com)

The occurrence of pharmaceuticals in the aquatic environment was mainly neglected in previous decades, although enormous quantities are used world-wide. In our laboratory, we therefore developed analytical procedures for a total of 84 drug-related analytes, enabling the simultaneous determination of polar drug residues belonging to different medicinal classes, such as lipid regulators, antibiotics, antiepileptics, anti-inflammatory agents, and estrogens in raw sewage, sewage treatment plant (STP) effluents, and river and drinking water. In a search for target analytes, 36 of 55 pharmaceuticals and 5 of 9 metabolites were quantified in at least one STP effluent. The highest concentration of drug residues were measured for the antiepileptic carbamazepine, with a maximum of 6.3 µg/L; X-ray contrast media were found in concentrations as high as 15 µg/L (iopamidol). In 40 German rivers and streams, 31 pharmaceuticals and five metabolites were quantified in at least one sample. Highest median values were detected for bezafibrate (0.35 µg/L) and carbamazepine (0.25 µg/L). Frequently, in small rivers and streams, much higher concentrations were detected than in higher-order rivers such as the Rhine or Main. Even in groundwater samples taken close to stream banks, sometimes relatively high concentrations of pharmaceuticals (e.g., up to 1.1 µg/L for carbamazepine) were detected. Hence, in the aquatic environment, drug residues are ubiquitously distributed. In distinct contrast, for drinking water only 10 of 69 target pharmaceuticals were found - always in the lower ng/L-range and in very few samples.

Introduction

While tons of individual pharmaceuticals have been used yearly in various countries over the last few decades, few investigations have been published about the assessment of their environmental relevance – a result of their inadvertent discharge from sewage treatment and land run-off. A survey of the current knowledge about exposure, effects, and environmental relevance is summarized in the reviews of Halling-Sørensen et al. (1), Daughton and Ternes (2), Jørgensen and Halling-Sørensen (3), and Ternes and Wilken (4). In Germany for instance, up to 100 t of individual drugs are prescribed every year. This amount underestimates the total usage of all drugs, many of which can be purchased without a pharmacy prescription and others procured illegally. Table I gives a rough estimate of the yearly tonnage of human medicines sold in Germany – both by prescription and "over the counter" (OTC). For example, approximately 100 t of ibuprofen was prescribed in 1997, with the remaining 80 t being OTC.

Table I: Estimation of quantities of selected pharmaceuticals sold for use in human medicine in Germany (5, 6)

Pharmaceuticals in human medicine	1997 in tons
Acetylsalicylic acid	> 500
Ibuprofen	180
Iopromide	130
Carbamazepine	80
Diclofenac	75
Sulfamethoxazole	60
Metoprolol	52
Bezafibrate	45
17α-Ethinylestradiol	0.050

Due to such high usage levels, detectable concentrations of drugs and their metabolites should not be unexpected in sewage. The overall concentrations, however, depend on each drug's pharmacokinetic behavior (half life, urinary/fecal excretion, metabolism, etc.). Many pharmaceuticals are excreted mainly as metabolites, and thus in addition to the non-metabolized compounds, principal metabolites have to be analyzed. However, metabolites formed by conjugation (e.g., with glucuronic acid or sulfate) within phase II reactions are likely cleaved during sewage treatment to yield the non-metabolized (free) pharmaceuticals, and hence may increase the relevant environmental concentrations. In diluted aerobic batch experiments, this phenomenon can be illustrated by the cleavage of two glucuronides of 17β-estradiol (17β-estradiol-17-glucuronide and 17β-estradiol-3-glucuronide) and of 4-methyliumbelliferyl-β-D-glucuronide (*10*).

Figure 1 gives two examples for phase I and phase II metabolism. Clofibric acid, which is the principal and active metabolite of three lipid regulators (clofibrate, etofyllin clofibrate [theofibrate], and etofibrate) is conjugated to clofibric-O-β-acylglucuronide (*11, 12*). More than 90% of the clofibric acid, administered in form of the pro-drugs clofibrate, etofyllin clofibrate, and etofibrate, is excreted as glucuronide. Another example is ibuprofen, which is first hydroxylated and then conjugated to ibuprofen-O-β-hydroxyglucuronide. Approximately 9% of ibuprofen is excreted as hydroxy-ibuprofen, 17% as the glucuronide of hydroxy-ibuprofen, and the remaining percentage is allocated to further metabolites and the non-metabolized ibuprofen (*11-13*).

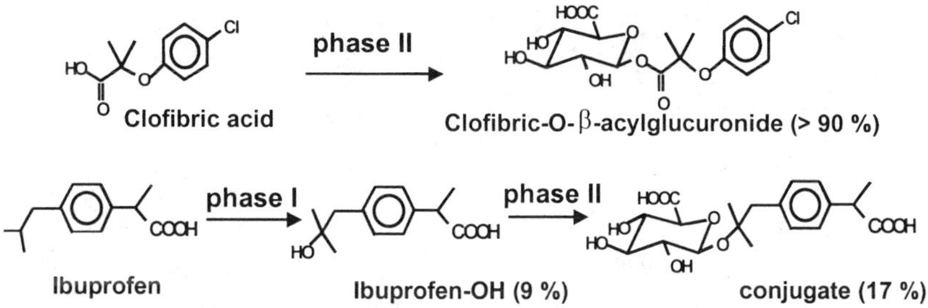

Figure 1. Metabolites of clofibric acid and ibuprofen excreted by humans

A prerequisite for determining the contamination of environmental matrices with metabolites is the availability of reference compounds. The availability of metabolite reference standards, however, is often problematic because most are not commercially available. Drug manufacturer research divisions can be another source from which to request reference compounds. When all attempts fail to procure metabolites, they must be synthesized, often at great time and expense.

The fates of veterinary and human drugs after urinary or fecal excretion are quite different. In general, excreted human pharmaceuticals pass through STPs prior to entering rivers or streams, whereas veterinary drugs are more likely to directly contaminate soil and groundwater (without previous sewage treatment) when liquid manure is used for top soil dressing (Fig. 2). After rainfall, surface waters can be polluted with human or veterinary drugs by run-off from fields (esp. agricultural) treated with digested sludge or livestock slurries; groundwater can also be contaminated.

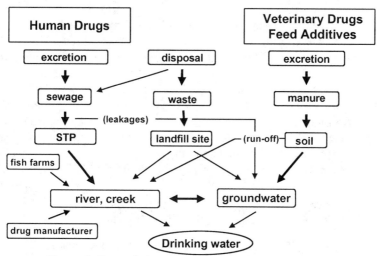

Figure 2. Fate of pharmaceuticals in the environment

The use of pharmaceuticals, especially antibiotics, in fish farms is a unique source for contamination of rivers and lakes, which receive the effluents of those ponds. Discharge of sewage from pharmaceutical manufacturing into surface waters is another point source (*8*). Transport of drugs via bank filtration from highly contaminated surface water into groundwater is also a possibility, as is the infiltration of sewage directly from sewer drain leakages. Drugs disposed together with domestic refuse can reach landfill sites, which can lead to groundwater contamination by leaching (*9*). The use/disposal of raw sewage or STP effluents by spray and broad irrigation on agricultural areas is yet another route of introduction.

Analytical Methods

In our laboratory, analytical procedures were elaborated for a total of 84 analytes, enabling the simultaneous determination of polar drug residues belonging to different medicinal classes such as lipid regulators, antibiotics, and estrogens, in sewage as well as drinking water (Fig. 3). Five methods were used for determining drugs and their metabolites in the lower ng/L-range. These incorporated solid phase extraction (SPE), derivatization, and detection by GC/MS and GC/MS/MS or LC-electrospray/MS/MS.

The first multi-residue method determines betablockers and β_2-sympathomimetics, as well as neutral drugs like diazepam or carbamazepine, at concentrations in the ng/L-range after solid phase extraction (*14*). A second multi-residue method allows the determination of acidic drugs possessing a carboxylic moiety, and additionally in some cases one or two hydroxy groups, down to the lower ng/L-range (*15*). Lipid regulators, antiphlogistics (anti-inflammatory agents), and three metabolites with carboxylic and hydroxy groups are included. In the third method, 18 antibiotics can be analyzed simultaneously after lypholization and LC-electrospray tandem MS detection (*16*). With the fourth method, X-ray contrast media were determined after SPE and LC-electrospray tandem MS detection down to 10 ng/L (*17*). Finally, contraceptives and natural estrogens were detected by GC/MS/MS after SPE extraction, silica gel clean-up, and silylation (*18*).

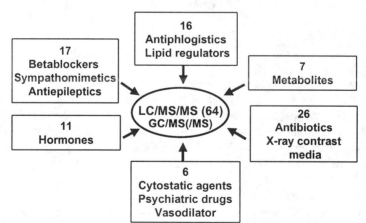

Figure 3. Number of analytes from different medicinal classes that can be determined in the lower ng/L-range with the analytical methods developed

Exposure of Sewage and the Aquatic Environment to Drug Residues

The very first reports of drug residues in STP effluents and the environment were mainly focused on clofibric acid, the active metabolite of three lipid regulators. Garrison et al. (19) and Hignite and Azarnoff (20) detected clofibric acid in the lower µg/L-range in treated sewage in the United States. Waggott (21) determined clofibric acid in the river Lee (Great Britain) at concentrations below 0.01 µg/L, and in Spain clofibric acid was even detected in groundwater samples by Galceran et al. (22). Stan et al. (23) and Heberer and Stan (24) identified clofibric acid up to 0.27 µg/L in Berlin tap water. On Iona Island (Vancouver/Canada), the two antiphlogistics, ibuprofen and naproxen have been identified in sewage by Rogers et al. (25). More systematic studies on the fate of drugs after excretion were reported by Van der Heide and Van der Pals (26) and Richardson and Bowron (8).

Recently, a multitude of pharmaceuticals and metabolites from distinctive medicinal classes such as anti-inflammatory agents, betablockers, β_2-sympathomimetics, antiepileptics, antibiotics, X-ray contrast media, lipid regulators, contraceptives, and antineoplastics were identified in sewage and rivers. Two recent reviews, Halling-Sørensen et al. (1) and Daughton and Ternes (2) summarize the literature in this new emerging field about the environmental relevance of pharmaceuticals.

Contamination of sewage with pharmaceuticals

Composite samples from a German municipal STP close to Frankfurt/Main were taken daily from the raw influent and the corresponding final effluent over a period of 6 days (7). In the influent, the concentration levels ranged up to 26 µg/L for acetaminophen and in the effluent up to 2 µg/L for carbamazepine and metoprolol. The elimination rates of the investigated drugs during passage through the STP ranged from negligible within in the standard deviation (appr. 7% for carbamazepine) to greater than 99% (salicylic acid) in the monitoring program presented in Table II. Generally more than 60% of the drug residues detected in the influent were removed. Only carbamazepine, clofibric acid, phenazone (antipyrine), and dimethylaminophenazone (aminopyrine) showed lower average removal rates. The antiepileptic carbamazepine was not significantly removed, thus nearly the same loads entering the STP were discharged into the receiving water. However, for other periods in the same STP much lower removal efficiencies for pharmaceuticals were observed, presumably influenced mainly by conditions of microbial activity in the

Table II. **Mean concentration and loads of selected pharmaceuticals in the effluent of a municipal STP close to Frankfurt/Main (investigation of corresponding 6 daily composite raw influent and 6 daily composite final effluents)**

Substance	Mean conc. influent in μg/L	Mean conc. effluent in μg/L	Load influent in g/d	Load effluent in g/d	Removal in %
Mean flow rate from 28 Jan. – 02 Feb. 1997: 55320 m³/d					
Acetylsalicylic acid	3.2 ± 1.2	0.62 ± 0.30	178 ± 66	34 ± 17	81 ± 12
Diclofenac	1.9 ± 0.2	0.58 ± 0.03	102 ± 11	32 ± 2	69 ± 4
Ibuprofen	4.4 ± 0.5	0.45 ± 0.13	241 ± 27	25 ± 7	90 ± 3
Naproxen	1.3 ± 0.1	0.45 ± 0.08	73 ± 8	25 ± 4	66 ± 7
Acetaminophen	26 ± 6	<LOQ[a]	1450 ± 320	<15	>99
Indometacine	0.29 ± 0.07	0.09 ± 0.02	16 ± 4	4 ± 1	83 ± 7
Bezafibrate	5.6 ± 1.6	0.96 ± 0.24	310 ± 88	53 ± 13	75 ± 9
Gemfibrozil	0.94 ± 0.11	0.29 ± 0.06	52 ± 6	16 ± 3	69 ± 7
Fenofibric acid	1.1 ± 0.2	0.38 ± 0.06	59 ± 9	21 ± 3	64 ± 8
Clofibric acid	1.2 ± 0.2	0.60 ± 0.09	69 ± 10	33 ± 5	51 ± 10
Carbamazepine	2.2 ± 0.4	2.0 ± 0.2	122 ± 20	114 ± 12	Non (7 ± 9)
Phenazone	0.25 ± 0.06	0.17 ± 0.03	14 ± 3	9 ± 2	33 ± 15
Mean flow rate from 24. March-30. March 1997: 63190 m³/d					
Dimethylamino-phenazone[b]	0.96	0.6	61	38	38
Mean flow rate from 24.-30 June 1996: 54300 m³/d					
Propranolol	10.0 ± 2.4	0.41 ± 0.07	520 ± 103	21 ± 5	95 ± 1
Metoprolol	7.2 ± 3.4	2.08 ± 0.36	374 ± 159	108 ± 22	67 ± 11
Bisoprolol	0.41 ± 0.08	0.14 ± 0.02	20 ± 3	5 ± 2	65 ± 9
Betaxolol	0.87 ± 0.15	0.17 ± 0.03	46 ± 6	9 ± 2	80 ± 5
Terbutalin	0.37 ± 0.11	0.11 ± 0.01	19 ± 6	6 ± 1	67 ± 18
Salbutamol	0.21 ± 0.13	n.d.[c]	9 ± 6	n.d.	>90
Carazolol	0.19 ± 0.03	0.07 ± 0.02	10 ± 1	3 ± 1	66 ± 8

[a] LOQ: Limit of quantification
[b] has only been detected in two influent and two corresponding effluent samples
[c] n.d.: not detectable

biological processes. For example, a major rainfall incident reduced the removal rates for some pharmaceuticals by more than 50% (7).

In a search for target analyses, 36 of 55 pharmaceuticals and 5 of 9 metabolites were quantified in at least one STP effluent. The highest concentrations of drug residues were measured for the *anti-epileptic* carbamazepine, with a maximum of 6.3 μg/L. *X-ray contrast media* were found in concentrations as high as 15 μg/L (iopamidol) and 11 μg/L (iopromide) (27). In addition to X-ray contrast media and carbamazepine, the *lipid regulators* bezafibrate and gemfibrozil, the *anti-inflammatory agents* diclofenac, ibuprofen, indometacine, naproxen, ketoprofen, and phenazone, and the *betablockers* metoprolol, sotalol, and propranolol were present with elevated concentrations in the majority of the 49 German municipal STP effluents investigated.

Metabolites of ibuprofen

Drug metabolites have high relevance as environmental contaminants since they are known to be the main excretion products. Figure 4 shows a GC/MS-total ion chromatogram of a raw sewage sample that has been analyzed after SPE (RP-C$_{18}$) followed by derivatization with diazomethane. Pharmaceutical residues are clearly the predominant peaks in the chromatogram, and certain metabolites are clearly of major significance. The determination of the two main metabolites of ibuprofen, 2-[4-(2-hydroxy-2-methylpropyl)phenyl]propionic acid (hydroxy-ibuprofen) and 2-[4-(2-carboxypropyl)phenyl]propionic acid (carboxy-ibuprofen) in municipal sewage, STP effluents, and even river water underlined the occurrence and relevance of metabolites in the environment (28). Since ibuprofen metabolites, just as with most of the other metabolites, could not be purchased commercially, they were extracted from the urine of a volunteer by solid phase extraction and semipreparative HPLC. Up to 6.7 μg/L of hydroxy-ibuprofen were quantified in raw sewage. While ibuprofen was eliminated nearly quantitatively during passage through a German STP, less than 20% of hydroxy-ibuprofen was removed. In 12 investigated German rivers, a median value of 0.34 μg/L hydroxy-ibuprofen was determined. The concentrations of the hydroxy metabolite were significantly higher in all water samples investigated than those of the parent drug ibuprofen.

Contamination of sewage with natural estrogens and contraceptives

In the German municipal STP close to Frankfurt/Main, the raw sewage was contaminated by 17β-estradiol and estrone, with average concentrations of 0.015 μg/L and 0.027 μg/L, respectively, yielding loads up to 1 g/d (18). The calculated removal rates were much lower than those obtained in the Brazilian STP (18). For instance, the loads of estrone and 17α-ethinylestradiol were not

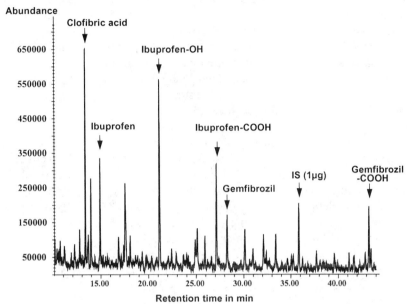

Figure 4. GC/MS total ion chromatogram of raw sewage for acidic drugs

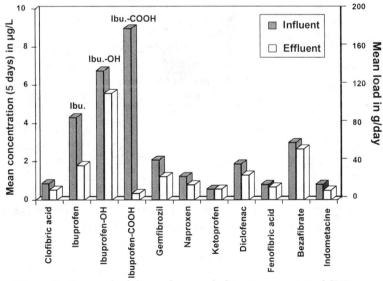

Figure 5. Removal of Ibuprofen metabolites in a municipal STP

appreciably reduced while passing through the German STP. Considering the standard deviation, no elimination rate could be observed. However, 16α-hydroxyestrone and 17β-estradiol were removed, with reduction in concentrations of 68% and about 64%, respectively. Because the efficiency of an STP for the elimination of drugs is influenced by several parameters such as microbial activity or rain events (7), only a long-term study can reveal whether the differences result from increased temperatures or whether other contributory factors are responsible. However, 17β-estradiol and 16α-hydroxyestrone were basically eliminated with a higher efficiency than 17α-ethinylestradiol and estrone.

In STP effluents, primarily the natural estrogens estrone, 17β-estradiol, 16α-hydroxyestrone, as well as the synthetically altered contraceptive 17α-ethinylestradiol could be found - in the lower ng/L-range (18). Mestranol (17α-ethinylestradiol-3-methyl ether) was only present in three samples and 17β-estradiol-17-valerate could not be detected at all (Table III). Estrone was determined with a maximum concentration of 0.070 μg/L and a median value of 0.001 μg/L. Additionally, median values could be calculated for 16α-hydroxyestrone as 0.001 μg/L. The contamination of STP effluents are within the concentration ranges found by Aherne and Briggs (29), Wegener et al. (30), Belfroid et al. (31), Desbrow et al. (32), and Routledge et al. (33).

Contamination of rivers and streams

Recent investigations reveal that more than 30 different pharmaceuticals belonging to nearly all important German medicinal classes could be found up to the μg/L-range in rivers and streams. In 40 German rivers and streams, a total of 31 pharmaceuticals and five metabolites were quantified. Highest median values were detected for bezafibrate (0.35 μg/L) and carbamazepine (0.25 μg/L). For instance, maximum concentrations were determined in German rivers up to 1.3 μg/L for carbamazepine (antiepileptic) and 1.2 μg/L for diclofenac (antiphlogistic). Frequently, in small rivers and streams, much higher concentrations were detected than in larger-order rivers like the Rhine or Main. Sometimes combined concentrations over 6 μg/L were quantified for the pharmaceuticals occurring in one sample (Fig. 6). These streams, located in the Hessian Ried, contain a high proportion of STP effluents, as indicated by their relatively high boron concentrations (ranging from 0.25 to 1.0 mg/L). Hence, the higher the proportion of treated sewage in the receiving water, the greater the potential for contamination by pharmaceutical residues (7).

Contamination of groundwater

In German groundwater samples taken close to stream banks in the Hessian Ried close to Frankfurt/Main, relatively high concentrations of pharmaceuticals

Table III: Concentrations of natural estrogens and contraceptives in German municipal STP effluents.

Substance	LOQ in µg/L	number of STPs	n>LOQ	median in µg/L	90-percentile in µg/L	maximum in µg/L
Estrone	0.001	38	20	0.001	0.021	0.070
17β-Estradiol	0.001	38	13	<LOQ	0.002	0.003
17β-Estradiol-17-valerate	0.004	38	0	<LOQ	<LOQ	<LOQ
17α-Ethinyl-estradiol	0.001	38	9	<LOQ	0.001	0.015
Mestranol	0.001	38	3	<LOQ	<LOQ	0.004
16α-Hydroxy-estrone	0.001	15	11	0.001	0.004	0.005

LOQ: Limit of quantification

(up to 1.1 µg/L) and X-ray contrast media (up to 2.4 µg/L) were sometimes detected. About 233 groundwater samples were analyzed for potential contamination (*34*). In 1% of the groundwater samples, mainly located close to contaminated small rivers or streams, the concentrations exceeded 1 µg/L, and 15% exhibited concentration levels between the limit of quantitation (LOQ) and 0.1 µg/L (Fig. 7). Irrigation with treated sewage was found to lead to groundwater contamination by diclofenac up to 3.5 µg/L indicating that indirect reuse of treated sewage may be responsible for appreciable pharmaceutical contamination of groundwater (Table IV). In a groundwater sample close to a domestic landfill site (another pathway for inadvertent environmental release) clofibric acid was detected as high as 11 µg/L and phenazon as 1 µg/L, indicating the contamination of an aquifer by leaking waste disposition sites.

Table IV. Pharmaceuticals in German groundwater (n = 233)

	n > LOQ	Maximum concentration in µg/L
Carbamazepine	22	1.1
Clofibric acid	79	11[a]
Diclofenac	30	0.93/(3.5)[b]
Bezafibrate	25	0.19

[a] groundwater influenced by a landfill site, [b] influenced by STP effluent irrigation

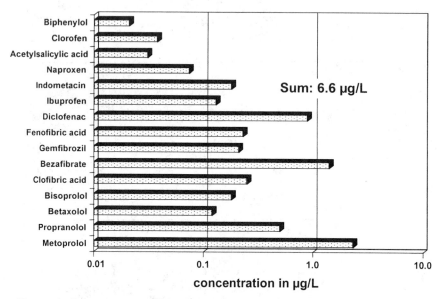

Figure 6: *Contamination of the Weschnitz, a small river in the Hessian Ried, Germany with pharmaceuticals and phenolic antiseptics (biphenylol = phenylphenol isomer; clorofen = chlorophene)*

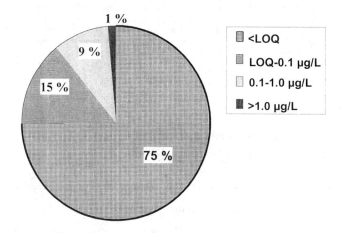

Figure 7. Concentration levels of selected pharmaceuticals found in German groundwater (Carbamazepine, Diclofenac, Clofibric acid, Bezafibrate)

Exposure of drinking water

In general, the target analytes could not be identified in drinking water. In a few cases, pharmaceuticals were found in drinking water but always due to the use of polluted surface water or groundwater contaminated with medicinal compounds by bank filtration. In drinking water, 10 of 69 target pharmaceuticals were found (Table V), and then, always in the lower ng/L-range (< 0.1 µg/L). Mainly clofibric acid and iodinated contrast media were detected above the LOQ.

Flocculation does not appreciably remove the selected pharmaceuticals (*35*). Applying ozone concentrations in the range between 1.0 and 2.0 mg/L, carbamazepine was totally eliminated, and diclofenac, bezafibrate, and clofibric acid were predominantly eliminated. Therefore, ozonation is a powerful technique for the removal of pharmaceuticals in water treatment. Additionally, in a waterworks the efficiency of granulated activated carbon (GAC) filtration for the removal of carbamazepine, diclofenac, and bezafibrate could be observed (*35*).

Conclusion

Pharmaceutical residues are widely distributed in the aquatic environment due to human use/consumption coupled with inadvertent discharge. Presumably in all countries with a developed medical care system, pharmaceuticals could be expected to be present in measurable (sometimes appreciable) quantities in sewage, STP effluents, and in receiving waters and groundwater. In addition to the original administered, unaltered parent compounds, a multitude of polar metabolites also can be expected in the environment. Since their removal is often incomplete in sewage treatment processes, one main route of drug residues into the environment are municipal STP effluents. Another important source is landfill sites, leading to groundwater contamination by leakages. Presumably, many additional drug residues are present in the environment, because many more compounds than those investigated in the current studies are approved in human and veterinary medicine by the regulatory bodies of Europe and nations worldwide. Hence, the elucidation of those pharmaceutical residues that may prove harmful to the environment will be a challenge for environmental scientists in the future. For meaningful risk assessment followed by appropriate risk management, exposure data and tailor-made toxicological test systems for pharmaceuticals in the environment are essential.

Table V: Selected pharmaceuticals in drinking water of different facilities in Germany

Drugs	LOQ in µg/L	Number of samples with conc. > LOQ	Number of samples with conc. > 0.010 µg/L	Median in µg/L	90- Percentile in µg/L	Maximum in µg/L
Antiphlogistics						
Diclofenac	0.001	8 of 30	0	<LOQ	0.002	0.006
Ibuprofen	0.001	3 of 30	0	<LOQ	0.001	0.003
Phenazone	0.010	1 of 12	1	<LOQ	<LOQ	0.050
Lipid regulators						
Clofibric acid	0.001	16 of 30	6	0.001	0.024	0.070
Fenofibric acid	0.005	1 of 30	1	<LOQ	<LOQ	0.042
Bezafibrate	0.025	1 of 30	1	<LOQ	<LOQ	0.027
Contrast media						
Iopamidol	0.010	4 of 10	4	<LOQ	0.070	0.079
Diatrizoat	0.010	5 of 10	5	0.021	0.075	0.085
Iopromid	0.010	1 of 10	1	<LOQ	<LOQ	0.086
Antiepileptics						
Carbamazepine	0.010	1 of 12	1	<LOQ	<LOQ	0.030

LOQ: Limit of quantification

References

1. Halling-Sørensen, B.; Nielsen, S. N.; Lanzky, P. F.; Ingerslev, F., Holten-Lützhøft, H. C.; Jørgensen S.E. *Chemosphere* **1998**, *36*, 357-393.
2. Daughton, C.; Ternes, T. A. Environ. Health Perspect. **1999**, *107* (suppl. 6), 907-938.
3. Jørgensen, S. E.; Halling-Sørensen, B. Drugs in the Environment. Ed.: Hutzinger O. **2000**, *40*, 691-793.
4. Ternes, T. A.; Wilken, R-D. (Eds.) Drugs and Hormones as Pollutants of the Aquatic Environment: Determination and Ecotoxicological Impacts **1998** *Sci. Total Environ.* 225, pp 176.
5. Schwabe, U.; Paffrath, D. Arzneiverordnungsreport '97 Aktuelle Daten, Kosten, Trends und Kommentare; Gustav Fischer Verlag, Stuttgart, Jena, 1998.
6. Steger-Hartmann, T.; Personal communication, Schering/Berlin, 1999.
7. Ternes, T. A. *Water Res. 1998*, *32*, 3245-3260.
8. Richardson, M. L.; Bowron, J. M. *J. Pharm. Pharmacol.* **1985**, *37*, 1-12.

9. Holm, J. H.; Rügge, K.; Bjerg, P. L.; Christensen, T. H. *Environ. Sci. Technol.* **1995**, *29*, 1415-1420.

10. Ternes, T. A.; Kreckel, P.; Müller, J. *Sci. Total Environ.* **1999**, *225*, 91-99.

11. Mutschler, E. Arzneimittelwirkungen - Lehrbuch der Pharmakologie und Toxikologie, Wissenschaftliche Verlagsgesellschaft mbH, Stuttgart, 1996.

12. Forth, W.; Henschler, D.; Rummel, W.; Starke, K. Allgemeine und spezielle Pharmakologie und Toxikologie. Wissenschaftsverlag Mannheim/Leipzig/Wien/Zürich, 7. Auflage, 1996.

13. Balfour, J. A.; McTarish, D., Heel, R. C. *Drugs* **1990**, *40*, 260-290.

14. Ternes, T. A.; Hirsch, R.; Müller, J.; Haberer, K. *Fres. J. Anal. Chem.* **1998**, *362*, 329-340.

15. Ternes, T. A.; Stumpf, M.; Schuppert, B.; Haberer, K. *Vom Wasser* **1998**, *90*, 295-309.

16. Hirsch, R.; Ternes, T. A.; Mehlig, A.; Ballwanz, F.; Kratz, K.-L.; Haberer, K. *J. Chromatogr.* **1998**, *815*, 213-223.

17. Hirsch, R.; Ternes, T. A.; Lindart, A.; Haberer, K.; Wilken, R.-D. *Fres. J. Anal. Chem.* 2000; 366, 835-841.

18. Ternes, T. A., Stumpf, M; Müller, J; Haberer, K; Wilken, R.-D.; Servos, M. *Sci. Total Environ.* **1999**, *225*, 81-90.

19. Garrison, A. W., Pope, J. D.; Allen, F. R. GC/MS Analysis of organic compounds in domestic wastewaters. In: Identification and Analysis of Organic Pollutants in Water. Eds. Keith C.H., Ann Arbor Science, Ann Arbor. 1976; 517-566.

20. Hignite, C.; Azarnoff D. L. *Life Sci.* **1977**, *20*, 337-41.

21. Waggott, A. Trace organic substances in the River Lee [Great Britain]. Chem. Water Reuse, **1981**, *2*, 55-99. Eds.: Cooper, William; J. Ann Arbor Sci.: Ann Arbor, Mich.

22. Galceran, M. T.; Rubio, R.; Rauret, G. *Water Supply* **1989**, *7*, 1-9.

23. Stan, H.-J.; Heberer, T.; Linkerhägner, M. *Vom Wasser* **1994**, *83*, 57-68.

24. Heberer, T.; Stan, H.-J. *Vom Wasser* **1996**, *86*, 19-31.

25. Rogers, I. H.; Birtwell, I. K.; Kruzynski, G. M. *Water Pollut. Res. J. Can.* **1986**, *21*, 187-204.

26. Van der Heide, E. F.; Hueck-Van der Plas, E. H. *Pharmaceutisch Weekblad* **1984**, *119*, 936-947.

27. Ternes, T. A.; Hirsch, R. W. Occurrence and behavior of iodinated X-ray contrast media in the aquatic environment. *Environ. Sci. Technol.*, **2000**; *34*, 2741-2748.

28. Stumpf, M.; Ternes, T. A.; Haberer, K.; Baumann, W. *Vom Wasser* **1998**, *91*, 291-303.

29. Aherne, G. W.; Briggs, R. *J. Pharm. Pharmacol.* **1989**, *41*, 735-736.

30. Wegener, G.; Persin, J.; Karrenbrock, F.; Rörden, O.; Hübner I. *Vom Wasser* **1999**, *92*, 347-360.

31. Belfroid, A. C.; Van der Horst, A.; Vethaak, A. D.; Schäfer, A.J.; Rijs, G. B. J.; Wegener, J. Cofino, W. P. *Sci. Total Environ.* **1999**, *225*, 101-108.

32. Desbrow, C.; Routledge, E.J.; Brighty, G.C.; Sumpter, J.P.; Waldock, M. *Environ. Sci. Technol.* **1998**, *32*,1549-1558.
33. Routledge, E. J.; Sheahan, D.; Desbrow, C.; Brighty, G. C.; Waldock, M.; Sumpter, J. P. *Environ. Sci. Technol.* **1998**, *32*, 1559-1565.
34. Ternes, T. A.; Berthold, G.; Seel, P.; Rückert, H.; Toussaint, B. Contamination of groundwater by pharmaceuticals. (unpublished data) in preparation for *Chemosphere*.
35. Ternes, T. A.; Meisenheimer M.; Wilken, R.-D.; Sacher, F.; Preuss, G.; Wilme, U.; Zulei-Seibert, N.; Brauch, H.-J.; Haist-Gulde, B. Behavior of selected pharmaceuticals during drinking water treatment. (unpublished data) in preparation for *Environ. Sci. Technol.*

Occurrence in the Environment

Chapter 3

Occurrence and Fate of Fluoroquinolone, Macrolide, and Sulfonamide Antibiotics during Wastewater Treatment and in Ambient Waters in Switzerland

Alfredo C. Alder, Christa S. McArdell, Eva M. Golet, Slavica Ibric, Eva Molnar, Norriel S. Nipales, and Walter Giger

Swiss Federal Institute for Environmental Science and Technology (EAWAG) and Swiss Federal Institute of Technology (ETH), Ueberlandstrasse 133, CH-8600 Dübendorf, Switzerland

The leading fluoroquinolone antibiotic in human medical use, ciprofloxacin, was measured in inflow and outflow of a municipal wastewater treatment plant (WWTP). Samples composited over 24 hours were collected and analyzed by HPLC/fluorescence. Ciprofloxacin concentrations in the inflow and outflow ranged from 220-370 and 60-75 ng/L, respectively. These results show that only 70-80% of ciprofloxacin is eliminated in this WWTP and that residual amounts reach the receiving surface waters. In addition, selected macrolide and sulfonamide antibiotics were detected by LC/MS in similar concentrations in the outflows of different WWTPs. Occurrences of antibiotics in ambient waters indicate different exposure routes based on the various uses of these drugs in human and veterinary medicine.

The occurrence of pharmaceutical chemicals in the aquatic environment has become a topic of substantial scientific and public concern (*1-4*). Pharmaceuticals, which are incompletely eliminated in wastewater treatment, may enter the aquatic environment. Little is known on the risk of low levels of drugs in the environment because pharmaceuticals are not properly addressed by current environmental risk assessment methodologies. Antibiotics are applied in human medicine and for veterinary purposes with different exposure routes for entering the aquatic environment, i.e., through municipal wastewaters and soil run-off (Fig. 1). Studies on the fate of antibiotics in wastewaters and surface waters are motivated by the question of whether trace concentrations in the environment may pose a risk for aquatic organisms and microbial populations. For an environmental risk assessment, it also must be evaluated whether the occurrence of antibiotics in the environment contributes to the increasing resistance of microorganisms towards antibiotics.

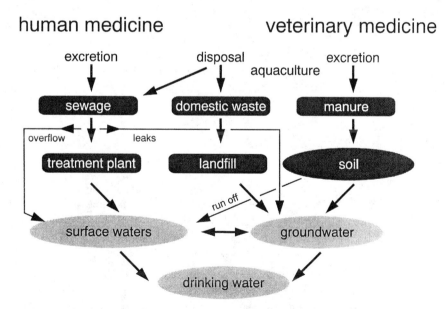

Figure 1. Different exposure routes of human and veterinary pharmaceuticals to the aquatic environment (adapted from (1)).

In Switzerland, the annual consumption of antibiotics in 1997 was about 80 tons (Fig. 2, 5), of which about 35% is used in human medicine and up to 65% by veterinarians. In human medicine, 20% of the antibiotics are applied in hospitals; the rest is domestic consumption. The annual consumption of antibiotics in human medicine remained roughly constant in the past few years.

Between 1992 and 1997, feed additives were reduced from 91 to 36 tons. This correlates with a reduction of the total consumption of antibiotics in these years from 125 to 80 tons. Within the same time frame the consumption of veterinary antibiotics for therapeutic purposes has doubled. Therefore, a shift from feed additives to therapeutic use has occurred. The application of antibiotics as growth promoters in feed additives has been forbidden in Switzerland since January 1999, which again creates a drastic change in the sales figures. Details on the use of antibiotics in veterinary medicine are difficult to obtain.

In human medicine, the most highly consumed antibiotics belong to the compound classes of penicillins, ampicillins, and cephalosporins (i.e., β-lactam antibiotics), macrolides and fluoroquinolones (Fig. 3). The tetracyclines are an additional important group of antibiotics. In veterinary medicine, the major antibiotic classes are again the β-lactam antibiotics as well as chloramphenicol. Sulfonamides, which are also used in considerable amounts, were not included in these statistics (5).

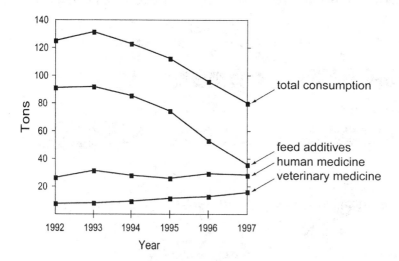

Figure 2. Annual consumption of antibiotics in Switzerland from 1992 to 1997 (5).

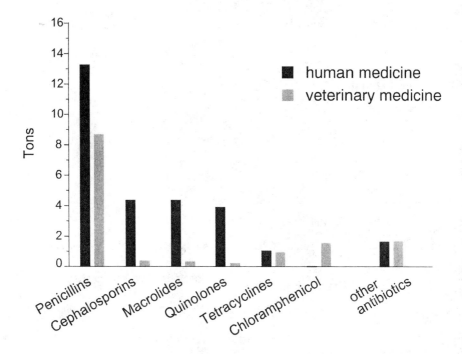

Figure 3. Annual consumption of different types of antibiotics in human and veterinary medicine in Switzerland in 1997 (5).

Our study emphasizes antibiotics belonging to the fluoroquinolone, macrolide, and sulfonamide groups. The chemical structures of the main antibiotics investigated are given in Figure 4. These antibiotics are widely used in human or veterinary medicine. Although β-lactam antibiotics are prescribed in the highest quantities, they are not very likely to occur in the aquatic environment because of their chemically unstable β-lactam ring, which is readily susceptible to hydrolytic cleavage (6). Therefore, β-lactam antibiotics were not included in our study.

Fluoroquinolone (FQ) antibiotics inhibit the bacterial DNA-unwinding enzyme, gyrase. These totally synthetic chemicals are active against many gram-negative and gram-positive pathogenic bacterial species (7). They are licensed as important broad-spectrum antibiotics for the use in human medicine (e.g., ciprofloxacin, norfloxacin, ofloxacin, levofloxacin) and veterinary applications (e.g., enrofloxacin, danofloxacin).

ciprofloxacin

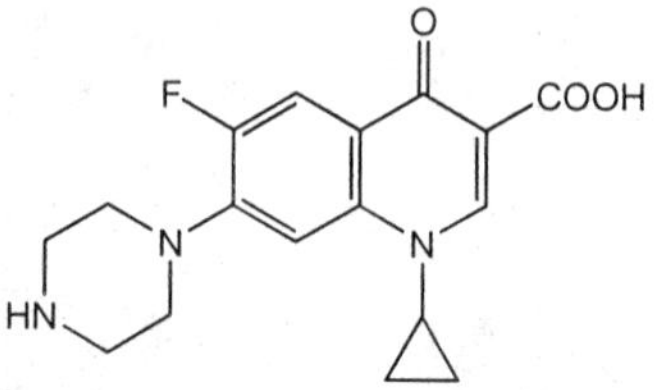

erythromycin
    R_1= O, R_2= H
clarithromycin
    R_1= O, R_2= CH_3
roxithromycin
    R_1= $NOCH_2O(CH_2)_2OCH_3$,
    R_2= H

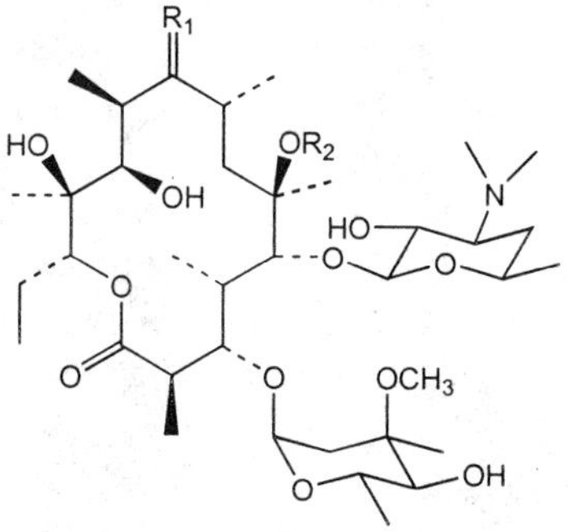

sulfamethoxazole R=

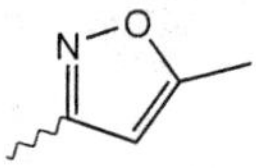

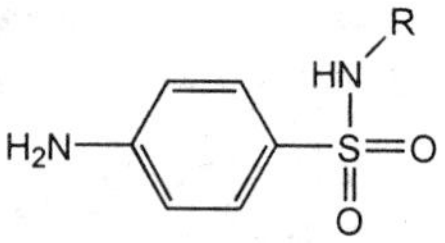

sulfamethazine   R=

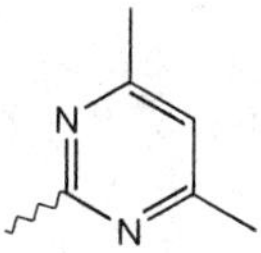

trimethoprim

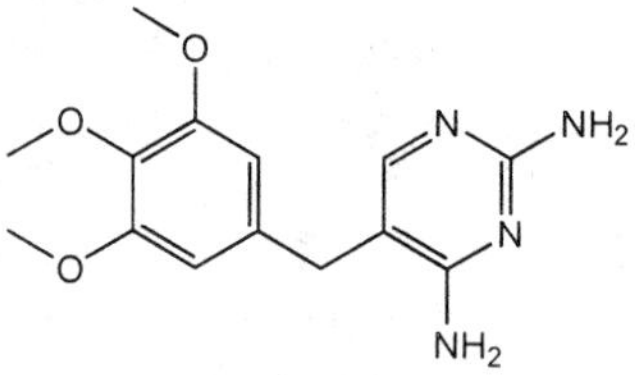

chloramphenicol

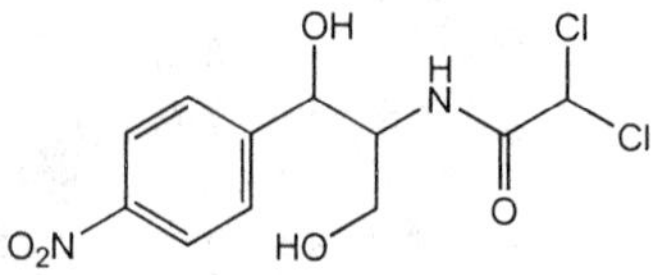

*Figure 4: Chemical structures and names of the main fluoroquinolone,
macrolide, and sulfonamide antibiotics investigated in this study.*

Macrolides are mainly active against gram-positive pathogenic bacterial species, and act as inhibitors of protein synthesis. They are the major alternative for patients who are allergic to β-lactam antibiotics.

Sulfonamide antibiotics are active against many gram-negative and gram-positive pathogenic bacterial species. They interfere in folic acid synthesis by competitive antagonism of p-aminobenzoic acid, resulting in a bacteriostatic effect on growing organisms. Nowadays, sulfonamides are mainly used in combination with a dihydrofolate reductase inhibitor such as trimethoprim. Sulfonamide antibiotics are applied in high quantities in human medicine (mainly sulfamethoxazole) and in veterinary medicine (mainly sulfamethazine). Chloramphenicol is a broad-spectrum antibiotic used in human medicine only in exceptional cases like meningitis. The veterinary use of this substance is forbidden in the European Union since 1995. It is also under discussion to be banned from the Swiss market, since it is believed to occasionally cause fatal aplastic anemia in humans due to its bone marrow damaging properties (8).

In preceding investigations, concentrations of the predominant FQ in human medicinal use, ciprofloxacin, were determined in hospital wastewaters by HPLC with fluorescence detection (9) and qualitatively confirmed by electrospray mass spectrometry (10). Concentrations ranged from 5 to 90 µg/L.

Here we report on field studies, which were performed in order to follow the input of several representative antibiotics of different groups into surface waters. Inflow and outflow concentrations of fluoroquinolone antibiotics were measured in municipal treatment plants in order to determine elimination rates. Daily variations of inflow and outflow were studied. For macrolide and sulfonamide antibiotics, first the results on measurements in WWTP outflows and in surface waters are presented. The focus of this investigation is to determine the occurrence of selected drug residues and to evaluate the different exposure routes of veterinary and human pharmaceuticals to the aquatic environment.

Experimental

Fluoroquinolones

Sampling

Samples were taken from the inflow and outflow of the municipal WWTP of Zurich-Werdhölzli. In this context, inflow means the effluent of the primary clarifier and outflow refers to the effluent of the WWTP after the secondary clarifier and flocculation filtration. In Switzerland, effluents of WWTPs are never disinfected. The WWTP of Zurich-Werdhölzli serves approximately a population of 450,000 inhabitants (wastewater discharge = 170,000 m^3/d). The wastewater has a residence time of 16-18 h (activated sludge system incl.

secondary clarifier) and the total sludge age in the biological treatment is 10-11 d (the aerobic sludge age is 7-8 d, the anoxic zone is 28% of the total activated sludge tank volume). The facility is operated with chemical phosphate precipitation and flocculation filtration.

Twenty-four hour flow proportional composite samples (1-L) of the inflow and outflow were collected over several days during a dry weather period. Samples were collected in plastic bottles containing 10 mL/L formaldehyde (37%) and were stored in the dark to avoid possible photodegradation. Samples were kept at 4°C until analysis, typically less than 24 h.

In newer studies, samples are collected in amber glass bottles and stabilized by lowering the pH to 3. No differences were observed between plastic and glass bottles. Formaldehyde is no longer used for preserving the samples because it may react with the amino group of the piperazynil group of ciprofloxacin to give the Schiff's base.

Analyses

The analytical methods used are summarized in Table I. Duplicate analyses were performed. Wastewater samples (200-400 mL) were adjusted to pH 4, filtered through a glass fiber filter with a nominal pore size of 0.7 μm, and enriched by solid-phase extraction using strong cationic exchanger Empore disks. Elution of the disk was performed by 3 mL 5% aqueous NH_3 (25%) in methanol. The extract was acidified with 200 μL phosphoric acid (85%) and subsequently analyzed by reversed-phase liquid chromatography with fluorescence detection (Ex: 278 nm, Em: 445 nm). An octadecylsilica column (Nucleosil 100, 5 μm, 250 x 3 mm) was operated at room temperature with a flow rate of 0.7 mL/min. The aqueous mobile phase (pH 2.4) was 20 mM KH_2PO_4, 30 mM ortho-phosphoric acid (85%) (eluent A), and 5 mM triethylamine. The organic modifier was acetonitrile (eluent B). Typically, elution started isocratically with 90% of eluent A for 1 min, followed by a linear gradient to 60% of eluent B in 15 min. Final conditions of 90% eluent B were reached in 2 min and held for 2 min before a 6-min linear gradient reestablished initial conditions, which were held for 4 min for column equilibration.

Quantification

Quantification was carried out using ciprofloxacin-HCl (Bayer AG, Wuppertal, Germany) as an external standard. The limit of quantification for wastewater was 50 ng/L (S/N = 10). However, this was limited by interferences of the wastewater matrix, not by instrumental sensitivity. Recoveries of ciprofloxacin from inflow and outflow were 40–50 and 50–70%, respectively.

Macrolides and Sulfonamides

Detailed descriptions of the analytical methods using LC/MS for the determination of macrolide and sulfonamide antibiotics in WWTP outflows and in surface waters will be published elsewhere (*11*). The method is a modified version of the procedure published by Hirsch et al. (*12*), who used LC/MS/MS for analyzing environmental samples. The limit of detection in groundwater samples was between 0.5 and 5 ng/L, except for spiramycin with a detection limit of 11 ng/L. Recoveries were in the range of 60 to 103%. Matrix interferences were small for macrolide antibiotics, but were of concern for sulfonamide antibiotics. Standard addition was therefore used for quantification if matrix effects needed to be considered.

Results and Discussion

Fluoroquinolones

Concentrations and mass flows of ciprofloxacin in wastewater

The concentrations of ciprofloxacin in the inflow and outflow are given in Figure 5. The concentrations are based on two replicate determinations of a single wastewater sample. Over a 14-day sampling period ciprofloxacin concentrations in the inflow of the WWTP Zurich-Werdhölzli ranged between 220 and 370 ng/L ($m = 284$, $s \pm 60$). Outflow concentrations of ciprofloxacin varied from 60 to 75 ng/L ($m = 70$, $s \pm 4$).

The mass flow calculations are based on an average wastewater discharge of 170,000 m^3/d. The mass flow of ciprofloxacin in the inflow varied from 37.4 to 63.6 g/d ($m = 48.0$, $s \pm 10.7$). In samples of the outflow, the mass flows were between 10.4 and 12.6 kg/d ($m = 11.9$, $s \pm 0.8$).

These results show that in this treatment plant ciprofloxacin is eliminated during activated sludge treatment with 70-80% efficiency, and that the residual amounts are discharged into ambient waters.

Ciprofloxacin degradation is likely to occur during composting procedures. It was demonstrated that ciprofloxacin is degraded by a brown rot fungus (*13*). At present, it is unknown whether ciprofloxacin is degraded during wastewater treatment. A recent study has shown that wastewater bacteria were not able to degrade ciprofloxacin in a closed bottle test (*14*). Elimination due to adsorption to sludge during activated sludge treatment is another plausible mechanism. Because the log K_{ow} of ciprofloxacin is approximately zero at pH 7 (*15*), hydrophobic partitioning into sludge cannot contribute significantly to the sorption. In wastewaters (pH 7-8), ciprofloxacin is present in the zwitterionic form with a positively charged amino group. Therefore, cationic exchange processes may dominate the sorption of ciprofloxacin to the negatively charged sludge.

Table I. Analytical methods for the determination of the antibiotics in wastewaters and ambient waters

Antibiotic	Analytical methods
Fluoroquinolones	**Enrichment** Solid phase extraction with strong cationic exchanger Empore cation exchange disks, pH 4 elution with 5% aqueous NH_3 in methanol **HPLC** Reversed-phase column, gradient elution with KH_2PO_4 (20 mM), H_3PO_4 (30 mM), $(C_2H_5)_3N$ (5 mM) and acetonitrile **Detection** Fluorescence detection (extinction: 278 nm, emission 445 nm)
Macrolides	**Enrichment** Solid phase enrichment with LiChrolute EN / LiChrolute C_{18} (2:5), pH 7 **HPLC** Reversed-phase column, gradient elution with NH_4Ac (10 mM) and acetonitrile **Detection** LC/MS, electrospray ionization
Sulfonamides	**Enrichment** Solid phase enrichment with Oasis HLB, pH 6 **HPLC** Reversed-phase column, gradient elution with NH_4Ac (1 mM) and acetonitrile **Detection** LC/MS, electrospray ionization

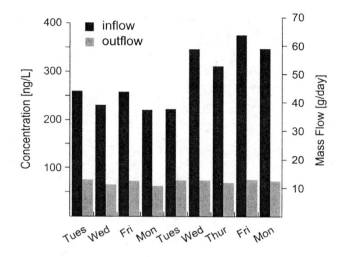

Figure 5. Daily variation of the concentration of ciprofloxacin in inflow and outflow of the wastewater treatment plant Zurich-Werdhölzli.

Plausibility estimation of the measured concentrations of ciprofloxacin

To make a plausibility estimation of the found concentrations, predicted concentrations (PEC) were compared with measured concentrations (MEC) of ciprofloxacin in wastewaters (Table II).

In a university hospital with 1000 beds, ciprofloxacin consumption is approximately 8 kg per year. Assuming a 50% excretion through the urine, and dividing this number by the mean annual wastewater discharge, an annual mean concentration of 8.7 µg/L can be predicted. This PEC corresponds well with the measured concentrations in grab samples of the hospital outflow. It must also be taken into account that most grab samples were collected in the morning when concentrations are assumed to be high.

Four hospitals with a total of 2000 beds are situated in the catchment area of the investigated WWTP. This corresponds to an estimated consumption of 16 kg ciprofloxacin per year. According to the Swiss market statistics, the domestic consumption (not including hospitals) in 1998 was approximately 1,400 kg corresponding to an annual consumption of 184 mg per person. With a population of 450,000 inhabitants in the catchment area of the WWTP, this yields a private consumption of 83 kg/y. Assuming an excretion of 50% as parent compound, a ciprofloxacin concentration of 0.8 µg/L in the raw sewage of the WWTP can be predicted. This value correlates well with the measured concentrations between 0.25-0.37 µg/L in the inflow, especially if we take into account that ciprofloxacin adsorbs to particles in the sewer and in the primary clarifier.

**Table II. Predicted (PEC) and measured (MEC) environmental
concentrations of ciprofloxacin**

Compartment	Ciprofloxacin consumption [kg/y]	Wastewater discharge [10^6 m^3/y]	PEC [$\mu g/L$]	MEC [$\mu g/L$]
1 Hospital (1000 beds)	8	0.459	8.7[a]	5–90 (grab samples, wastewater)
4 Hospitals in catchment area (2000 beds)	16 (estimated)			
Domestic consumption in catchment area	83[b]	62.05	0.8[a]	0.25–0.37 (24 h samples, inflow)

[a]Assumption: 50% excreted as parent compound in urine.

[b]Annual private consumption in Switzerland (1998): 1.4 t, corresponding to 184 mg per person; population in the WWTP catchment area: 450,000.

Macrolides and Sulfonamides

Macrolide and sulfonamide antibiotics were determined in the outflow of four wastewater treatment plants (24-h composite and grab samples) and in several surface water grab samples from various lakes and rivers in Switzerland. WWTP inflow samples were also collected. However, quantification proved impossible with the currently used method because of matrix effects. The ranges of measured concentrations in WWTP outflows and in surface waters are summarized in Figure 6. Minimum concentrations correspond to the lowest concentrations measured or to the limit of detection if nothing was detected. From the group of macrolide antibiotics, erythromycin, clarithromycin, roxithromycin, tylosin, and spiramycin were determined. Of these five antibiotics, only clarithromycin and roxithromycin were found in the samples, at concentrations of up to 131 and 35 ng/L, respectively. These findings are not surprising, considering that in Switzerland clarithromycin is consumed in the highest quantities of all macrolide antibiotics used in human medicine. Erythromycin was not detected in its original form but as a degradation product with an apparent loss of water (erythromycin-H_2O). This degradation product no longer exhibits antibiotic properties (*16*). Spiramycin and tylosin were not detected. Both antibiotics are used in veterinary medicine, whereas erythromycin, roxithromycin, and clarithromycin are only used for humans. The three latter macrolide antibiotics were also found in the surface water samples. Concentrations were between 2 and 18 ng/L. The measurements indicate that these human-use antibiotics reach the surface water by the outflow of the WWTP and are found in ambient waters in concentrations corresponding to a dilution factor of two to ten.

Results from measurements of sulfonamide antibiotics (sulfamethoxazole, sulfamethazine, sulfaguanidine, and sulfadiazine), trimethoprim, and chloramphenicol are also summarized in Figure 6. Sulfamethoxazole, which is mainly used in human medicine, was found in relatively high concentrations of up to 473 ng/L in WWTP outflows. A dilution factor of approximately 15 fold was found for the concentrations in surface waters. The same behavior was observed for trimethoprim, which is normally used in combination with sulfamethoxazole in a ratio of 1:5.

A significantly different picture arises from the measurements of chloramphenicol and sulfamethazine, which are mainly used in veterinary medicine. Both antibiotics were found in higher concentrations in surface waters than in WWTP outflows. Chloramphenicol was detected in only one WWTP outflow sample at a level of 22 ng/L, and only two surface water samples showed concentrations of 10 and 30 ng/L, respectively. The relatively high concentration in one WWTP outflow sample may have been the result of a point input of chloramphenicol applied in human treatment. In general, chloramphenicol did not occur regularly in the aquatic environment. Sulfamethazine was also found in only one WWTP outflow sample in a concentration of 4 ng/L, which may have been caused by improper disposal of veterinary drugs into sewage. However, sulfamethazine was present in nearly all surface water samples in concentrations of up to 54 ng/L. Sulfamethazine is used in the highest quantities of all sulfonamides applied in veterinary medicine, but exact amounts are not available. Sulfaguanidine and sulfadiazine, both mainly used in veterinary medicine, were not found at detectable concentrations.

These results indicate that the expected exposure route of veterinary antibiotics reaching surface water is mainly by run-off and field drains from manure application and animal excretion on soils.

68

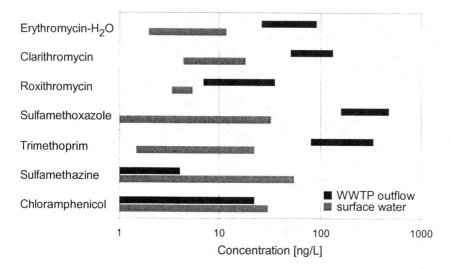

Figure 6. Concentration ranges of antibiotics found in WWTP outflows and in ambient waters.

Conclusions and Outlook

Antibiotics are ubiquitous trace contaminants in the aquatic environment. Due to their physico-chemical properties, their elimination in WWTP is not complete. Thus, residual amounts can reach the aquatic environment. This is the predominant exposure route for antibiotics used in human medicine. Antibiotics employed in veterinary medicine can enter surface waters by run-off and field drains from soils where they were applied by agricultural activity. Studies on several sulfonamide antibiotics used in human and veterinary medicine confirm these different exposure routes.

The presented work shows the occurrence of selected macrolide and sulfonamide antibiotics in WWTP outflows and in ambient waters. Further research will focus on the elimination of these antibiotics in WWTPs depending on different treatment processes and will incorporate studies on degradation and sorption behavior. However, LC/MS/MS is needed to overcome problems with quantifying antibiotics in WWTP inflow samples due to matrix effects.

Our findings indicate that ciprofloxacin occurs in treated wastewaters in concentrations that, in combination with other fluoroquinolone antibiotics, might have effects on some of the most susceptible aquatic organisms. Ciprofloxacin was used as a compound to investigate the daily input variations of human antibiotics in surface waters. However, it has to be considered that fluoroquinolone antibiotics comprise a group of individual chemicals. Therefore, for an exposure assessment for these antibiotics, the environmental fate of other fluoroquinolones also needs to be studied (*17*).

Acknowledgments

We thank Bayer AG (Wuppertal, Germany) for supplying the reference compound ciprofloxacin and financial support. The ICSC-World Laboratory is gratefully acknowledged for the Wilhelm Simon fellowship provided to N.S. Nipales. We thank Abbott GmbH (Wiesbaden, Germany) for the supply of clarithromycin. We appreciate the comments on the manuscript by Erika Vye.

References

1. Ternes, T. *Water Res.* **1998**, *32*, 3245-3260.
2. Halling-Sørensen, B.; Nors Nielsen, S.; Lansky, P.F.; Ingerslev, F.; Holten Lützhøft, H.C.; Jørgensen, S.E. *Chemosphere* **1998**, *36*, 357-393.
3. Stumpf, M.; Ternes, T.A.; Wilken, R.-D.; Rodrigues, S.V.; Baumann, W. *Sci. Total. Environ.* **1999**, *225*, 135-141.
4. Daughton, C.G.; Ternes, T.A. *Environ. Health Perspec.* **1999**, *107*, 907-938.
5. *Annual Report of Swiss Importers of Antibiotics*, Bern, Switzerland, 1998.
6. Hirsch, R.; Ternes, T.A.; Haberer, K.; Kratz, K.L. *Sci. Total. Environ.* **1999**, *225*, 109-118.
7. *Quinolone Antibacterials;* Kuhlmann, J.; Dalhoff, A.; Zeiler, H.-J. Eds.; Handbook of Experimental Pharmacology, Springer-Verlag: Berlin, Germany, 1998; Vol. 127.
8. *Allgemeine und spezielle Pharmakologie und Toxikologie;* Forth, W.; Henschler, D.; Rummel, W.; Strake, K. Eds.; Spektrum Akademischer Verlag: Heidelberg, Germany, 1996.
9. Hartmann, A.; Alder, A.C.; Koller, T.; Widmer, R.M. *Environ. Toxicol. Chem.* **1998**, *17*, 377-382.
10. Hartmann, A. Ph.D. Thesis No. 12762, Swiss Federal Institute of Technology (ETH), Zurich, Switzerland, 1998.
11. McArdell, C.S.; Nipales, N.S.; Molnar, E.; Giger, W. *In preparation,* **2000**.
12. Hirsch, R.; Ternes, T.A.; Haberer, K.; Ballwanz, F.; Kratz, K.L. *J. Chromatogr. A* **1998**, *518*, 213-223.
13. Wetzstein, H.-G.; Stadler, M.; Tichy, H.-V.; Dalhoff, A.; Karl, W. *Appl. Environ. Microbiol.* **1999**, *65*, 1556-1563.
14. Al-Ahmad, A.; Daschner, F.D.; Kümmerer, K. *Arch. Environ. Contam. Toxicol.* **1999**, *37*, 158-163.
15. Takács-Novák, K.; Józan, M.; Hermecz, I.; Szász, G. *Int. J. Pharm.* **1992**, *79*, 89-96.
16. *Macrolides; chemistry, pharmacology and clinical uses.* Bryskier, A.J.; Butzler, J.-P.; Neu, H.C.; Tulkens, P.M., Eds.; Arnette Blackwell: Paris, 1993.
17. Golet, E.M. Ph.D. Thesis, Swiss Federal Institute of Technology (ETH), Zurich, Switzerland. *In preparation.*

Chapter 4

Occurrence of Pharmaceutical Residues in Sewage, River, Ground, and Drinking Water in Greece and Berlin (Germany)

Th. Heberer[1], B. Fuhrmann[1], K. Schmidt-Baumler[1], D. Tsipi[2], V. Koutsouba[2], and A. Hiskia[3]

[1]Institute of Food Chemistry, Technical University of Berlin, Sekr. TIB 4/3-1, Gustav-Meyer-Allee 25, 13355 Berlin, Germany
[2]General Chemical State Laboratory, 16, An. Tsoha, 11521, Athens, Greece
[3]Institute of Physical Chemistry, NCSR Demokritos, 15310, Athens, Greece

A number of pharmaceuticals used in human medical care are not completely eliminated in the municipal sewage treatment works. They are discharged as persistent contaminants into the aquatic environment. Due to their polar structures, some of these residues are not significantly adsorbed in the subsoil and may, under unfavorable conditions, also leach into the groundwater aquifers from the contaminated surface waters. Especially in conurbations such as Berlin (Germany), with high municipal sewage water outputs and low surface water flows, there is a potential risk of drinking water contamination when groundwater recharge is used in drinking water production. In 1998 and 1999, in the framework of a German-Hellenic project, the occurrence of drug residues in the aquatic environment in Berlin and in different cities in Greece was investigated and compared. The results demonstrate the extent and variety of surface water contamination by drug residues in the aquatic environment.

Introduction

When applying pharmaceuticals to humans or livestock, many of these compounds are excreted without being metabolized in the target organism. They are excreted only slightly transformed or even unchanged and often conjugated to polar molecules (e.g., as glucoronides). Considering human pharmaceuticals, the excreted drugs and drug metabolites are passed into the sewage system and will be processed by municipal sewage treatment works (STWs). In the STWs the drug conjugates are easily cleaved and then discharged into the receiving waters (1-9). Due to their polar structures, many of these persistent residues are not significantly adsorbed in the subsoil (10,11). In case of influent conditions, they may also leach into the ground water aquifers from the contaminated surface waters (12-15). Especially in conurbations such as Berlin (Germany) with high municipal sewage water outputs and low surface water flows, there is a potential risk of drinking water contamination when groundwater recharge is used in drinking water production (13-15). In 1992, the drug metabolite clofibric acid (2-(4)-chlorophenoxy-2-methyl propionic acid) was first found in Berlin groundwater samples and in ground water samples collected from former sewage irrigation fields near Berlin (10,11,14,16). Clofibric acid is the active metabolite of the drugs clofibrate, etofyllin clofibrate, and etofibrate, which are used as blood lipid regulators in medical care. Meanwhile, the occurrence and fate of pharmaceutical residues in the aquatic environment has become a subject of public concern (1,2,4,5). More than 40 pharmaceuticals have been identified in sewage effluents and surface waters up to the μg/L-level (1-6,9), and a few pharmaceutical residues also show up in ground and drinking water (13-15).

Most investigations concerning pharmaceutical residues have been carried out in Germany, but recently several findings have also been reported from Brazil, Canada, Denmark, Italy, Switzerland, The Netherlands, and the USA (2,8,14,17-23). In the terms of a Hellenic-German Scientific Cooperation entitled "Investigation and Determination of New Environmental Contaminants in Greek Surface Water" (24), the occurrence of several drug residues in the aquatic system in different regions in Greece and in Berlin (Germany) was investigated and compared. Additionally, in the years between 1994 and 1999, sewage, surface water, ground water, and drinking water samples from the Berlin area were analyzed for several drug residues. Over the years, the spectrum of the target compounds was extended to more than 30 pharmaceutical residues (25). Several results from these investigations are presented here. They show the impact of sewage discharges on the surface water quality of large conurbations such as Berlin. Moreover, the results show that several pharmaceutical compounds are cycled from human application via human excretions, STWs, surface waters, ground water recharge, and back to human drinking water.

Discharges of Persistent Drug Residues from Municipal Sewage Treatment Plants in Greece and Berlin

In terms of a German-Hellenic cooperation (*24*), the occurrence of drug residues in influents and effluents of STWs from different regions in Greece and in Berlin was investigated and compared. This survey was the first study of the occurrence of drug residues in sewage effluents in Greece. Sometimes it was difficult to encourage the operators of the sewage treatment plants to participate in this study. Thus, only random sewage effluent samples were provided and for the second major survey in 1999 only three Greek STWs supplied us with samples. Nevertheless, the analyses of these samples gave some interesting results and indications.

Table I. Results from the first screening of influents and effluents of several STWs in Greece for clofibric acid, diclofenac and propyphenazone in 1998.

STWs (date of sampling)	influent/ effluent[a]	clofibric acid ng/L	diclofenac ng/L	propyphenazone ng/L
Psittalia (Athens)	in	n.d.[b]	85	n.d.
11/09/98	out	n.d.	100	n.d.
Psittalia (Athens)	in	n.d.	115	n.d.
11/11/98	out	n.d.	80	n.d.
Metamorphossi	in	n.d.	560	10
11/16/98	out	n.d.	10	n.d.
Metamorphossi	in	n.d.	35	n.d.
11/17/98	out	5	50	< 1
Thessaloniki	in	n.d.	105	n.d.
11/18/98	out	n.d.	100	n.d.
Thessaloniki	in	n.d.	120	n.d.
11/19/98	out	n.d.	365	n.d.

[a] random samples; [b] n.d.: not detected

In a first screening in 1998, influents and effluents from several STWs located in the major conurbations of Greece such as Athens and Thessaloniki were sampled and analyzed for different inorganic and organic contaminants such as polar pesticides, nitrate, nitrite, ammonia, phosphate, boron, chemical oxygen demand (COD), and total organic carbon (TOC). Additionally, the samples were analyzed for three pharmaceutical residues that have been detected at higher concentrations in German STWs, namely clofibric acid and the analgesics diclofenac and propyphenazone. Table I compiles the concentrations of these drug residues measured in the influents and effluents of three Greek STWs. Diclofenac was found as a pharmaceutical residue in the effluents of all STWs at concentrations up to 560 ng/L. Clofibric acid, which has been found in sewage, surface, and drinking water in Berlin, excepting one single sample was not detected in Greece (5 ng/L in Metamorphossi effluent on 11/17/98). Propyphenazone was also only found at trace-level concentrations in two sewage samples. Additional to the pharmaceuticals, several important acidic pesticides were also analyzed but mecoprop (MCPP) was the only pesticide which was also found in a single sewage sample at a concentration of only 70 ng/L (Metamorphossi influent 11/16/98).

Table II. Comparison of positive findings for several drug residues detected in the effluents of STWs in Greece and Berlin (25)[a].

	Berlin (Germany)[b] ng/L	Greece[c] ng/L
clofibric acid	415 – 510	n.d.[d]
diclofenac	210 – 1110	200 – 340
gemfibrozil	n.d.	n.d. – 150
ketoprofen	n.d.	270 – 820
mefenamic acid	n.d.	80 – 220
naproxen	n.d. – 120	n.d.
primidone	n.d. – 880	n.d.
propyphenazone	n.d. – 740	n.d.
(salicylic acid)	n.d. – 65	640 – 2000

[a] sampling series in May-July 1999; [b] STWs in Berlin: Ruhleben, Schönerlinde, Waßmannsdorf (mixed samples: 24hours); [c] STWs in Greece: Psittalia (Athens), Ioannina and Herakleon (all random samples); [d] n.d.: not detected

The investigations of the STWs in Greece were repeated and extended in 1999 to include a broader spectrum of drug residues. In parallel, sewage effluents (mixed samples, 24 hours) from three STWs in Berlin were analyzed for these residues. As shown in Table II, significant differences in the spectrum of drug residues were observed for the sewage effluents from Greece and Germany, which may be accounted for by different prescription behavior for pharmaceuticals. Again, clofibric acid, which was found in sewage effluents in Berlin at concentrations between 415 and 510 ng/L, was not detected at all in sewage effluents from Greece. Only diclofenac and salicylic acid were found both in Berlin and in Greece. Salicylic acid may, however, not only originate from drug residues as a metabolite of aspirin, because it is also used as keratolytic, dermatice, and preservative of food and may also be formed naturally (*3*). The lipid regulator gemfibrozil and the analgesics ketoprofen and mefenamic acid were only detected in Greece, whereas the analgesics naproxen and propyphenazone and the antiepileptic drug primidone were only detected in Berlin's STWs.

Due to the limited set of data and due to the fact that only random samples could be collected from the STWs in Greece, no final conclusions can be drawn from these results. However, the investigations strongly indicate that there are significant differences in the spectrum of drug residues used in human medical care in Germany and Greece. In Greece, the sewage effluents are mainly discharged directly into the Mediterranean Sea where the sewage effluents are highly diluted. Thus, in two random samples from the Bay of Athens (Elefisina Port), no drug residues were detected. The investigations in Greece can, however, only be seen as a first screening and should be repeated and intensified in the future.

The Berlin Situation

In Berlin, municipal wastewater contamination is of great relevance because of the high proportion of bank filtrate and artificial ground water enrichment of approximately 75% used in drinking water production in Berlin (*13*). High concentrations of sewage contaminants can be expected in the receiving waters with regard to the high contributions of STWs effluents to the surface waters resulting from low surface water flows and large amounts of raw sewage produced by its population of around 3.5 million people. Additionally, the total drinking water usage per inhabitant in Germany of 127 liters per day is low compared with as much as 300 liters per day in the USA.

Figure 1 shows a map of Berlin and its important lakes, rivers, and canals. The map also shows the locations and drainages of the STWs (A-G), the Berlin water works (I-XIV) and the sampling sites of a surface water monitoring carried out in 1996 (*26*). Berlin's most important rivers, the Spree and the Havel, are typical shallow and slow-flowing lowland rivers. The river Spree enters Berlin from the southeast, passes through the whole city, and discharges into the river Havel in the northwestern districts of Berlin. The river Havel enters Berlin from the northwest and leaves the town in southwest direction.

The Teltowkanal, a canal located in the south of Berlin (see Fig. 1), was built between 1901 and 1906. The canal was used as drainage for rainwater and industrial waste water from districts formerly located outside of Berlin. Additionally, it was used as a shipping canal for industrial supply and as a short cut for the shipping routes between the rivers Oder and Elbe. The canal has a length of approximately 35 kilometers and connects the rivers Dahme and Havel. Today, it is characterized by high proportions of sewage effluents discharged into the canal by Berlin's two largest STWs (Ruhleben and Waßmannsdorf). Additionally, the STWs in Stahnsdorf and Marienfelde (phased out in September 1999) discharge their effluents into this canal.

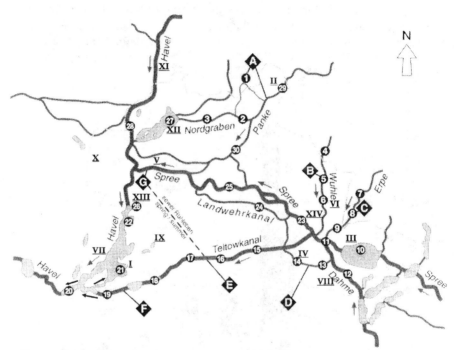

Figure 1. Map of Berlin showing the locations of the STWs (A: Schönerlinde, B: Falkenberg, C: Münchehofe, D: Waßmannsdorf, E: Marienfelde, F: Stahnsdorf and G: Ruhleben), the water works (I-XIV: for affiliations refer to Table IV) and 30 sampling locations of a surface water monitoring in Berlin in September 1996 (26). Adapted with permission from reference 26. Copyright 1998 John Wiley & Sons.

Drug Residues In Surface Waters

Surface water monitoring for several organic and inorganic sewage contaminants was carried out in Berlin in 1996 (26). Analyzed for several environmental pollutants such as phenols, synthetic musks, some polar pesticides, pharmaceutical residues, and other polar organic compounds were 30 surface water samples representative of the Berlin area with respect to possible contaminations by STW discharges. The monitoring of the surface water samples provided an overall picture of the degree of contamination of the Berlin waters by drug residues and other organic contaminants. Pharmaceuticals and pharmaceutical metabolites were found up to the µg/L-level at all sampling sites located downstream from the municipal STWs (26).

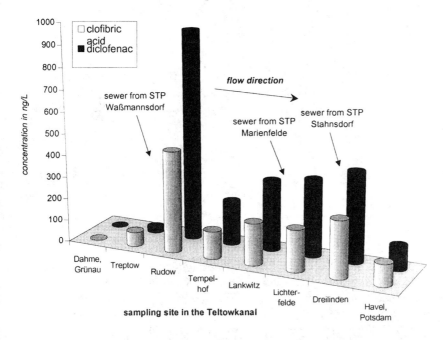

Figure 2. Concentration profiles for clofibric acid and diclofenac in the Teltowkanal in Berlin. Selected results from a surface water monitoring in September 1996 (26).

The water quality of the Teltowkanal is of public importance because its surface water is used for bank filtration in drinking water production. As already mentioned, the surface water in the Teltowkanal is characterized by high proportions of treated municipal sewage. During the sampling series in September

1996, three STWs deposited their effluents into the canal. In spring and summer time, the canal is fed by a force main with additional effluents from Berlin's largest STW in Ruhleben. Figure 2 shows the concentration profiles for diclofenac and for the drug metabolite clofibric acid (26). The final sampling site labeled as Havel, Potsdam shows the concentration of clofibric acid and diclofenac that is detected in the river Havel downstream of Berlin, containing all discharges from Berlin's STWs. Both compounds occur in the surface water samples at maximum concentration levels of approximately 500 and 1000 ng/L, respectively. With both compounds, peak concentrations were observed in the canal where sewage effluents were discharged by the three STWs. The impact of sewage effluents on the surface water quality was found to be much more significant for these pharmaceutical contaminants than for any of the common chemical parameters (26).

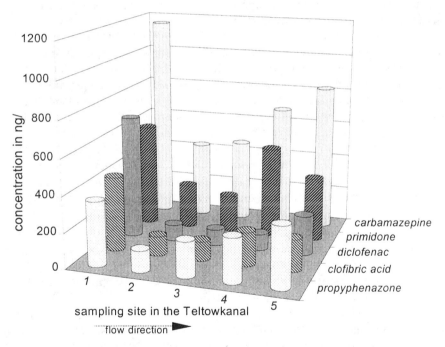

Figure 3. Concentration profiles for carbamazepine, clofibric acid, diclofenac, primidone, and propyphenazone in the Teltowkanal in Berlin. Selected results from investigations in July 1999 (25). Sampling locations (1-5) at the downstream of the STWs of Wassmannsdorf (1), Marienfelde and Ruhleben (4), and Stahnsdorf (5).

Figure 3 shows the concentration profiles for several pharmaceuticals that have been found in the Teltowkanal in terms of another surface water survey in July 1999 (25). In this survey, only sampling sites located downstream of the STWs were collected. Again, individual concentrations of up to more than 1 µg/L (for the antiepileptic drug carbamazepine) were detected, and again, peak concentrations were observed in the canal where sewage effluents are discharged by the three STWs. With the exception of primidone, the concentration profiles for all residues are almost identical. The decrease in the concentrations from sampling sites 1 to 2 can be explained by dilution (drainage, run-off). At sampling site 1, the canal consists mainly of discharges from the STWs in Waßmannsdorf.

Occurrence of Drug Residues in Groundwater Aquifers

The polar contaminants that were found in the surface waters may also leach through the subsoil into the groundwater aquifers whenever contaminated surface water is used for groundwater recharge.

In investigations of 17 groundwater wells in the catchment area of a Berlin drinking water plant, several drug residues were found up to the µg/L-level (15). This particular drinking water plant is located at a canal downstream of STWs and an abandoned pharmaceutical production plant. It uses a high proportion of bank filtrate in drinking water production. Independent of the contamination source, the results showed that several drug residues are not eliminated on their way through the subsoil and leach easily into the groundwater aquifers of the drinking water plant. A compilation of these results and of additional results from investigations carried out in Berlin in 1999 (25) is given in Table III.

Occurrence of Clofibric Acid and other Organic Residues in Berlin Drinking Water

In 1993, clofibric acid was detected in Berlin tap water samples at concentrations up to 165 ng/L (12). Further investigations of drinking water samples from 14 water works in Berlin (13,14) compiled in Table IV show the correlation between bank filtration (BF), artificial groundwater enrichment (GWE), and the level of drinking water contamination. The concentrations of clofibric acid found in tap water samples of the individual Berlin water works correlate well with the proportions of groundwater recharge used by the particular water works in drinking water production. In this survey, clofibric acid and the polar contaminant N-(phenylsulfonyl)-sarcosine (not a drug residue) were found at maximum concentrations of 270 ng/L and 105 ng/L, respectively (13,14). The geographical situation of the water works and the relative contamination of the neighboring watercourses used as a source for groundwater recharge have been identified as the factors determining the concentration levels of clofibric acid and N-(phenylsulfonyl)-sarcosine in the drinking water samples (*14*).

Table III. Concentrations of drug residues found in Berlin groundwater wells located near contaminated surface waters. Results of investigations carried out in 1996 (15) and 1999 (25).

drug residues	concentration range in ng/L
clofibric acid	n.d.[a] – 7300
diclofenac	n.d. – 380
fenofibrate	n.d. – 45
gentisic acid	n.d. – 540
gemfibrozil	n.d. – 340
ibuprofen	n.d. – 200
ketoprofen	n.d. – 30
phenazone	n.d. – 1250
primidone	n.d. – 690
propyphenazone	n.d. – 1465
(salicylic acid)	n.d. – 1225
clofibric acid derivative	(n.d. - 2900)[b]
N-methylphenacetin	(n.d. – 470)[b]

[a] n.d.: not detected; [b] concentrations were only estimated, because standards were not commercially available

As can be seen exemplarily in Figure 4, several polar organic compounds appear at trace-level concentrations in drinking water samples in Berlin. In this case, the drug residues clofibric acid (40 ng/L), propyphenazone (80 ng/L), and diclofenac (<10 ng/L) were identified in this sample. Additionally, N-(phenyl-sulfonyl)-sarcosine and the polar DDT metabolites o,p'- and p,p'-DDA (bis-(chlorophenyl)acetic acid) (*27*) were also detected at trace-level concentrations in this sample.

Conclusions

Drug residues are discharged into the receiving waters by municipal STWs. The spectrum of pharmaceuticals varies significantly, however, between Greece and Germany. In Greece, the sewage effluents of the big cities are discharged directly into the Mediterranean Sea where the sewage effluents are highly diluted. In Berlin (Germany), they are found at concentrations up to the µg/L-level in surface water samples collected from several canals, lakes, and rivers. When

Table IV. Clofibric acid and N-(phenylsulfonyl)-sarcosine in drinking water samples collected from the Berlin water works (13,14). Adapted with permission from reference 14. Copyright 1997 Gordon and Breach Publishers.

Water works	no.	BF^a in %	GWE^b in %	clofibric acid in ng/L^e	NPS^c in ng/L^e	SWC^d
Beelitzhof	I	66	7	60	105	+
Buch	II	0	0	< 5	-	+
Friedrichshagen	III	82	0	10	25	o
Johannisthal	IV	62	0	170	105	++
Jungfernheide	V	52	43	45	-	+
Kaulsdorf	VI	0	0	-	-	-
Kladow	VII	68	0	6	17	+
Köpenick	VIII	74	0	6	8	o
Riemeisterfenn	IX	17	0	11	70	+
Spandau	X	30	48	-	-	-
Stolpe	XI	?	+	13	-	(-)
Tegel	XII	54	27	25	-	+
Tiefwerder	XIII	61	0	55	35	+
Wuhlheide	XIV	58	0	75	-	+

[a] BF : bank filtration; [b] GWE : artificial ground water enrichment; [c] NPS : N-(phenylsulfonyl)-sarcosine; [d] SWC : degree of contamination of the neighboring watercourse; [e] - : below the limit of detection of 1 ng/L for clofibric acid or 2 ng/L for N-(phenylsulfonyl)-sarcosine.

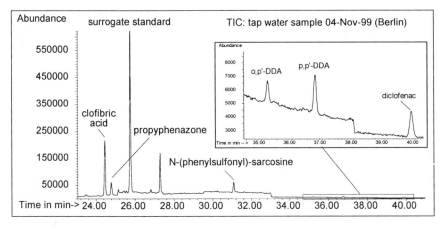

Figure 4. Multiple ion detection (MID) chromatogram recorded with GC-MS in selected ion monitoring (SIM) mode of a derivatized extract of a Berlin tap water sample. Reprinted from reference 25.

groundwater recharge is used in drinking water production, these contaminants can leach from the contaminated watercourses into the groundwater aquifers and appear at trace-level concentrations in drinking water.

References

1. Stan, H.-J.; Heberer, Th. "Pharmaceuticals in the Aquatic Environment." In (coord.: Suter, M.J.F.) Dossier Water Analysis. *Analusis* **1997**, *25*, M20-23.
2. Halling-Sørensen, B.; Nielsen, N.; Lansky, P.F.; Ingerslev, F.; Hansen Lützhøft, H.C., Jørgensen, S.E. "Occurrence, fate and effects of pharmaceutical substances in the environment. A review." *Chemosphere* **1998**, *36*, 357-394.
3. Ternes, T.A. "Occurrence of drugs in german sewage treatment plants and rivers." *Water Res.* **1998**, *32*, 3245-3260.
4. Heberer, Th., Stan, H.-J. "Arzneimittelrückstände im aquatischen System." *Wasser & Boden* **1998**, *50* (4), 20-25.
5. Daughton, C.G.; Ternes, T.A. "Pharmaceuticals and personal care products in the environment: agents of subtle change?" *Environ. Health Perspect.* **1999**, *107* (suppl. 6), 907-938.
6. Möhle, E.; Horvath, S.; Merz, W.; Metzger, J.W. "Bestimmung von schwer abbaubaren organischen Verbindungen im Abwasser - Identifizierung von Arzneimittelrückständen." *Vom Wasser* **1999**, *92*, 207-223.
7. Hirsch, R.; Ternes, Th.A.; Haberer, K.; Kratz, K.L. "Occurrence of antibiotics in the aquatic environment," *Sci. Total Environ.* **1999**, *225*, 109-118.

8. Ternes, T.A; Stumpf, M.; Mueller, J.; Haberer, K.; Wilken, R.-D.; Servos, M. "Behavior and occurrence of estrogens in municipal sewage treatment plants - I. Investigations in Germany, Canada and Brazil." *Sci. Total Environ.* **1999**, *225*, 81-90.

9. Wilken, R.-D.; Ternes, T.A.; Heberer, Th. "Pharmaceuticals in Sewage, Surface and Drinking Water in Germany." In (Deininger, R.A. et al. eds.) *Security of Public Water Supplies;* Kluver Academic Publishers: Dodrecht, 2000; pp. 227-240.

10. Scheytt, T.; Heberer, Th.; Stan, H.-J. "Pharmaceuticals in Groundwater: Clofibric acid beneath sewage farms south of Berlin, Germany," in *Pharmaceuticals and Personal Care Products in the Environment: Scientific and Regulatory Issues*; Daughton, C.G. and Jones-Lepp, T., Eds.; American Chemical Society: Washington, DC (see Chapter in this book).

11. Heberer Th. "Identification and Quantification of Pesticide Residues and Environmental Contaminants in Ground and Surface Water Applying Capillary Gas Chromatography - Mass Spectrometry" (in German), Wissenschaft & Technik Verlag: Berlin, 1995; 437p.

12. Stan, H.-J.; Heberer, Th.; Linkerhägner, M. "Vorkommen von Clofibrinsäure im aquatischen System - Führt die therapeutische Anwendung zu einer Belastung von Oberflächen-, Grund- und Trinkwasser?" *Vom Wasser* **1994**, *83*, 57-68.

13. Heberer, Th.; Stan, H.-J. "Vorkommen von polaren organischen Kontaminanten im Berliner Trinkwasser." *Vom Wasser* **1996**, *86*, 19-31.

14. Heberer, Th.; Stan, H.-J. "Determination of clofibric acid and n-(phenylsulfonyl)-sarcosine in sewage, river and drinking water." *Int. J. Environ. Anal. Chem.* **1997**, *67*, 113-124.

15. Heberer, Th.; Dünnbier, U.; Reilich, Ch.; Stan, H.-J. "Detection of drugs and drug metabolites in ground water samples of a drinking water treatment plant." *Fresenius Environ. Bull.* **1997**, *6*, 438-443.

16. Stan, H.-J.; Linkerhägner, M. "Identifizierung von 2-(4-Chlorphenoxy)-2-methyl-propionsäure im Grundwasser mittels Kapillar-Gaschromatographie mit Atomemissionsdetektion und Massenspektrometrie." *Vom Wasser* **1992**, *79*, 75-88.

17. Buser, H.-R.; Müller, M.D.; Theobald N. "Occurrence of the pharmaceutical drug clofibric acid and the herbicide mecoprop in various Swiss lakes and in the North Sea," *Environ. Sci. Technol.* **1998**, *32*, 188-192.

18. Buser, H.-R.; Poiger, Th.; Müller M.D. "Occurrence and fate of the pharmaceutical drug diclofenac in surface waters: Rapid photodegradation in a lake," *Environ. Sci. Technol.* **1998**, *32*, 3449-3456.

19. Buser, H.-R.; Poiger, Th.; Müller M.D. "Occurrence and environmental behavior of the chiral pharmaceutical drug ibuprofen in surface waters and in wastewater," *Environ. Sci. Technol.* **1999**, *33*, 2529-2535.

20. Stumpf, M.; Ternes, Th.A.; Wilken, R.-D.; Rodriguez, S.V.; Baumann, W. "Polar drug residues in sewage and natural waters in the state of Rio de Janeiro, Brazil," *Sci. Total Environ.,* **1999**, *225,* 135-141.

21. Seiler, R.L.; Zaugg, S.D.; Thomas, J.M.; Howcroft, D.L. "Caffeine and pharmaceuticals as indicators of waste water contamination in wells." *Ground Water* **1999**, *37,* 405-410.

22. Eckel, W.P.; Ross, B.; Isensee, R.K. "Pentobarbital found in ground water." *Ground Water* **1993**, *31,* 801-804.

23. Belfroid, A.C.; Van der Horst, A.; Vethaak, A.D.; Schäfer, A.J.; Rijs, G.B.J.; Wegener, J.; Cofino, W.P. "Analysis and occurrence of estrogenic hormones and their glucoronides in surface water and waste water in the Netherlands." *Sci. Total Environ.,* **1999**, *225,* 101-108.

24. Hellenic-German Scientific Cooperation (1997-1999) *Investigation and Determination of New Environmental Contaminants in Greek Surface Water*; European Union Project GRI-161-97.

25. Heberer, Th.; Fuhrmann, B.; Dünnbier, U.; *J. Hydrogeol.,* in preparation.

26. Heberer, Th.; Schmidt-Bäumler, K.; Stan, H.-J. "Occurrence and distribution of organic contaminants in the aquatic system in Berlin. Part I: Drug residues and other polar contaminants in Berlin surface and ground water." *Acta Hydrochim. Hydrobiol.* **1998.**, *26,* 272-278.

27. Heberer, Th.; Dünnbier, U. "DDT metabolite bis(chlorophenyl)acetic acid (DDA): The neglected environmental contaminant." *Environ. Sci. Technol.* **1999**, *33,* 2346-2351.

Chapter 5

Pharmaceuticals in Groundwater: Clofibric Acid beneath Sewage Farms South of Berlin, Germany

Traugott Scheytt[1], Susanne Grams[1], Elzbieta Rejman-Rasinski[1], Thomas Heberer[2], and Hans-Jürgen Stan[2]

[1]Institute of Applied Geosciences II, Technical University of Berlin, Berlin, Germany
[2]Institute of Food Chemistry, Technical University of Berlin, Berlin, Germany

One of the first findings of pharmaceuticals in groundwater was reported from the sewage irrigation farms south of Berlin. Clofibric acid, the active metabolite of blood lipid regulators, was measured in concentrations up to 4.2 µg/L in groundwater. The fate of clofibric acid was investigated within the project "Sewage Farms South of Berlin" and with laboratory experiments. The distribution of clofibric acid in the unsaturated and saturated zone of the sewage irrigation farms reflects the high mobility and persistence of this substance. Results from laboratory batch experiments with sediments from the vicinity of the sewage irrigation farms show generally low sorption of clofibric acid. There is clear indication for the input of clofibric acid via sewage irrigation although no uniform spatial distribution pattern was found. Concentrations of clofibric acid in groundwater decrease due to the cessation of sewage farming operations.

Occurrences of pharmaceuticals in surface water and sewage water have been widely reported *(1-3)*. Pharmaceuticals have been analyzed in German Rivers (e.g., Rhine, Danube), in the Italian river Po, in Swiss lakes, and even in the North Sea *(4)*. Compared with those occurrences, the reported incidence of drugs in groundwater samples is much rarer.

The processes leading to the input of pharmaceuticals into aquifers need further investigation, and the behavior of these substances, once within an aquifer, is almost unknown.

Groundwater input paths

In the 1980s, the environmental behavior and the risks from pharmaceutical chemicals in the aquatic environment were assessed by Richardson and Bowron *(5)*. The first analysis of a human medical compound, clofibric acid, in groundwater samples of Berlin area was reported in 1991 *(6)*.

Meanwhile, occurrences of pharmaceuticals in groundwater have not only been reported from the Berlin area, but also from several other places worldwide. Investigations from the state of Nevada (U.S.) show the presence of human pharmaceuticals (i.e., chlorpropamide, phensuximide, and carbamazepine) in groundwater *(7)*. These investigators made use of these drugs, together with caffeine and elevated nitrate levels in groundwater, as indicators of recharge from domestic wastewater. The authors pointed out, however, that the usefulness of human pharmaceuticals as indicators of groundwater contamination by domestic wastewater is limited due to the unpredictable presence of drugs.

Organic compounds originating from pharmaceutical manufacturing waste were identified in the groundwater down gradient of a landfill in Grindsted (Denmark). Among others, sulfonamides, barbiturates, and propyphenazone were the primary constituents detected in groundwater *(8)*. Many of the identified organic compounds appeared at high concentrations in the sampling points (wells) close to the landfill. Sulfanilic acid (up to 6.5 mg/L) and propyphenazone (up to 4.0 mg/L) for example, were present in high concentrations. Almost all pharmaceuticals were no longer measurable within a distance of 115 m downstream of the landfill. The authors *(8)* argued that this could not be explained by dilution alone, when taking into consideration the chloride ion, which acts as a conservative tracer in this part of the aquifer. Most of the compounds were attenuated in the anaerobic zone characterized as methanogenic/sulfate-reducing, and generally the largest reduction seemed to take place under more reducing redox conditions.

In the 1970s Garrison et al. *(9)* and Hignite and Azarnoff *(10)* detected clofibric acid in sewage effluent. Clofibric acid is the active metabolite of the blood lipid regulators like clofibrate, etofibrate, and etofyllin clofibrate. Their investigations showed that only a small percentage of clofibric acid was eliminated by sewage water treatment. If surface water is contaminated by pharmaceuticals, these substances can reach groundwater in case of influent conditions (losing stream) and bank infiltration. Furthermore, groundwater contamination through sewage water is likely to be due to leaky sewage and septic systems. Irrigation with effluent water is another path by which drugs can be input into groundwater, as shown by Heberer and Stan *(11)*.

In Germany, the use of blood lipid regulators increased in the early 1960s. The daily doses of, e.g., clofibrate amounts to 1.5 g with about 29 million day-doses prescribed annually in 1993 in Germany. On the total, 198 million day-doses of clofibric acid derivatives and analogs were prescribed in 1993 with an estimated amount of about 15-21 t of clofibric acid released in 1993 with the sewage in Germany (Figure 1).

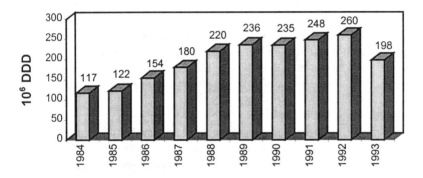

Figure 1. Prescriptions of clofibric acid derivatives and analogs in Million "Defined Day Doses" (DDD) from 1984 to 1993 in Germany; source: (12).

Recent investigations revealed that not only clofibric acid but a whole series of drugs and drug metabolites were found at concentrations up to the μg/L-level in groundwater samples of Berlin waterworks. Among these pharmaceuticals detected in groundwater are phenazone, propyphenazone, diclofenac, ibuprofen, fenofibrate, and several other compounds *(13)*. Clofibric acid was detected in groundwater and drinking water with maximum concentrations of up to 7,300 ng/L *(14)*.

The effect of sewage farm irrigation with the target of optimizing the reuse of sewage water is investigated at the sewage irrigation farms in Cairo, Egypt.

Sewage farms are designed to irrigate sewage effluent with the aim of improving the fertility of the soil. Even in the groundwater beneath these well-engineered sewage irrigation farms, a pharmaceutical was detected. Clofibric acid was found in concentrations of 40 ng/L to 75 ng/L in groundwater samples from sewage irrigation farms Gabal Al-Asfar (north of Cairo) in Egypt *(15)*. These findings indicate that the occurrence of drugs in groundwater is not a local phenomenon.

The fate of clofibric acid was investigated within the project "Sewage Farms South of Berlin" and with laboratory experiments. The objectives of these studies were to determine the environmental and transport behavior of clofibric acid, especially in the unsaturated and saturated zone.

Sample Preparation and Analysis of Clofibric Acid

The water samples collected at the sewage irrigation farms south of Berlin had a volume of at least 1 liter. Samples were collected from sewage effluent, from surface water, from the unsaturated zone, and from the saturated zone in different depths. After collecting, the samples were transported directly to the laboratory. The sample extracts prepared by solid-phase extraction (SPE) were derivatized with different reagents such as pentafluorobenzyl bromide to enable gas chromatographic detection of the acidic analytes. All sample extracts were analyzed applying capillary gas chromatography-mass spectrometry (GC-MS) with selected ion monitoring (SIM). The detection limit was about 1 ng/L, and the limit of determination was 10-25 ng/L. More detailed information on analysis of clofibric acid is provided elsewhere *(11, 14)*.

Sewage Irrigation Farms South of Berlin

The sewage irrigation farms in the south of Berlin have been investigated within a research project that started in 1992. These farms are located about 10 km south of Berlin city limits. Sewage farming took place on an area of about 16 km² and started in 1894. About 80 years later, the area that was irrigated by sewage effluent decreased gradually until 1994, when irrigation with sewage effluent stopped completely. During the period of operation, the sewage irrigation farms were flooded with sewage effluent at an average of 1,400 mm to 3,000 mm per year *(16)*. Within these farms, there are interconnected drainage channels. After irrigating with sewage water, only a part of the water infiltrated so deep as to recharge groundwater. A large amount of sewage water infiltrated into the unsaturated zone and was drained by the drainage channels after passing via a short soil path (ca. 2 m). The run-off from the sewage irrigation farms, as well as drainage water, flowed to the rivers, an unquantified portion of which reached the river Spree upstream of Berlin.

Geology and Hydrogeology

Geology beneath the sewage irrigation farms consists of quaternary sediments. These sediments belong to three major glaciations: the Elsterian, the Saalian, and Weichselian Glaciation. Tertiary Sediments are buried under these younger Pleistocene series. The quaternary sediments have a total thickness of 50 m in morphological highs, and up to 300 m in subsurface channels. These sediments consist of mostly glaciofluviatile fine-to-coarse grained sands with tills of differing thicknesses in between. Furthermore, the glacial sequence is often glaciotectonically disturbed and exhibits erosional patterns.

From a hydrogeological point of view, these quaternary sediments can be subdivided into four aquifers (1: uppermost aquifer, 4: deepest quaternary aquifer) and 3 aquitards in between. The deepest, fifth aquifer belongs to the Tertiary. The aquifers are composed of sands, whereas the aquitards are composed of the tills. Hydraulic conductivities range from 1×10^{-3} m/s to 4×10^{-5} m/s for the aquifers, and the aquitards are characterized by K values between 1×10^{-5} m/s to 9×10^{-8} m/s. Due to these relatively high conductivities, even of the aquitards, and due to glaciotectonics, the aquifers are not isolated by the aquitards. Furthermore, the aquifers show a high number of interconnections, especially due to geological windows within the aquitards. The deepest, fifth aquifer belongs to the Tertiary and consists of fine-grained sand.

Figure 2 shows a map of the sewage irrigation farms with the location of wells. Groundwater elevation of the 1 aquifer is portrayed by isolines of piezometric levels. Due to sewage irrigation, piezometric levels are high in the part of the sewage irrigation farms where sewage effluent was irrigated until the end of the operation. Piezometric levels decrease towards the edges of the study region. The groundwater flow direction is vertical in the 1 aquifer and radial in the 2 aquifer.

Groundwater Ages

Transport of groundwater constituents (including these pharmaceuticals) is mainly based on groundwater flow. Knowledge about groundwater ages can be used to infer groundwater flow paths. Groundwater ages were determined from tritium analysis of 84 groundwater samples. With these results, ages of up to 45 years can be dated.

Generally, groundwater in the uppermost aquifer has ages less than 10 years. Commonly, groundwater ages as young as 10 years can be found beneath the sewage irrigation farms within the **2** aquifer (Figure 3). This is due to the high recharge at the sewage irrigation farms and an almost vertical groundwater flow in the subsurface beneath the sewage irrigation farms. Vertical groundwater flow is almost completely limited to the **1** aquifer, whereas the predominant flow direction in the deeper aquifers is horizontal. Outside of the area of sewage farming operation, only the uppermost aquifer (**1** aquifer) has groundwater with ages less than 10 years.

Ages of groundwater increase (up to 40 years) in deeper aquifers. The deepest, tertiary aquifer generally has ages of more than 45 years *(18)*.

Groundwater Chemistry

The state of soil and groundwater contamination was analyzed, including heavy metals and organic contaminants, with a total of up to 845 groundwater samples for most of the constituents. Inorganic constituents analyzed were the main cations (Na^+, K^+, Ca^{2+}, Mg^{2+}, $Fe^{2+/3+}$, Mn^{2+}, NH_4^+) and anions (HCO_3^-, NO_3^-, NO_2^-, PO_4^{3-}, Cl^-, SO_4^{2-}). The physico-chemical parameters (temperature, specific conductance, pH, redox potential, O_2) were measured in the field. The organic compounds that were analyzed include pesticides, herbicides, PCBs, halogenated aliphatics, and halogenated aromatics.

To discriminate groundwater that is influenced by sewage farming from naturally recharged groundwater, the relationship between sodium and calcium was found to be valuable. Ratios of concentrations (mg/L) of sodium to calcium higher than 0.4 proved to be the best indicators of groundwater influenced by sewage farming. Ratios of Na/Ca in groundwater without the influence of sewage farming vary between 0.1 and 0.2. Groundwater from the upper two aquifers beneath the sewage irrigation farms have typically Na/Ca ratios between 0.6 and 0.9. This classification coincides well with groundwater ages of this region as the groundwater in the two uppermost aquifer has generally ages less than 10 years due the high recharge through sewage irrigation.

Because of the above-mentioned vertical groundwater flow in the upper aquifer, the variations of groundwater constituents exhibit a remarkable spatial variation in the **1** aquifer. In the deeper aquifers, concentration variations decrease due to hydrodynamic dispersion and diffusion.

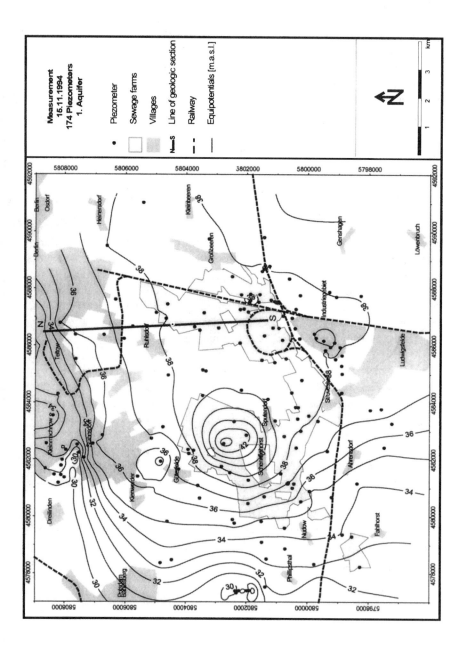

Figure 2. Map of equipotentials of piezometric levels of the 1 Aquifer in the area of the sewage farms south of Berlin; also location of cross section. (Reproduced from reference 17. Copyright 1998 Springer Verlag.)

92

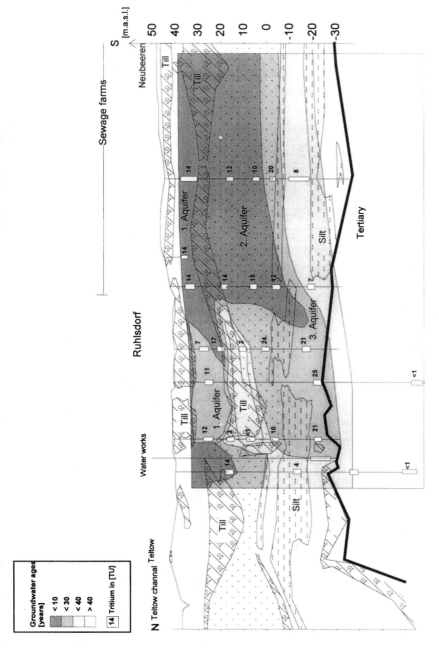

Figure 3. Groundwater ages beneath the sewage farms (18)

Distribution of Clofibric Acid

Clofibric acid was measured in sewage effluent (10 samples), surface water, and in water from drainage channels (82 samples). Clofibric acid concentrations in these samples varied between values below the detection limit (1 ng/L) to 4,550 ng/L. The mean value for water from drainage channels was 1,100 ng/L. This average was even higher in places with effluent groundwater conditions. This suggests that after irrigation with sewage water, which subsequently infiltrated into the subsurface, a certain amount flows into the drainage channels, leading to these high concentrations. Surface water from the same area that is not directly influenced by sewage farm operation exhibits mean values ranging from 125 ng/L to 310 ng/L.

Water samples from different depths within the unsaturated zone were collected, and the content of clofibric acid was measured. The average concentrations in pore water from the unsaturated zone declined from higher values near ground surface (1,430 ng/L) to lower values near the groundwater table (65 ng/L).

The distribution of clofibric acid in groundwater was based on the analysis of 597 samples. Generally higher amounts of clofibric acid were measured in the upper aquifers and lower concentrations were detected in the deeper aquifers. However, the concentrations in most of the groundwater samples were below the detection limit. The maximum concentration in groundwater was found to be 4,175 ng/L below the sewage irrigation farms.

Box plots are used to show the distribution of clofibric acid in sewage water, surface water, and the five aquifers (Figure 4). Upper and lower quartiles define the top and bottom of the box, and the median is marked by a horizontal line inside the box. The horizontal lines ("whiskers") connected to the upper and lower box sides are "adjacent values", i.e., the most extreme values that are not outliers. The observations that exceed the adjacent values are "outside values" or "outliers" and are marked as dots. This statistical analysis was done according to standard statistical procedures with the software package SPSS.

About 80% of 690 samples had clofibric acid concentrations that were below the detection limit. Decreasing concentrations from sewage effluent to surface water, and in the subsurface with increasing depth, can be seen clearly in Figure 4. There are, however, a high number of outside values.

Statistical analysis revealed that there was no clear correlation between clofibric acid concentration and most of the groundwater constituents or with different redox or pH environments. There was only a slight correlation between the concentration of clofibric acid and PO_4^{3-}, DOC, and COD as well as Na/Ca ratio; these parameters and ratios were found to be indictors of sewage recharge. A typical spatial distribution pattern for a conservative tracer could be found in the aquifers of the sewage irrigation farms for chloride and was expected for clofibric acid. However, no such "conservative" distribution was observed for clofibric acid: there were samples from wells of the same aquifer that contained high amounts adjacent to wells with no clofibric acid at all.

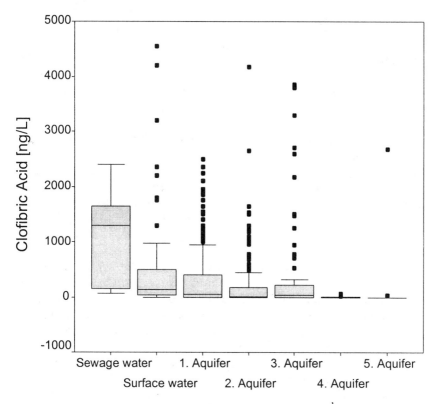

Figure 4. Boxplots of clofibric acid concentrations [ng/L] in sewage effluent, surface water, and in the five aquifers; number of samples: Sewage water: 10; Surface water: 82; 1 Aquifer: 325; 2 Aquifer: 169; 3-5 Aquifer: 103. (Reproduced from reference 17. Copyright 1998 Springer Verlag.)

The wells were sampled several times since 1993, during which time a decrease of clofibric acid concentrations was observed in almost all wells, due to the cessation of sewage farming. Table I shows an almost continuous decrease of concentrations in well Rb1/78 over time. Although no homogeneous distribution was found in groundwater, the relative concentrations of clofibric acid remained about the same over time. High concentrations in groundwater from certain wells were found to also yield high concentrations during the next sampling campaign. In well Rb1/78 (Table I), level MP2 at a depth of 12.9 m below ground surface exhibited the highest concentrations compared with the other levels for all sampling campaigns, although the values decreased markedly from the sampling campaign in fall 1993 to sampling campaign in summer 1999.

Table I. Concentration of clofibric acid in groundwater from multi-level well Rb1/78; level screenings are from top to bottom

Well Name	Aquifer	Concentration of clofibric acid [ng/L]			
		Fall 93	Fall 94	Spring 95	Summer 99
Rb1/78 OP	1	50	14	0	dry well
Rb1/78 MP1	1	30	10	7	0
Rb1/78 MP2	1	950	270	316	160
Rb1/78 MP3	1	450	220	230	25
Rb1/78 UP	2	135	30	40	15

Laboratory Batch Experiments

The decrease of clofibric acid concentration within the unsaturated zone and the lack of a uniform (i.e., "conservative") distribution in the groundwater at the sewage irrigation farms might be the effect of degradation or sorption. Therefore, batch experiments were conducted to investigate the sorption behavior of this substance. Medium-grained sand of the unsaturated zone was used as the sediment material. The samples were collected to the north of the sewage irrigation farms to (1) obtain sediment material similar to that from the sewage irrigation farms, and (2) to reduce background contamination; the location from which the sediment was obtained had never been in use as a sewage irrigation farm. To perform the experiments, a special fluid circulation device was used to better simulate natural conditions. This device operates with undisturbed soil samples and a permanent flow through the samples to enable "dynamic" batch experiments. The major advantage is that soil samples remain undisturbed compared with normal batch experiments.

Clofibric acid was added to deionized water to give concentrations of 300 ng/L, 1.0 µg/L, 3.0 µg/L, and 10.0 µg/L of clofibric acid in the solution. Dry weight of the samples (medium-grained sand) was about 250 g, and the amount of clofibric acid / water solution in the cycle about 175 mL. The experiment was conducted at two temperatures — 8 °C and 19°C; these temperatures were chosen because they correspond to the natural average temperature and maximum natural temperature in soils in the region of the sewage irrigation farms. The duration of each run was 24 hours. Set-up for all runs was kept the same except for temperature and clofibric acid concentration. Immediately after adding clofibric acid to the solution (about 1 hour) the actual initial concentration was measured.

In general, the observed sorption was low (Table II). The concentrations after the run typically showed that sorption was low at lower concentrations (0.3 µg/L and 1.0 µg/L) and at the higher temperature (19°C). Sorption increased at higher concentrations (10 µg/L) and the lower temperature (8 °C).

Table II. Results of "dynamic" batch tests with the fluid-circulation device; concentration values are clofibric acid in the solution (water)

Mixing concentration [µg/L]	Actual initial concentration [µg/L]	8 °C Concentration after experiment run [µg/L]	19°C Concentration after experiment run [µg/L]
0.3	0.34	0.26	0.40
0.3	0.34	-	0.34
1.0	1.01	0.95	0.84
1.0	1.01	0.70	1.06
3.0	2.40	2.13	2.46
3.0	2.40	-	2.00
10.0	6.86	5.72	6.30
10.0	6.86	5.82	6.45

Interestingly, there was a difference between intended initial concentration and actually measured initial concentration. A comparison of these two concentrations reveals that although there was almost no difference at 0.3 µg/L and 1.0 µg/L, increasing differences were found at the higher concentrations (3 µg/L and 10 µg/L). Compared with the sorption results, these differences exceeded the amount of clofibric acid sorbed on the sediment material. This difference between intended initial solution at a concentration of 10 µg/L and the actually measured initial concentration not only occurred at these batch tests but also at a laboratory degradation experiment and column experiment *(15)*.

Results of a laboratory column experiment *(15)* with medium-grained sand show that clofibric acid is transported with the velocity of the conservative tracer (lithium). It was also interesting that the redox potential decreased remarkably from + 500 mV to -250 mV after adding clofibric acid and increased again after flushing the column with pure groundwater. Again, no correlation between clofibric acid and pH, oxygen content, or the concentration of groundwater constituents was found, whereas a possible dependence of the redox potential on clofibric acid was observed.

Discussion and Conclusion

Data analysis from the sewage irrigation farms clearly indicates the input of clofibric acid by sewage irrigation. The input path of clofibric acid can be traced from sewage effluent and drainage water into the unsaturated zone and saturated zone. Analysis of groundwater ages shows that the older the groundwater, the lower the mean concentration of clofibric acid. Clofibric acid can consequently be used as an indicator for anthropogenic input sources. Significant amounts of clofibric acid even in the deeper aquifers demonstrate the high persistence of this contaminant.

A typical spatial distribution pattern for a conservative tracer (chloride) was found in the aquifers of the sewage irrigation farms. This was also anticipated for clofibric acid because a previously conducted laboratory column experiment showed a transport behavior comparable to a conservative tracer. No such distribution, however, was observed for clofibric acid at the sewage irrigation farms. Instead, groundwater samples from wells of the same aquifer that contained high levels were adjacent to wells with no detectable clofibric acid. These variations may be due to varying input of clofibric acid as the concentrations in sewage effluent itself varied. According to Landesumweltamt Brandenburg *(16)*, the mean annual precipitation (1951 - 1980) is 641 mm and the potential evapotranspiration is 664 mm in the study region. The climatic water balance is slightly negative. Therefore, varying evapotranspiration may, for example, lead to an increased concentration in summer when evapotranspiration is high. In conclusion, the input function is unpredictable to some extent and with this the distribution of clofibric acid in the unsaturated and saturated zone.

Because there was no uniform concentration distribution of clofibric acid in groundwater, degradation and/or sorption processes were assumed. Additionally, the decrease in the clofibric acid concentrations in soil water from the unsaturated zone with increasing depth also suggests possible degradation and/or sorption processes. Results of the sorption experiments with the fluid circulation device exhibited almost no sorption at lower concentrations or higher temperatures, and only slightly higher sorption tendency at higher concentrations and lower temperatures.

The laboratory experiments exhibited an interesting difference between the intended initial concentration and the actually measured initial concentration. The actually measured initial concentration was much lower, and this phenomenon was also observed in several other experiments. We conclude that degradation processes, in addition to sorption, lower the concentration of clofibric acid in the study site. One of these degradation processes might be microbial metabolism.

Despite the lack of a uniform distribution, clofibric acid concentrations in groundwater did exhibit certain consistent patterns. Groundwater samples with higher concentrations at certain wells in one sampling campaign showed the higher concentrations during other sampling campaigns again. Relative concentrations remain about the same over time. Concentrations showed almost no dependence on predominant physicochemical conditions or groundwater constituents. Groundwater flow paths seemed to determine the concentration of clofibric acid in groundwater. Furthermore, there has been a general decrease in concentration since 1993, coinciding with the end of sewage farm operations.

References

1. Sacher, F.; Lochow, E.; Bethmann, D.; Brauch, H.-J. *Vom Wasser* **1998**, *90.*
2. Heberer, Th.; Schmidt-Bäumler, K.; Stan, H.-J. *Acta Hydrochim. Hydrobiol.* **1998**, *26*, 272-278.
3. Ternes, T.A. *Wat. Res.* **1998**, *32*, 3245 - 3260.
4. Buser, H.-R.; Müller, M.D. *Environ. Sci. Technol.* **1998**, *32* (1), 188-192.
5. Richardson, M.L.; Bowron, J.M. *J. Pharm. Pharmacol.* **1985**, *37*, 1-12.
6. Stan, H.-J.; Linkerhägner, M. *Vom Wasser* **1992**, *79*, 85-88.
7. Seiler, R.L.; Zaugg, S.D.; Thomas, J.M.; Howcroft, D.L. *Ground Wat.*, **1999**, *37* (3), 405-410.
8. Holm, J.V.; Rügge, K.; Bjerg, P.L.; Christensen, Th.H. *Environ. Sci. Technol.*, **1995**, *29* (5), 1415-1420.
9. Garrison, A.W.; Pope, J.D.; Allen, F.R. In: *Identification & Analysis of Organic Pollutants in Water*; Keith, C.H., Ed.; Ann Arbor Science Publishers Inc.: Ann Arbor , Michigan, 1976; pp 517-556.
10. Hignite, C.H.; Azarnoff, D.L. *Life Sci.*, **1977**, *20*, 337-342.
11. Heberer, Th.; Stan, H.-J. *Int. J. Environ. Anal. Chem.* **1997**, *67*, 113-124.

12. Klose, G.; Schwabe, U. In *Arzneimittel-Report 94*; Schwabe, U.; Pfaffrath, D., Eds.; Gustav Fischer Verlag: Stuttgart, 1994; pp 271-277.

13. Heberer, Th.; Fuhrmann, B.; Schmidt-Bäumler, K.; Tsipi, D.; Koutsouba, V.; Hiskia, A. In *Pharmaceuticals and Personal Care Products in the Environment: Scientific and Regulatory Issues*; Daughton, C.G.; Jones-Lepp, T., Eds.; ACS book series; ACS/Oxford University Press: Washington, DC, 2000.

14. Heberer, Th.; Dünnbier, U.; Reilich, Ch.; Stan, H.-J. *Fresenius Environ. Bull.* **1997**, *6*, 438-443.

15. Scheytt, T.; Grams, S.; Fell, H. In *Gambling With Groundwater - Physical, Chemical, and Biological Aspects of Aquifer-Stream Relations;* Brahana, J.V., Eckstein, Y., Ongley L.K., Schneider, R., Moore, J.E., Eds.; IAH/AIH Proceedings Volume: St. Paul, 1998; pp 13-18.

16. Landesumweltamt Brandenburg, *Rieselfelder südlich Berlins - Altlast, Grundwasser, Oberflächengewässer*; Studien- und Tagungsberichte Bd. 13/14, Potsdam, 1997; pp 297.

17. Scheytt, T.; Grams, S.; Fell, H. *Grundwasser* **1998**, *2*, 67-79.

18. Asbrand, M. Ph.D. thesis, Technical University Berlin, Berlin, 1997.

Chapter 6

Phenazone Analgesics in Soil and Groundwater below a Municipal Solid Waste Landfill

Marijan Ahel[1] and Ivana Jeličić[2]

[1]Center for Marine and Environmental Research, Ruđer Bošković Institute, Bijenička 54, HR-10000 Zagreb, Croatia
[2]Faculty of Food Technology and Biotechnology, University of Zagreb, Pierottijeva 6, HR-10000 Zagreb, Croatia

Analgesics of the phenazone type, including propyphenazone, aminopyrine, and antipyrine, were determined in solid waste and leachate from the main landfill of the city of Zagreb, Croatia, as well as in soil and groundwater below the landfill. All structural identifications and quantitative analyses were performed using high-resolution gas chromatography/mass spectrometry. The analyses of solid waste revealed that propyphenazone and aminopyrine were among the most abundant specific anthropogenic compounds in the landfill, while antipyrine was detected only in trace concentrations. Since the landfill does not include any protective barrier, heavily contaminated landfill leachate plume penetrates rapidly into the underlying soil, which is composed of highly permeable alluvial sediments, and eventually reaches groundwater aquifer. As a consequence, propyphenazone and aminopyrine were determined in high concentrations (up to 67µg/L) in groundwater below the landfill. However, aminopyrine seems to be strongly attenuated in the aquifer section near the landfill, while propyphenazone showed a rather high persistence and mobility in the leachate plume.

In the last few years, the number of reports in the literature expressing concern about possible environmental effects of various pharmaceutical chemicals is rapidly increasing (see ref. *1* for review). A wide variety of these highly biochemically active chemicals, including lipid regulating agents, antiphlogistics, betablockers, β_2-symphatomimetics, psychiatric drugs, antiepileptic drugs, synthetic estrogens, and antibiotics, is used in human and veterinary medicine and subsequently released into the environment (*1,2,3*). The percentage of manufactured pharmaceuticals that are eventually excreted into the aquatic environment is dependent on their pharmacokinetics and therefore is very specific for individual drugs. The direct input of pharmaceuticals into the environment, such as wastewater discharge from manufacturing units and disposal of unused medications, can also be significant, especially at some hot-spots.

Antiphlogistic drugs (analgesics-antipyretics) belong to the most popular pharmaceutical chemicals. Several reports indicated recently that analgesic compounds are rather common constituents of municipal wastewaters (*2-5*). Analgesics of the phenazone type were detected in most of the examined German municipal sewage effluents and rivers with the median concentration values of 0.16 and 0.024 µg/L, respectively (*3*). Heberer et al. (*6*) indicated a possible impact of sewage effluents, containing propyphenazone and antipyrine, on groundwater from wells situated downstream of sewage treatment plants. The concentrations of phenazone analgesics in sewage effluents and groundwater were of the same order of magnitude, suggesting their efficient infiltration into groundwater.

Another important route of input of the pharmaceutical chemicals into groundwater is the release from landfills. Since many of the disposal sites are open dumps without protective barriers and leachate collection systems (*7*), the infiltration of heavily contaminated leachates could seriously jeopardize the quality of groundwater near landfills (*8, 9*). Landfill leachates were shown to contain complex assemblages of various contaminants (*10*) but until recently pharmaceutical chemicals have been neglected as possibly important groundwater contaminants (*9, 10*). However, studies by Eckel et al. (*11*) and Holm et al. (*12*) indicate that pharmaceutical chemicals, including analgesics, are among the most prominent anthropogenic constituents in landfill-leachate polluted groundwater. The aim of this study was to determine the impact of phenazone analgesics (Fig. 1) from the municipal solid waste landfill of the city of Zagreb (Jakuševec landfill), Croatia, on underlying soil and groundwater below and near the landfill.

Materials and Methods

Field site

This study was conducted on the main landfill of the city of Zagreb (Jakuševec landfill), Croatia (Fig. 2), which has been operated since 1965 and covers an

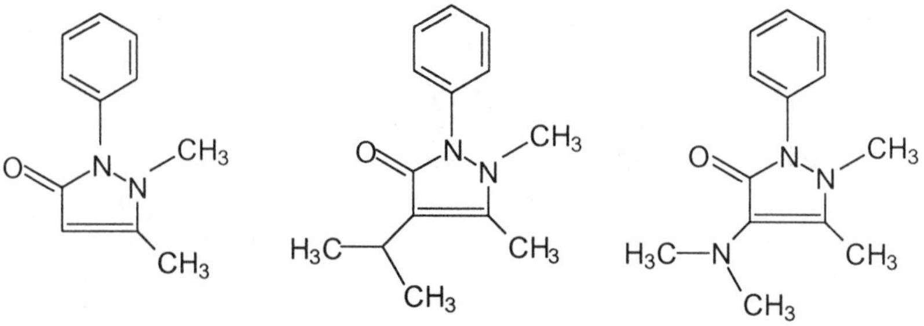

ANTIPYRINE   PROPYPHENAZONE   AMINOPYRINE

Figure 1. Chemical structures of phenazone analgesics

area of 800,000 m^2. The landfill contains presently about 5-7 million tons of solid waste (*13*). The waste is mostly of domestic origin, but significant contributions of commercial and industrial waste were also detected (*14*). Until 1997, the whole landfill, which is now under reconstruction, was operated as an open dump and posed a major risk to groundwater resources situated downstream (*15, 16*). The prevailing direction of groundwater flow is parallel with the Sava River bed (toward southeast). The waste has been disposed directly onto highly permeable alluvial sediments, while the permanently saturated zone of the aquifer is situated only 2 to 4 m below the landfill bottom. Consequently, the penetration of landfill leachates into the groundwater is very fast and efficient, resulting in a large leachate plume, which can be detected at distances longer than 1.5 km (*16*). This study was carried out in the southeastern part of the landfill, which consisted of two different sectors depicted as A and B in Figure 2. The height of filling in sector A was highly variable, ranging from 0-1 m (boreholes JRP-2, JRP-3, JRP-6) to 9-12 m (boreholes JRP-4, JRP-5, JRP-7 and JRP-9). Its average age at the moment of sampling was <10 years, with significant portions of the waste being in a highly active acetogenic phase of the waste stabilization (*14*). The deeper layers of the landfill, however, were sulfate- reducing and/or methanogenic (*14*). The sector B (boreholes JRP-11 to JRP-17) is an abandoned part of the landfill containing waste older than 15 years. The thickness of the waste layer in this section is generally lower than 2m.

Sample collection

Samples of solid waste and soils were obtained by drilling, which was carried out in such a manner that the original vertical structure of the core (i.d., 10 cm) was kept undisturbed. The analyses of solid waste were performed on

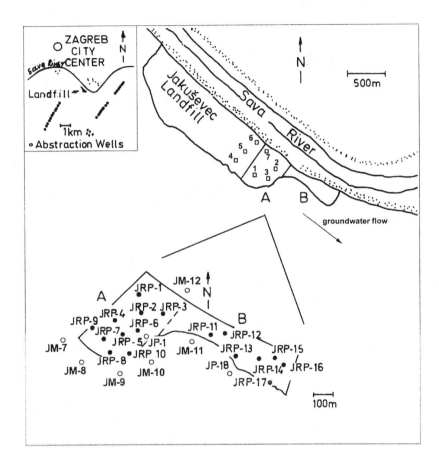

Figure 2. Map of the main landfill of the city of Zagreb (Jakuševec landfill) indicating locations of boreholes for solid phase sampling (filled circles; numbers assigned with JRP), piezometers for groundwater sampling (open circles) and locations of the leachate collection (squares). The boreholes JRP-9 and JRP-10 served also as piezometers.

composite samples, which were obtained by mixing subsamples from different depths, representing discrete 1-m layers. Each soil sample, however, represented a distinct 1-m subsurface soil layer situated immediately below the landfill bottom. All analyses of solid samples were performed on their size-fractions smaller than 2 mm, which were obtained by sieving from air-dried original samples. Groundwater samples were collected from several observation wells (Fig. 2), which were made of perforated high-density polyethylene (i.d. 105 mm). The sampling was performed by suction, using a Grunfos MP-1 pump equipped with a PTFE tubing. Before sample collection, each piezometer was purged by pumping groundwater for at least 20 min. Leachate samples were collected by hand using a glass beaker from small streams emanating from the landfill after a rain event.

Analysis

Solid samples (50 g of soil and 30 g of solid waste) were extracted with dichloromethane for 45 minutes using an ultrasonic bath. After the collection of the dichloromethane extract, the same procedure was repeated with methanol and the two extracts were combined. The total extract was passed through anhydrous sodium sulfate and evaporated to dryness using rotary evaporation. The concentrated extract was subsequently subjected to liquid chromatographic separation using a column filled with deactivated silica (15% organic-free water) (17). Nonpolar and moderately polar compounds were eluted with hexane and dichloromethane, respectively, and collected as separate fractions. Phenazone analgesics were subsequently eluted in the most polar fraction using methanol. The aqueous samples (5 L of groundwater and 1-2.5 L of leachate) were extracted with dichloromethane in 2.5-L glass bottles by vigorous shaking for 45 min. on a shaker. The organic phase was separated in a separatory funnel and processed further in the same manner as the solid sample extracts. The methanol fractions from the silica separations were evaporated to dryness, reconstituted in dichloromethane, and subjected to gas chromatography/mass spectrometry (GC/MS) operated in full scan mode to elucidate the structures of detected compounds. The identification of analgesics was performed using authentic standards of individual compounds as a reference. Quantitative GC/MS determinations were carried out in selected ion monitoring mode (SIM) after addition of deuterated phenanthrene into concentrated extracts as an internal standard. The molecular ions of analyzed compounds (m/e 188, 230, and 231) were used for the SIM-GC/MS detection. A Hewlett Packard system consisting of a gas chromatograph (Series 5800) and mass selective detector (MSD 5971A) was used for all analyses. The apparatus was equipped with splitless injector and a 0.25-μm HP5 fused-silica column (20 m x 0.2 mm).

The reproducibilty of the analytical procedure was 5-15% and the detection limits of individual compounds, based on 1 L of aqueous sample and 50 g of solid sample, were around 10 ng/L and 2 ng/g, respectively. The recovery of a duplicate extraction of propyphenazone, aminopyrine, and antipyrine from aqueous samples was 90%, 64%, and 25%, respectively. The final concentrations were corrected for the recovery.

Results and Discussion

Phenazone analgesics belong to the most abundant classes of anthropogenic compounds in the main landfill of the city of Zagreb (14, 18). Qualitative analyses of solid waste and landfill leachates from the southeastern part of the landfill by GC/MS provided positive identification of three phenazone chemicals, including propyphenazone (PRO; 4-(1-methylethyl)-2,3-dimethyl-1-phenyl-3-pyrazolin-5-one), aminopyrine (AM; 4-(dimethylamino)-2,3-dimethyl-1-phenyl-3-pyrazolin-5-one), and antipyrine (ANT; 2,3-dimethyl-1-phenyl-3-pyrazolin-5-one). Figure 3 shows SIM-GC/MS chromatogram of a typical landfill leachate extract, containing high levels of propyphenazone and aminopyrine, while antipyrine was detected only at trace levels (below 50 ng/L). A similar feature of the distribution of individual phenazone compounds was found in solid waste, soils, and groundwater, and therefore, for sake of clarity, antipyrine was omitted from Figures 4-8.

Figure 4 shows the distribution of propyphenazone and aminopyrine in solid waste. Propyphenazone was present in all analyzed samples from sector A in detectable concentrations, ranging typically from 0.05 to 0.65 mg/kg, while a hot-spot at borehole JRP-10 reached up to 20 mg/kg. Aminopyrine levels were significantly lower (0.005-0.02 mg/kg). Extension of the study into sector B confirmed a widespread presence and heterogeneous spatial distribution of propyphenazone and aminopyrine in the landfill. In the older part of the landfill (sector B) propyphenazone was also much more abundant than aminopyrine, reaching high levels of 1.6 and 22 mg/kg in the boreholes JRP-11 and JRP-15, respectively. In fact, the ratio between the two phenazone compounds was very similar in most of the moderately contaminated cores from the sectors A and B, possibly indicating a similar origin of the pharmaceutical waste disposed in different periods during the 30-year history of the landfill site (Table I). However, the concentration ratio between propyphenazone and aminopyrine was dramatically enhanced in the cores from the hot-spot boreholes (Table I), suggesting a larger disposal of medications, which contained only propyphenazone. The results may indicate a persistence of phenazone compounds in the landfill over a significant time scale of several decades. This is in agreement with some other suggestions that under anoxic conditions supposedly unstable drugs persist over long time-periods (11).

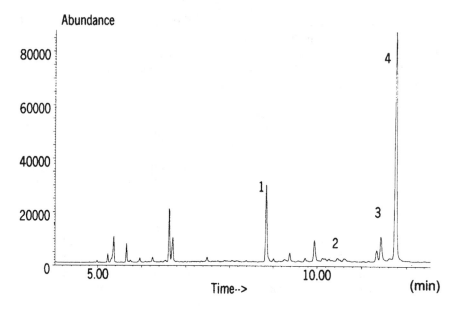

Figure 3. SIM-GC/MS chromatogram of a landfill leachate sample collected at the edge of the landfill of the city of Zagreb; 1 - deuterated phenanthrene (internal standard), 2 - antipyrine, 3 - aminopyrine, 4 - propyphenazone.

The amount of phenazone compounds in the southeastern part of the landfill was estimated assuming that each core was representative of the zone situated around the given borehole and taking into account the thickness of the solid waste layer in each zone (Table I). Since the total area of the southeastern part of the landfill (sectors A and B) was 155,000 m², the area associated with each of 17 boreholes was 9000 m². The estimate shows that this part of the landfill contains about 800 kg of propyphenazone and 3 kg of aminopyrine.

Propyphenazone and aminopyrine are relatively polar compounds, which reveal high solubility in water (*19*) and moderate log K_{ow} values of 2.32 (*12*) and 0.80 (*20*), respectively. As a consequence, both compounds could be easily remobilized from the solid waste by infiltrating leachates. The distribution of propyphenazone and aminopyrine in landfill leachate samples collected at different locations around the southeastern part of the landfill are presented in Figure 5. All leachates contained combined phenazone analgesics with total concentrations ranging from 4.3 to 67 µg/L. Most of the leachate samples were collected after strong rain events and are therefore more dilute compared with those leachates that are formed by a slower percolation of infiltrating water during the drier periods. Propyphenazone was the clearly predominant individual compound in all landfill leachates (3.7-60 µg/L), while the concentration of aminopyrine was lower (0.06-16 µg/L). However, the relative

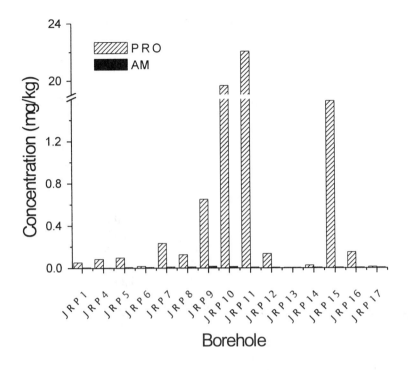

Figure 4. Distribution of phenazone analgesics in solid waste from the main landfill of the city of Zagreb. PRO – propyphenazone; AM – aminopyrine.

contribution of aminopyrine was somewhat higher than in the solid waste. This is probably a consequence of a comparatively higher solubility of aminopyrine in the aqueous phase. Moreover, leachate integrates the contaminants over a larger area than the solid core, and its composition may be more representative of an average situation.

The water balance for the southeastern part of the landfill predicts that highly reducing leachate infiltrates into the aquifer at an average rate of about 100 m^3 per day (*21*). Taking an average concentration of 25 μg/L of propyphenazone and 3 μg/L of aminopyrine (Fig. 5), the total amount of phenazone analgesics released from this part of the landfill is estimated to be about 1 kg/year. Considering that the landfill contains about 800 kg of phenazone analgesics (Table I), it further suggests that the amount present should be sufficient to maintain their high levels in groundwater for more than 100 years.

Table I. Estimation of the total amounts of phenazone analgesics in solid waste disposed of in the sectors A and B (Fig. 2) of the southeastern part of the Jakuševec landfill (PRO=propyphenazone; AM=aminopyrine).

Borehole	Waste thickness (m)	Amount[a] (kg)		Ratio
		PRO	AM	PRO/AM
JRP1	1.7	0.56	0.054	10.4
JRP2	0	0	0	-
JRP3	0	0	0	-
JRP4	10	5.60	0.203	27.6
JRP5	9	3.69	0.269	13.7
JRP6	0.6	0.13	0.057	2.3
JRP7	10	15.7	0.608	25.7
JRP8	4	3.38	0.243	13.9
JRP9	12	28.08	0.691	40.6
JRP10	2.1	408	0.291	1407
JRP11	2.0	308	0.098	3143
JRP12	1.6	1.54	0.069	22.3
JRP13	0	0	0	-
JRP14	0.6	0.18	0.035	5.0
JRP15	1.2	14.9	0.057	266
JRP16	1.5	1.66	0.067	24.8
JRP17	1.2	0.180	0.091	2.0
Total	-	791.72	2.83	-

[a]The total area of the sectors A and B = 155 000 m^2 (area/borehole = 9000 m^2); average density of the analyzed waste = 2.5 g/cm^3.

The analyses of soil from the layer situated immediately below the landfill bottom (0-1 m) showed that phenazone analgesics were present in all samples in detectable concentrations with minima of 3 ng/g found in the boreholes JRP-13 and JRP-17 (Fig. 6). This can be explained by the fact that borehole JRP-13 was situated at the edge of the landfill and had no visible layer of solid waste, while solid waste from borehole JRP-17 contained very low concentration (0.012 mg/kg) of phenazone analgesics. The concentration of propyphenazone in soil was much lower than that in solid waste and varied in a wide range from 0.003 to 2.9 mg/kg. The spatial distribution reflected the general situation in the overlying waste, showing a maximum at the borehole JRP-10. Aminopyrine was also detected in the majority of the analyzed samples, but its concentration was low and varied in a relative narrow range of 0.003 to 0.007 mg/kg.

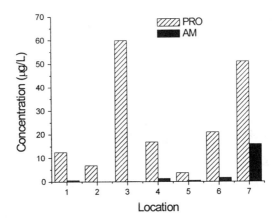

Figure 5. Distribution of phenazone analgesics in leachate samples collected on the main landfill of the city of Zagreb. PRO – propyphenazone; AM – aminopyrine.

Taking into account the concentration of phenazone analgesics in soils at different locations of the landfill, it was estimated that their total amount sorbed in a 1-m soil layer below the landfill equals about 60 kg. This is only about 10% of the amount present in the landfill (Table I). Still, considering that the release of phenazone analgesics into the aquifer by infiltrating leachates at a rate of 1 kg/year resulted in a significant contamination of underlying groundwater, the phenazone pool in the polluted soils below the landfill must be considered significant. Therefore, it was recommended that remediation of the Jakuševec landfill should include not only removal of solid waste, but also the excavation of at least 1 m of contaminated soils below the landfill.

The analyses of groundwater samples collected from the piezometers within (JRP 9-10, JP1) and near (JM 7-12, JP18) the landfill (Fig. 2) in 1995 and 1996 revealed that phenazone analgesics were present in significant concentrations (Fig. 7 and Fig. 8). Propyphenazone was the prevailing compound in most of the analyzed samples, which clearly reflected its predominance in the landfill leachate. The concentration of propyphenazone in piezometers situated within the landfill was found typically in the range of 5-50 µg/L, which is the same order of magnitude as in the collected leachates. The concentration of aminopyrine varied widely from <0.050 to 36 µg/L. The maximum concentration of aminopyrine was determined in groundwater from the piezometer JRP-10 situated below heavily contaminated waste (Fig. 1). This concentration was even higher than the maximum concentration in analyzed leachate samples (16 µg/L). Since the dilution factor of the leachate plume in the

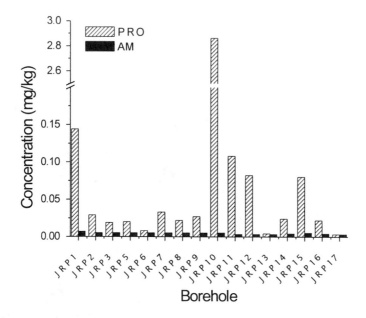

Figure 6. Distribution of phenazone analgesics in soils situated below the main landfill of the city of Zagreb. PRO – propyphenazone; AM – aminopyrine.

aquifer below the landfill, which was estimated using chloride as an conservative tracer, ranged from 3 to 5 (*22*), elevated concentrations of phenazone analgesics in groundwater indicate the existence of other hot-spots outside sectors A and B. They also suggest that the vertical transport of both phenazone analgesics into the subsurface by highly reducing leachates is very efficient.

The concentration of propyphenazone and aminopyrine gradually decreased with distance from the hot-spots mentioned above (JRP-10 and JRP-11), but the spatial distribution was also dependent upon direction and the sampling date. Sampling conducted in May and July 1995 (Fig. 7) represent time-points immediately before starting a major reconstruction of the Jakuševec landfill. The reconstruction had begun with excavations of the waste in sector A and its translocation to the neighboring part of the landfill in the vicinity of the borehole JRP-7. As can be seen, the concentration and the composition of phenazone analgesics in groundwater below the landfill (piezometers JRP-9 and JRP-10) were relatively constant in the period before the excavations started with slightly lower concentrations found in July 1995 than May 1995.

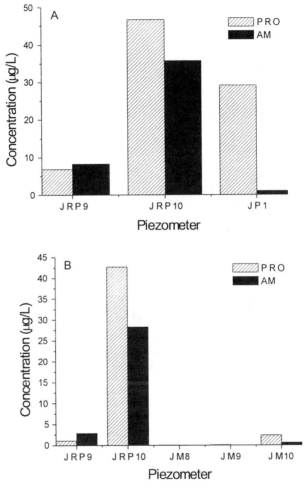

Figure 7. Distribution of phenazone analgesics in groundwater below and near the main landfill of the city of Zagreb before remediation of its southeastern part: A) May 1995 and B) July 1995; PRO - propyphenazone; AM – aminopyrine.

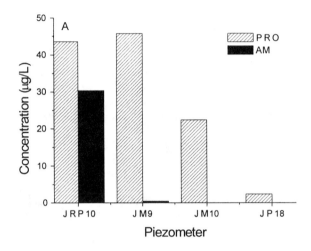

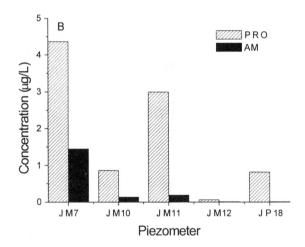

Figure 8. Distribution of phenazone analgesics in groundwater below and near the main landfill of the city of Zagreb during remediation of its southeastern part: A) September 1995 and B) January 1996; PRO - propyphenazone; AM – aminopyrine.

The distribution of analgesics in July showed that the plume was spreading mainly in the direction of the piezometer JM-10, which is typical of medium levels of groundwater (*16*), while piezometers JM-8 and JM-9 were almost unaffected, with concentrations of phenazone analgesics below 50 ng/L. The situation observed in September 1995 (Fig. 8), characterized by a rather high concentration of propyphenazone in piezometer JM-9, is very interesting because it illustrates the effect of hydrological conditions following continued heavy rains and the effect of the excavation on the distribution of contaminants in the adjacent aquifer. Enhanced groundwater levels divert the main direction of the plume spreading towards the south (*16*), while the partial translocation of the waste resulted in a smaller thickness of the waste layer, allowing faster and more efficient penetration of the leachate into the subsurface. After the completion of the major part of the excavations (January 1996), the concentration of phenazone analgesics in all piezometers slightly decreased, except in piezometer JRP-7, which became a new hot-spot due to the translocation of the contaminated waste near that location (Fig. 8). Enhanced concentration, found in the piezometer JM-11, reflects probably the impact of the hot-spot at the borehole JRP-11 (Table I).

Behavior of aminopyrine and propyphenazone in the leachate plume seem to be different. Namely, the concentration decrease of aminopyrine in groundwater with increasing distance from the landfill was much faster than the concentration decrease of propyphenazone. Enhanced concentrations of aminopyrine were observed only in groundwater below the landfill, while in the most distant piezometer, JP18, the concentration was below the detection limit (20 ng/L). The prevailing redox conditions in the aquifer section between the two piezometers were shown to change from methanogenic or sulfate-reducing to manganese/nitrate reducing conditions (*16*). Propyphenazone displayed apparently a very persistent behavior in the aquifer and can be regarded as an excellent tracer of the leachate plume. Similar observations on the occurrence of propyphenazone in a leachate plume were made by Holm et al. (*12*).

Conclusions

Analgesic compounds of the phenazone type, notably propyphenazone and aminopyrine, can occur in significant concentrations in municipal solid waste landfills. Due to their high solubility in water, they exhibit a rather high mobility in the landfill, which results in an efficient transport into the subsurface by landfill leachates. The retention of phenazone analgesics by soils below the landfill seems to be very limited. Thus, lacking a protective barrier, it can be predicted that the largest portion of these compounds released from the landfill will reach the groundwater aquifer. However, the leaching is not a flushing-like process but proceeds at a rather moderate rate (1 kg/year), which results in a chronic contamination of the aquifer by phenazone analgesics at concentrations exceeding 10 µg/L over long time periods. Comparison of concentration levels of individual analgesic compounds in solid waste, soil, landfill leachate, and groundwater suggests that aminopyrine is probably eliminated by yet unknown biotic and/or abiotic transformations in the aquifer zone near the landfill. In contrast, propyphenazone showed a very high mobility and persistence in the leachate plume. Following such estimates of the possible long-term effects, the Jakuševec landfill is currently subject of a major remediation project, which includes building a double protective barrier and leachate collection system.

Acknowledgments. This work was carried out within a project aimed at assessing strategies for the remediation of the Jakuševec landfill that was funded by the city of Zagreb. Support by the Ministry of Science and Technology of the Republic of Croatia is also acknowledged. We thank Nataša Tepic for technical assistance.

References

1. Daughton, C.G.; Ternes, T.A. *Environ. Health Perspec. Suppl.* **1999**, *107* (suppl. 6), 907-938.
2. Stan, HJ.; and Heberer, T. *Analusis* **1997**, *25*(7), 20-23.
3. Ternes, T.A. *Wat. Res.* **1998**, *32*, 3245-3260.
4. Rogers, I.H.; Birtwell, I.K.; Kruzynski, G.M. *Water. Poll. Res. J. Can.* **1986**, *21*, 187-204.
5. Buser, HR., Poiger, T.; Müller M. *Environ. Sci. Technol.* **1999**, *33*, 2529-2535.
6. Heberer T.; Dünnbier, U.; Reilich, C.; Stan, HJ. *Fresenius Envir. Bull.* **1997**, *6*, 438-443.
7. Stegmann, R. In *Proceedings Sardinia 95, 5th International Landfill Symposium*, Christensen, T.H.; Cossu, R.; Stegmann, R., Eds.; CISA: Cagliari, Italy, 1995; Vol. 1, pp 2-12.
8. Christensen, T.H.; Kjeldsen, P.; Jansen, J.C. In *Landfilling of Waste: Leachate*; Christensen, T.H.; Cossu, R.; Stegmann, R., Eds.; Elsevier Applied Science: London, Great Britain, 1992; pp 29-49.
9. Kerndorff, H. *Int. J. Environ. Anal. Chem.* **1995**, *60*, 239-256.
10. Christensen, T.H. In *Landfilling of Waste: Leachate*; Christensen, T.H.; Cossu, R.; Stegmann, R., Eds.; Elsevier Applied Science: London, Great Britain, 1992; pp 441-483.
11. Eckel, W.P.; Ross, B.; Isensee, R.K. *Groundwater* **1993**, *31*, 801-804.
12. Holm, J.V.; Rugge, K.; Bjerg, P.L.; Christensen, T.H. *Environ. Sci. Technol.* **1995**, *29*, 1415-1420.
13. Olujic, M. *Gospodarstvo i okoliš* **1995**, *III*(5), 288-295.

14. Ahel, M.; Mikac, N.; Cosovic, B. Prohic, E.; Soukup, V. *Wat. Sci. Tech.* **1998**, *37*(8), 203-210.
15. Ahel, M. *Bull. Environ. Contam. Toxicol.* **1991**, *47*, 586-593.
16. Mikac, N.; Cosovic, B.; Ahel, M.; Andreis, S.; Toncic, Z. *Wat. Sci. Tech.* **1998**, *37*(8), 37-44.
17. Ahel, M.; Giger, W. *Kem. Ind.* **1985**, *34*, 295-309.
18. Ahel, M.; Jelicic, I. In *Proceeding of the 3rd IWA Specialized Conference on Hazard Assessment and Control of Environmental Contaminants. ECOHAZARD'99*; Matsui, S., Ed.; Kyoto University: Kyoto, Japan, 1999; pp 497-504.
19. *The Merck Index*; Budavari, S., Ed.; An Encyclopedia of Chemicals, Drugs, and Biologicals, Twelfth Edition; Merck Research Laboratories: Whitehouse Station, NJ, 1996; pp 1741.
20. Leo, A.; Hansch, C.; Elkins, D. *Chem. Rev.* **1971**, *71*, 525-616.
21. Milanovic, Z.; Matasovic, M.; Jurkovic, J.; Shalabi, M.; Galic, I. In *Proceedings of 2nd Croatian Conference on Waters*; Gereš, D., Ed.; Hrvatske vode: Zagreb, Croatia, 1999; pp 157-165 (in Croatian).
22. Ahel, M.; Mikac, N.; Cosovic, B.; Prohic, E. In *Proceedings Sardinia 99, Seventh International Waste Management and Landfill Symposium*; Christensen, T.H.; Cossu, R.; Stegmann, R., Eds.; CISA: Cagliari, Italy, 1999; Vol. 4, pp 141-148.

Chapter 7

Pharmaceuticals and Personal Care Products in the Waters of Lake Mead, Nevada

Shane A. Snyder[1,4], Kevin L. Kelly[2], Andrew H. Grange[3], G. Wayne Sovocool[3], Erin M. Snyder[4], and John P. Giesy[4]

[1]Southern Nevada Water Authority, Las Vegas, NV 89153
[2]United States Bureau of Reclamation, Denver, CO 80225
[3]U.S. Environmental Protection Agency, ORD/NERL, Las Vegas, NV 89119
[4]Department of Zoology, Michigan State University, East Lansing, MI 48824

Several recent reports have indicated that a variety of pharmaceuticals and personal care products are entering the aquatic environment via wastewater discharges. The Boulder Basin of Lake Mead, Nevada, receives a significant input of highly treated wastewater from the Las Vegas Wash. The objectives of the studies presented here were to determine which compounds entering Lake Mead may be able to induce endocrine disruptive effects on aquatic life and to identify previously unreported xenobiotic compounds. Various locations in the Las Vegas Wash and Boulder Basin were investigated between 1997 and 1999. Certain pesticides, polychlorinated biphenyls (PCBs), alkylphenols (APs), polycyclic aromatic hydrocarbons (PAHs), and steroids were target compounds. In the first tier of this research, genetically engineered cell lines were used to screen for estrogenic compounds in organic extracts of Lake Mead water samples. Radioimmunoassay was used to identify and quantitate natural and synthetic estrogens. Bioassay-directed fractionation and mass-balance were used to determine which xenoestrogens exhibited the greatest potency in an estrogen-sensitive cell line. This work demonstrated that natural and synthetic estrogens within the mixture of organic compounds extracted from Lake Mead water samples were responsible for the greatest induction in the estrogen-responsive cellular bioassay. Detectable concentrations ranged from 0.2 – 0.5 and 0.2 – 3 ng/L for ethynylestradiol (EE2) and 17β-estradiol (E2), respectively. The second tier of analysis involved the extraction of 100 L of water followed by gas chromatography with mass selective detection. This methodology was used to identify and quantify known organic compounds and to screen for previously unreported compounds. High-resolution mass spectrometry was used to identify and verify several polar organic compounds that have not previously been reported to be present in Lake Mead. Compounds identified include APs, pharmaceuticals, caffeine, nicotine, flame-retardants, and insect repellants. Detectable concentrations of pharmaceuticals and personal care products ranged from 1 – 1500 ng/L. Concentrations of xenobiotics were greater in samples collected near the confluence of the Las Vegas Wash with Lake Mead at the Las Vegas Bay.

Lake Mead, Nevada is the largest reservoir on the Colorado River by volume (approximately 36.7×10^9 m^3) and covers approximately 593 km^2 of surface area. Lake Mead was formed in 1935 with the impoundment of the Colorado River by the Hoover Dam. The Colorado River provides approximately 97% of Lake Mead's water. The remaining water originates from the Virgin and Muddy Rivers ($\approx$ 1.5%), which discharge into the Overton Arm, and the Las Vegas Wash ($\approx$ 1.5%), which discharges into the Boulder Basin (Figure 1). More than 22 million people depend on water from Lake Mead and the lower Colorado River for domestic and agricultural usage (*1*).

Tertiary-treated wastewater, storm water, urban runoff, and shallow groundwater enter the Las Vegas Bay of the Boulder Basin of Lake Mead via the Las Vegas Wash. The Las Vegas Wash is a 19-km channel that drains the entire 4100 km^2 Las Vegas Valley hydrographic basin. Approximately 5.8×10^8 L of water per day (excluding storm water) flow through the Las Vegas Wash.

The drinking water intakes for southern Nevada are approximately 10 km down stream of the Las Vegas Wash confluence. The Las Vegas Bay exhibits frequent eutrophication due to nutrient loading from the Las Vegas Wash (*1-3*). Secchi readings of less than 0.5 m and chlorophyll *a* concentrations of greater than 300 mg/m^3 have been reported (*1*). Suspended solids are elevated in the Las Vegas Bay because the Las Vegas Wash flows rapidly and erodes its channel, picking up solids before entering the lake. Water from the Las Vegas Wash is warmer and has greater conductivity, turbidity, and suspended and dissolved solids than main-body Lake Mead water. Although Las Vegas Wash water is warmer, this water has a greater density, which often results in an underflow that extends until density equilibrium is met. This intrusion then can form an interflow or overflow above the hypolimnion in the main body of Lake Mead. At certain times of the year, the intrusion of the Las Vegas Wash can be detected at various depths to Hoover Dam (approximately 16 km down stream) (*1*).

In 1996, feral carp from the Las Vegas Wash and Las Vegas Bay were reported to have significantly different plasma sex steroid and vitellogenin levels than carp collected from a reference site in Callville Bay (*4*). These types of endocrine disruptive effects have been associated with estrogenic substances found in wastewater effluents (*5-10*). Concentrations of some synthetic organic chemicals were found to be greater in water, sediment, and fish tissues from the Las Vegas Wash and Las Vegas Bay when compared with similar samples from the Callville Bay reference site (*4*). For these reasons, various studies were initiated to determine what compounds may be impacting the aquatic environment of the Las Vegas Bay and whether the effects documented in feral fish could be reproduced in fish caged at the same sites.

In 1997, Michigan State University (MSU) was invited to use toxicity identification and evaluation (TIE) methodology to screen the waters of Lake Mead for estrogenic compounds. The analytical methodology has been reported previously (*11*). Briefly, organic compounds were extracted from 5 L of water using solid-phase extraction (SPE). The corresponding organic extracts were separated into three fractions based on polarity. Each extract and fraction was tested for estrogenicity using the MVLN (MCF-7-luc) *in vitro* gene expression assay. This cellular bioassay is extremely sensitive to estrogen (*12, 13*). Once extracts or fractions exhibiting estrogenic activity were identified, they were separated into finer fractions and again tested using the bioassay. This bioassay-directed fractionation and subsequent instrumental analysis allowed for mass-balance evaluation. Radioimmunoassay also was applied to determine concentrations of the natural estrogen 17β-estradiol (E2) and the synthetic estrogen ethynylestradiol (EE2), which is used in oral contraceptives.

From 1998 through 1999, MSU began to extract organic compounds from 100-L samples of Lake Mead to increase analytical sensitivity and generate greater quantities of extract for additional testing. The method was designed to identify and quantify compounds detected in the 1997 study and to identify other yet undetected organic compounds. Extracts were again separated into three fractions based on polarity. Gas chromatography with quadrupole mass spectrometry (GC/MS) was used to identify and quantify these target compounds and to screen for previously unreported organic compounds. Previously undetected compounds were identified using the National Institute of Standards and Technology (NIST) mass spectral library. Structural confirmation was achieved using authenticated standards when available. Quantitation was accomplished using the GC/MS in selected ion monitoring (SIM) mode with calibration against authentic standards.

Gas chromatography with high-resolution mass spectrometry (GC/HRMS) was used for structural verification of those compounds never before reported in the waters of Lake Mead. Mass Peak Profiling from Selected Ion Recording Data (MPPSIRD) and a Profile Generation Model (PGM) were used in conjunction with GC/HRMS to identify trace concentrations of organic compounds, including pharmaceuticals. These methods are effective for determining elemental compositions of unknown organics present in complex environmental extracts (*14-16*). The data presented for previously unreported compounds are semi-quantitative due to the methodologies applied.

Toxicity Identification and Evaluation of Estrogenic Compounds

Since estrogenic-type effects were documented in fish from the Las Vegas Bay of Lake Mead, a project was initiated to aid in determining what estrogenic compounds might be present in these waters. To screen for estrogen-mimicking compounds in a more comprehensive fashion, a TIE method was developed. TIE schemes combine bioanalytical endpoints with analytical chemistry data to identify specific chemicals or groups of chemicals responsible for observed toxicity. Following bioassay-directed chemical fractionation coupled with subsequent instrumental analyses, mass-balance could be performed to ascertain whether all observed bioactivity could be accounted for by the compounds identified. This method was applied at one site in the Las Vegas Wash, two sites in the Las Vegas Bay, and two reference sites, all within the Boulder Basin of Lake Mead (Figure 1).

Analytical method

A detailed description of the analytical method has been published previously (*11*). Briefly, 5-L water samples were extracted at each field site using solid-phase extraction (SPE) Empore™ disks. Organic extracts from these SPE disks were separated into three fractions based on polarity using normal-phase high-performance liquid chromatography (NP-HPLC). These fractions were called F1, F2, and F3, in order of increasing polarity. In some cases, F2 and F3 extracts were further fractionated with reverse-phase HPLC (RP-HPLC) with fractions collected every 3 min to better isolate bioactive compounds (Figure 2). Chromatographic conditions for fine fractionation were described previously (*17*). NP-HPLC and RP-HPLC separations were accomplished using silica and C_{18} analytical columns, respectively. Instrumental analyses included HPLC with fluorescence detection, GC/MS, and competitive radioimmunoassay (RIA) (*11*).

In vitro bioassay

The MCF-7 human breast carcinoma cell line, stably transfected with an estrogen receptor-controlled luciferase reporter gene construct (MVLN or MCF-7-luc cells), was used to screen organic extracts for estrogenic compounds. This cell line has an endogenous human estrogen receptor (ER) and is able to detect as little as 50 pM E2 (*13*). Extracts were dissolved in culture media to yield a final concentration of 1.0% extract. A 3-fold dilution of each extract also was prepared, yielding a concentration of 0.33% extract. Test wells were dosed with 125 µL 1.0% or 0.33% extract in media to yield final in-well concentrations of 0.50% and 0.165% extract. Solvent control wells were dosed with 125 µL of media spiked with 1.0% of the appropriate solvent to yield a final in-well concentration of 0.50% solvent. Blank wells received 125 µL of the appropriate media. A 5-point E2 calibration curve was run on approximately every other plate using standards prepared in the same manner as the sample extracts. Dosed cells were exposed for 72 h at standard incubation conditions. Light production as a function of ER activity was measured using a plate-scanning luminometer. Light production from sample wells was compared with the maximum light production from the E2 standard curve as %E2-max.

Results

Several classes of xenobiotic compounds were detected in Lake Mead water samples. Of particular interest were the alkylphenols (APs), octylphenol (OP)

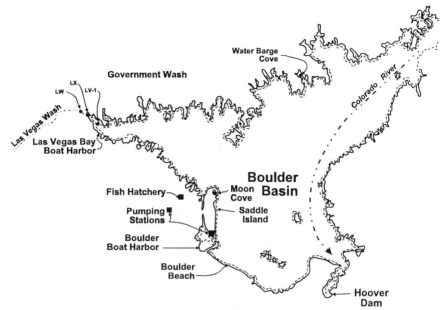

Figure 1. Boulder Basin Study Area of Lake Mead, Nevada

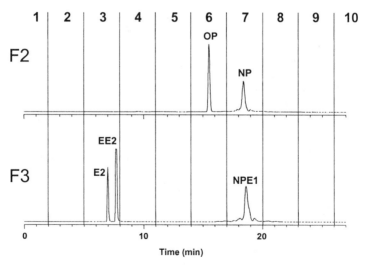

Figure 2. HPLC/fluorescence fine fractionation of spike standards. OP=octylphenol; NP=nonylphenol; E2=17β-estradiol; EE2=ethynylestradiol; NPE1=nonylphenol monoethoxylate.

and nonylphenol (NP), which are estrogenic degradation products of the widely used nonionic alkylphenol polyethoxylate (APE) surfactants used in several personal care products. Concentrations of these APs ranged from 27 to 1140 ng/L in the Las Vegas Wash and Las Vegas Bay and were not detected at the reference sites. Although E2 and EE2 were not detected using conventional analytical instruments, these estrogens were detected and quantified using RIA. Concentrations of E2 and EE2 were detectable only in the Las Vegas Wash and Las Vegas Bay at levels ranging from 0.19 – 2.7 and 0.25 – 0.52 ng/L, respectively (*11*).

Results of the bioanalyses indicated that E2 and EE2 were responsible for all estrogenic activity within the water extracts. Only the third, most polar, fraction of the extracts was able to induce significant estrogenic activity. This indicates that concentrations of non-polar and moderately-polar compounds, such as halogenated aromatic hydrocarbons (HAHs) and APs, were not great enough to induce a detectable response in the MVLN cellular bioassay. When the F3 extracts were fractionated further, and each resulting fine fraction was analyzed using the MVLN assay, the maximum induction closely correlated to the HPLC elution times of E2 and EE2 (Figure 3). Conversely, no induction was observed at the elution time of APEs, although these were readily detectable using instrumental analyses. Likewise, F2 extracts were finely fractionated to determine whether interfering compounds were antagonizing the cellular responses to the instrumentally detectable concentrations of the known estrogen agonists, OP and NP. Instrumentally determined concentrations of estrogenic compounds and their corresponding estrogenic potencies (as determined by the MVLN cellular bioassay) were compared with observed MVLN cellular responses to the sample extracts in order to determine mass-balance. It was concluded that the extract concentrations of E2 and EE2 were sufficient to induce the observed responses. Concentrations of OP, NP, and APEs were not great enough to induce a detectable cellular response. These data suggest that E2 and EE2 are the most potent estrogenic compounds (as determined by the MVLN assay) in water samples from the Las Vegas Wash and Las Vegas Bay. Although it is not appropriate to infer that E2 and EE2 are solely responsible for estrogenic effects observed in feral carp captured from these waters, it is certain that these compounds contribute to the overall estrogenicity of these waters. Furthermore, laboratory studies have shown that E2 and EE2 in low ng/L concentrations can induce certain estrogenic effects in fish (*6, 18-20*).

Large Volume Water Extraction and Mass Spectrometry

Following the results of the 1997 study where endogenous and synthetic estrogens were detected in the Las Vegas Wash and Las Vegas Bay of Lake Mead, a new analytical strategy was developed. During the 1998 to 1999 study, a 100-L volume of water was extracted to perform more thorough instrumental analyses. Initially low-resolution mass spectrometry (LRMS) (quadrupole technology) was used in conjunction with gas chromatography to identify selected target compounds such as pesticides, APs, APEs, HAHs, PAHs, and steroidal estrogens. Further, LRMS was used in scan mode to identify previously unreported compounds by using spectral matching software. When "new" compounds were tentatively identified, authentic standards were procured and spectra compared. The Environmental Protection Agency's National Environmental Research Laboratory (EPA-NERL) in Las Vegas, Nevada, analyzed some extracts using high-resolution mass spectrometry (HRMS) to determine molecular formulas of organic compounds in the extracts and to verify the composition of new compounds detected by LRMS. Using these technologies, several pharmaceuticals and personal care products were identified.

Water Extraction

Water samples were collected from five sites in the Boulder Basin of Lake Mead, namely LW, LX, LV-1, Water Barge Cove (WB), and Moon Cove (MC) (Figure 1). Samples were filtered at the sampling location by pumping water through a Pentaplate parallel filtration system (Micro Filtration Systems) equipped with one to five 30-cm glass fiber filters (GFFs). Approximately 130 L of filtered water was collected in a Teflon™-lined high-density polyethylene tank. The sample was transported to the Southern Nevada Water System (SNWS) laboratory at Lake Mead. Organic compounds were concentrated from the water samples by solid-phase extraction (SPE) using a stainless-steel preparative HPLC column packed with approximately 55 g of styrenedivinylbenzene resin (Bond Elut ENV™, 125-μm particle size, Varian, Harbor City, CA). Water was pumped through the SPE column at 0.1 L/min until a total volume of 100 L had been extracted (approximately 17 h). The SPE column was then capped and stored at 4 °C until elution.

Organic compounds isolated on the SPE column were eluted by passing various solvents through the column using an HPLC pump. Acetone, dichloromethane (DCM), and hexane were passed through the SPE column and collected in a 1-L separatory funnel containing 100 mL of a 25% sodium chloride solution. The resulting mixture was liquid/liquid extracted (LLE) and the organic layer retained in a separate container. The aqueous phase was extracted twice again with DCM. The combined organic phases were dried over anhydrous sodium sulfate and concentrated. This extract was fractionated using a gravimetric silica gel column into three fractions based on relative polarity. The first fraction (F1) was eluted with 100 mL of 20% DCM in hexane. A second and third fraction (F2 and F3) were eluted with 100 mL of 75% DCM in hexane and 100 mL 50% DCM in methanol, respectively. F1 contained the compounds of least polarity, such as PAHs, HAHs, and many organochlorine (OC) pesticides. F2 contained nonylphenol (NP), octylphenol (OP), and greater polarity OC pesticides. F3 contained the compounds of greatest polarity, including triazine herbicides; pharmaceuticals; N,N-diethyl-m-toluamide (DEET); oxybenzone; flame-retardants; steroids; and APEs. Each fraction was concentrated separately to 1 mL for instrumental analyses.

Low-Resolution Mass Spectrometry

LRMS analyses used a HP 5890 series II GC with a HP 5972 mass selective detector (MSD). A detailed description of this system has been published (*11, 12*). Initially, the mass spectrometer was operated in scan mode to identify new organic compounds in the water extracts. The NIST mass spectral database was used to tentatively identify unknown chromatographic peaks from the total ion chromatograms (TICs). Various columns and temperature programming were necessary to resolve multiple peaks from these complex chromatograms. Once tentative identifications were made, authentic standards were obtained and analyzed using the same GC/MS system. Tentative spectral identification was "verified" first by comparing the spectrum of a standard with the spectrum of the suspect peak from the water extract TIC (Figure 4). To quantify these compounds, and for further validation, the extracts were analyzed by operating the MSD in SIM mode with 2 – 3 ions per compound (Figure 5). Identifications were further confirmed by retention times and ion ratios, and quantitation was performed using HP ChemStation software. Spike recovery studies using this method with both laboratory and matrix water resulted in consistent recoveries greater than 70%. Some of these spike recoveries were done with standard concentrations much greater than the concentrations found in the natural water samples. This indicates that break-through due to column saturation was not a significant issue.

Several pharmaceuticals and personal care products were detected in water samples from Lake Mead (Table I). With the exception of DEET, oxybenzone, and o-homomenthyl salicylate, pharmaceuticals and personal care products were detected only at the Las Vegas Bay (LW, LX, LV-1) sites. Compounds marked

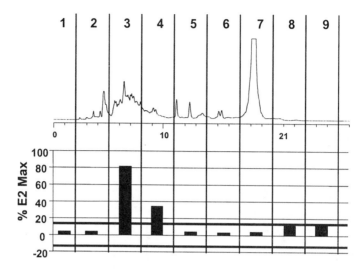

Figure 3. HPLC fine fractionation and corresponding estrogenicity of Las Vegas Wash water extract.

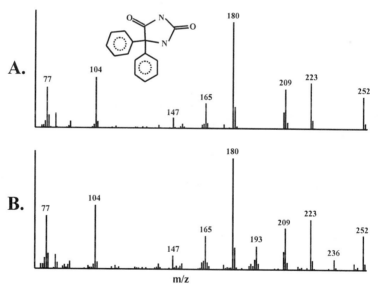

Figure 4. Mass spectra of phenytoin (Dilantin): (A) authentic standard; (B) LV Bay water extract

126

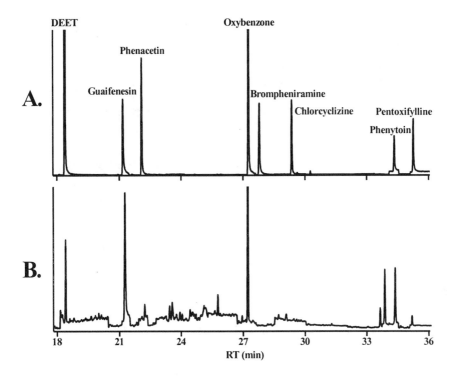

Figure 5. Mass spectral SIM chromatograms I: (A) authentic standards; (B) LV Bay water extract

Table I. Pharmaceuticals and Personal Care Products Detected in the Las Vegas Bay of Lake Mead.

Compound	Usage	Concentration Range (ng/L)
Phenytoin (Dilantin)	seizure treatment	53 – 261
Phenobarbital (Luminal)	seizure treatment	11 – 39
Carbamazepine (Tegretol)	seizure treatment	14 – 35
Primidone (Myidone)	seizure treatment	11 – 130
Hydrocodone	pain medication	6 – 13
Codeine	pain medication	17 – 123
Diazepam (Valium)	depression	3 – 62
Guaifenesin (Robotusin)	expectorant	21 – 52
Pentoxifylline (Trental)	Blood thinner	5 – 50
Meprobamate	antianxiety	Qualitative
Carisoprodol	muscle spasms	Qualitative
Methocarbamol	muscle spasms	Qualitative
Mequinol	skin pigments	Qualitative
Fulvicin	antifungal	Qualitative
Triclosan	antibacterial	Qualitative
Acetophenone	Fragrance	Qualitative
o-methoxyacetophenone	Fragrance	Qualitative
Vanillin	Fragrance	Qualitative
Octylphenol	surfactant (degradation)	2 - 23
Nonylphenol	surfactant (degradation)	5 - 308
Bisphenol A	plasticizer	2.4
DEET	insect repellant	3 – 268
Oxybenzone	Sunscreen	30 – 364
Homomenthyl salicylate	Sunscreen	qualitative
Acridine	Dye	qualitative
p-Chloroaniline	Dye	qualitative
Caffeine	coffee, tea, etc.	10 – 47
Nicotine	tobacco products	5 – 23

as qualitative were identified by NIST mass spectral database matches of greater than 90% certainty and/or by HRMS.

High-Resolution Mass Spectrometry

The GC/HRMS system used a CTC Analytics Model A220S autosampler (Zwingen, Switzerland) to make 1-μL, splitless injections onto a 30-m, 0.25-μm ID, 0.25-μm film, J&W Scientific (Folsom, CA), DB-5 GC column. Analytes eluted from the column in the oven of an HP 6890 GC into the ion source of a Finnigan MAT 900 S-trap hybrid mass spectrometer (Bremen, Germany). Only the double focusing MS stage was used in these studies. Full-scan, low-resolution (1000) mass spectra over mass ranges of 50-300 Da and 100-400 Da were acquired at 1.8 s/cycle and 1.1 s/cycle, respectively. Elemental compositions of specific ions observed in background-subtracted mass spectra were determined using MPPSIRD to acquire data and the PGM to interpret the data automatically. The SIM mode provided 31 m/z ratios. Five m/z ratios monitored 40% of the mass range of a lock mass peak profile and another five m/z ratios similarly monitored a calibration profile. The masses of these two profiles bracketed the mass of analyte profiles. Up to three experiments were required to determine each ion's composition. First, the remaining 21 m/z ratios monitored a 2000-ppm wide mass range centered on an estimated mass from a low resolution mass spectrum using 3000 resolution. This survey data was examined to ensure further study would focus on the most abundant analyte ion at its nominal mass and provided an exact mass estimate based on only three points across the profile. Next, this value was used as the center mass for a second experiment for which 10 m/z ratios at 10-ppm mass increments monitored the analyte's full profile with 10,000 resolution. When multiple compositions remained possible for the ion, a third experiment used three sets of seven m/z ratios to monitor 60% of the mass range of three partial analyte profiles again at 10,000 resolution or with 4-ppm mass increments using 25,000 resolution. Historically, a maximum error of 6 ppm for one exact mass determination made with 10,000 resolution was established using a VG70-SE double focusing mass spectrometer (16). Although 6 ppm overestimates the error when both a lock mass and a calibration mass are used, the presence of interferences for some low abundance ions can increase the errors observed. An error limit of ±3 ppm was used for exact masses obtained from profiles monitored with 25,000 resolution. A monitoring time of 20 ms for each m/z ratio and a 10-ms transition time between m/z ratios provided a 1-s cycle time. The secondary electron multiplier voltage was 1.60 kV, 1.85 kV, and 2.00 kV, at resolutions of 3000, 10,000, and 25,000, respectively. No quantitation by GC/HRMS was performed.

The F3 extracts contained hundreds of compounds, some of which continually degraded the GC column as injections were made. The most polar compounds within the extract provided wide chromatographic peaks. For instance, the width of the chromatographic peak for phenytoin (e.g., Dilantin) increased from 0.75 min to 1.5 min during the study. The complexity of the extract and the poor chromatography made background subtraction to obtain clean mass spectra difficult unless a large amount of a particular analyte was present or unless few coeluting interferences were evident. Background-subtracted mass spectra often contained extraneous ions or lacked low-abundance ions that should have been present. Only for a small fraction of compounds were clean mass spectra with numerous abundant ions obtained that provided convincing matches with library mass spectra and compelling tentative identifications.

Matches for poor quality mass spectra can lead to mis-identifications. Two such examples are shown in Figure 6 for compounds initially suspected to be talbutal, a barbiturate sedative, and acetaminophen, an over-the-counter analgesic. Figures 6b and 6d are the background-subtracted mass spectra. Figures 6a and 6c are the NIST library mass spectra for these two drugs. Both drugs provide mass spectra with few ions having large abundances. To increase confidence in tentative identifications or to minimize mis-identifications, elemental compositions of apparent molecular ions, and if necessary, fragment ions were determined from HRMS data using MPPSIRD and the PGM.

The data plotted in Figure 7a, revealed at least three ions with a nominal mass of 209. The ion chromatograms for the maxima of the three profiles displayed nearly steady signals for two ions, indicative of ever-present compounds such as calibrants or column bleed, and a chromatographic peak for the analyte. The estimated exact mass of 209.13889 Da for this low-abundance analyte ion was determined as the weighted average of the top three points defining its profile. This value was then used as the center mass to acquire the data plotted in Figure 7b. The analyte profile in Figure 7a was resolved into two. Only the ion chromatogram for the maximum of the higher-mass profile displayed a chromatographic peak. A much better estimate of the analyte ion's exact mass was obtained from the weighted average of the top six points that delineated its baseline-resolved profile.

The PGM provided a list of possible compositions containing C, H, O, N, P, F, S, or Si atoms with exact masses of 209.15451 ± 6 ppm (1.25 mmu for this mass). The possible compositions were $C_3H_{19}N_9Si$, $C_8H_{20}N_3O_2F$, $C_9H_{24}NO_2P$, and $C_{13}H_{21}O_2$. To prepare a SIM descriptor to monitor partial profiles, the most plausible composition, $C_{13}H_{21}O_2$, was chosen as the hypothetical composition. A hypothetical composition was required to calculate predicted masses for the m/z 209, +1 (m/z 210), and +2 (m/z 211) partial profiles. The +1 and +2 profiles arise from heavier isotopes of the elements in the monoisotopic m/z 209

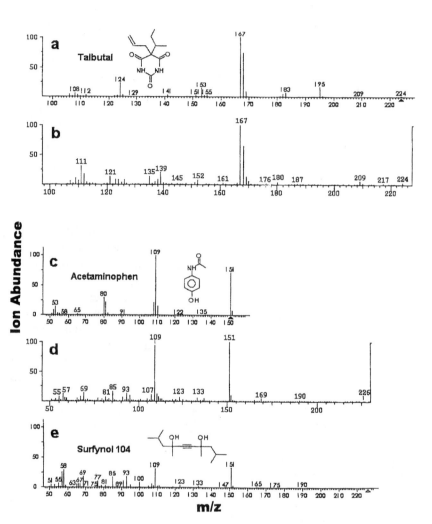

Figure 6. (a,c,e) NIST library mass spectra of Talbutal, acetaminophen, and Surfynol 104, respectively and (b,d) background subtracted mass spectra for a lake water extract.

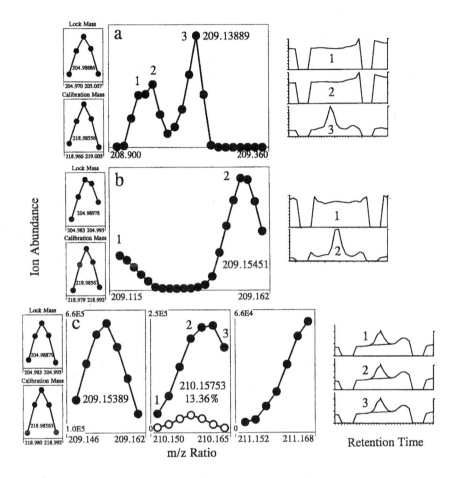

Figure 7. (a) Mass peak profiles plotted with 100-ppm mass increments using 3000 resolution over a 2000-ppm mass range near m/z 209. At the right are ion chromatograms for the labeled points on the profiles. (b) Integrated chromatographic peak areas at 10-ppm increments across a 200-ppm mass range about the mass obtained from (a) acquired with 10,000 resolution. The ion chromatograms for the points labeled 1 and 2 are at the right. (c) Partial profiles for hypothetical composition, $C_{13}H_{21}O_2$, plotted with 10-ppm steps using 10,000 resolution. Ion chromatograms for three points on the +1 partial profile are at the right.

ion (in this case, ^{13}C, ^2H, ^{17}O, and ^{18}O). Data was again acquired using a mass increment of 10 ppm with 10,000 resolution. The partial profiles obtained are displayed in Figure 7c. The partial profile that provided an exact mass of m/z 209.15389 was centered within the mass range monitored and had a symmetrical shape, which suggested a lack of interferences. The +1 partial profile was not well centered and was wider than expected. The three ion chromatograms for the first (210.15127), middle (210.15757), and last (210.16386) m/z ratios used to monitor the +1 profile displayed a chromatographic peak rising above a high background level. Data system integration provided the areas between the two excursions to the baseline that were induced by reversing the polarity of an ion source voltage for 5 s before and after the analyte eluted. The plotted areas provided the distorted +1 partial profile delineated by solid circles. Manual integration of the peak areas above the baselines provided the partial profile defined by the open circles. This profile provided 210.15753 Da as the exact mass of the +1 partial profile and 13.36% for its abundance relative to the m/z 209 partial profile. The relative abundance was determined as the ratio of the sum of the seven chromatographic peak areas used to plot the +1 partial profile to the same sum for the m/z 209 partial profile times 100%. Overwhelming interference rendered the +2 partial profile useless. This was commonly the case, since the relative abundance of the +2 profile is less than a few percent unless S, Si, Cl, or Br atoms are present. When Cl was present, abundant +2 partial profiles were easily monitored, but provided only the number of Cl atoms present. The large isotopic abundance of ^{37}Cl (24.2%) so dominated the +2 profiles that information about the numbers of C, O, or N atoms in the ion was obscured.

The measured exact masses of the m/z 209 and +1 partial profiles and the relative abundance of the +1 profile were entered into the PGM to provide the first list of possible compositions in Table II. A composition in addition to those from which the hypothetical composition was selected was consistent with the lower exact mass obtained from the partial profiles. But only the last composition had theoretical values in agreement with the three experimental values within their error limits. This composition was not consistent with loss of CH_3 from the molecular ion of talbutal, $C_{11}H_{16}N_2O_3^+$ (m/z 224) to yield $C_{10}H_{13}N_2O_3^+$ (m/z 209). No partial profile was observed for the molecular ion of talbutal using 10,000 resolution, which further refuted the tentative identification of this compound as talbutal. The observed exact mass for a full profile using 10,000 resolution was 224.17717, which is consistent with composition 5 plus an additional methyl group, $C_{14}H_{24}O_2^+$ (m/z 224.17763). The m/z 168 ion in Figure 6a was identified unequivocally as $C_{10}H_{16}O_2^+$. Clearly, this compound was not talbutal, but probably, a compound that provided a molecular ion of $C_{14}H_{24}O_2^+$ having 3 rings and double bonds, which lost CH_3 and C_4H_8 to yield the other two ions investigated.

Table II. Possible Compositions, Mass Defects, and M+1 or F+1 Relative Abundances

m/z 209.15389 ± 6 ppm   Elements: C H N O P F S Si   Res: 10000

#	RDB Range	Composition	209	+1	%+1 (%+1 Range)	
1	0.0	$C_3H_{19}N_9Si$	.15327	.15329 X	4.89(2.27- 8.07)	X
2	0.0 1.0	$C_8H_{20}N_3O_2F$	.15396	.15687	9.66(7.57-11.79)	X
3	-1.0 0.0	$C_9H_{24}NO_2P$	.15447	.15776	10.80(8.87-12.80)	X
4	4.0	$C_{11}H_{19}N_3O$	.15281	.15583 X	11.90(9.13-14.58)	X
5	3.5	**$C_{13}H_{21}O_2$**	.15415	.15757	14.85(12.47-17.25)	

Experimental Values: .15389   .15753   13.36

m/z = 252.08992 ± 3 ppm   Elements: C H N O P F S Si   Res: 25000

#	RDB Range	Composition	252	+1	%+1 (%+1 Range)
30	3.0	$C_9H_{15}N_2OF_3Si$	.09057	.09278	14.21(11.12-17.60)
35	6.0	$C_{13}H_{18}PFSi$	.08994	.09248	17.94(15.00-21.06)
36	11.0 12.0	**$C_{15}H_{12}N_2O_2$**	.08988	.09324	17.03(14.41-19.70)

Experimental Values: .08992   .09296   17.37

m/z 180.08106 ± 6 ppm   Elements: C H N O P F S Si

#	Err (mmu & ppm)		RDB Range	Composition	
1	-0.1	-0.8	2.5	$C_3H_8N_7F_2$	X
2	-0.6	-3.4	0.5 1.5	$C_4H_{14}N_3O_3Si$	X
3	+0.4	+2.1	1.5	$C_4H_{12}N_5PF$	X
4	+0.9	+4.9	0.5	$C_5H_{16}N_3P_2$	X
5	-0.9	-4.8	0.0	$C_6H_{17}OFSi_2$	X
6	+0.7	+4.0	0.0	$C_6H_{16}O_4Si$	X
7	+0.1	+0.4	6.0	$C_8H_9N_4F$	X
8	+0.6	+3.2	5.0	$C_9H_{13}N_2P$	X
9	+0.3	+1.5	9.5	**$C_{13}H_{10}N$**	

Superimposed ion chromatograms acquired with 10,000 resolution, free of interferences from other ions with the same nominal mass, would provide overlapping chromatographic peaks for ions associated with the same compound. The elemental compositions of the most abundant ion (m/z 167) and of ions related to it could be determined. Then, only a limited number of possible isomers for the compound would remain, and a literature search might indicate which isomers are produced commercially (*14*).

For the molecular ion of phenytoin (m/z 252), 25,000 resolution with 4-ppm mass increments was used to determine its composition. The three compositions in the second list in Table II were possible after comparing the theoretical and measured values for the mass defects of the m/z 252 and +1 partial profiles and the relative abundance of the +1 profile. Again, the +2 partial profile was not useful. Consideration by the PGM of the sums of atomic masses, isotopic abundances, and elemental valences alone could not reveal the correct composition. However, the first two compositions contained Si in combination with other heteroatoms and were unlikely. In addition, the low resolution mass spectrum in Figure 3 displayed several high-abundance ions consistent with phenytoin. Finally, an exact mass was obtained for the m/z 180 ion in Figure 3. When the third list of possible compositions in Table II based on the exact mass of 180.08106 Da was compared with the list above it for the parent ion, only the last composition, $C_{13}H_{10}N$, could be a fragment ion formed from one of the possible compositions for the parent ion. The exact mass difference between the m/z 252 and m/z 180 ions was 72.00886 Da. The PGM provided $C_2H_2NO_2$ (72.00855 Da), the composite neutral loss between $C_{15}H_{12}N_2O_2$ and $C_{13}H_{10}N$, as the only possible composition for this exact mass, further confirming the compositions of the parent and fragment ions. Thus, the parent ion composition was unequivocally that of phenytoin, $C_{15}H_{12}O_2N_2$.

The number of GC/HRMS analyses was limited by the volume of sample extract. Up to 32 different ions were investigated for each 1-μL injection. However, each ion required multiple data collections. Ions observed in numerous low resolution mass spectra were not investigated before the F3 extract was consumed. Thus, the tentatively identified compounds in Table III are only a partial list of the compounds in the extract. In addition, elemental compositions were determined for 83 ions formed from compounds for which no tentative identifications were made. Many of these ions containing only C, H, and O atoms probably were formed from normal and branched chain isomers of APEs used in detergent formulations and known to be present in these extracts. Their low resolution mass spectra generally provided only a few abundant ions, and standards would be required before identifications could be made.

The m/z 151 and m/z 190 ions from Figure 6d and listed in Table III were determined to be $C_{10}H_{15}O^+$ and $C_{14}H_{22}^+$, respectively. A search of the NIST library provided the mass spectra in Figures 6c and 6e for acetaminophen and 2,4,7,9-tetramethyl-5-decyn-4,7-diol, also known as Surfynol 104, a wetting agent. The m/z 151 and m/z 190 ions were consistent with losses of $C_4H_{11}O$ and two water molecules, respectively, from $C_{14}H_{26}O_2^+$, the unobserved molecular ion of Surfynol 104. The m/z 151 ion was not $C_8H_9NO_2^+$, the molecular ion of acetaminophen, and the possible match for this drug was rejected.

For the low-abundance, m/z 209 ion, Figure 7 illustrated how MPPSIRD and the PGM were used to determine elemental compositions of ions in an extract containing hundreds of compounds. Resolving analyte and interference ions with the same nominal mass provided an orthogonal separation after imperfect chromatography. The large-volume sample collection, extraction strategy, and HRMS detection of ions used in this study provided a sensitive and specific methodology for identifying trace-level compounds in Lake Mead.

The ions in Table III and the 83 additional ions for which elemental compositions were determined indicate that further study of Lake Mead water extracts is warranted. With enough sample extract, many more compounds should be tentatively identifiable using MPPSIRD and the PGM. Of course, confirmation of compound identities will require purchase of standards for retention time and mass spectral comparison.

Conclusions

The detection of trace concentrations of pharmaceuticals and personal care products in the Las Vegas Bay of Lake Mead is not surprising considering the volume of highly treated wastewater effluents entering the system. Several of these compounds have been detected in sewage effluents since the 1980s or earlier in certain cases (*11, 12, 21-23*). Sewage treatment facilities are unable to completely remove all the compounds entering the system. The ability to detect these types of trace contaminants has improved rapidly with the increased usage of more sensitive and reliable instrumentation such as mass spectrometry. Future studies of Lake Mead should utilize state of the art technology such as liquid chromatography with mass spectral detection to better quantitate compounds previously detected, and to look for new compounds which have not yet been reported. However, little is currently known regarding toxicity of trace concentrations of these compounds. Further, it should be emphasized that these compounds were only detected in the Las Vegas Bay and do not appear to be widely distributed in Lake Mead.

The toxicological relevance of the findings of pharmaceuticals and personal care products in Lake Mead is unknown at this time. Further study should be directed at determining whether these compounds are capable of producing adverse effects in the aquatic environment at environmentally relevant

Table III. Tentatively Identified Compounds and Elemental Compositions, Exact Masses, and Relative Abundances of Ions

Tentatively Identified Compound	Ions	M	M+1	M+2	%M+1	%M+2
methyl pyruvate	$C_4H_6O_3^+$	102.03174 (+0.4)				
thiazole or iso-thiazole	$C_3H_3NS^+$	84.99865 (+0.4)				
methylmethoxybenzene	$C_8H_{10}O^{+a}$	122.07324 (+0.6)	123.07661 (+0.5)			
methylcyclopentanol	$C_6H_{10}^+$ (M-H_2O)	82.07828 (+0.4)				
ethylfuraldehyde	$C_7H_8O_2^+$	124.05215 (−2.3)				
trimethylcyclohexenedione or dimethyloxocyclohexenecaboxaldehyde	$C_9H_{12}O_2^+$	152.08348 (−1.6)	153.08766 (+3.4)		11.2	
	$C_5H_4O_2^+$	96.02162 (+0.5)			No N tail	
methylethylmaleimide or methoxymethylpyridinol	$C_7H_9NO_2^+$	139.06342 (+0.7)	140.06723 (+3.8)		8.7	
C_4-alkylphenol (3 isomers)	$C_{10}H_{14}O^+$	150.10450 (+0.2)	151.10785 (0)		9.8	
p-cymenol	$C_{10}H_{14}O^+$	150.10441 (−0.4)	151.10869 (+5.5)		11.1	
1-(4-ethylphenol)-ethanone	$C_{10}H_{12}O^+$	148.08886 (+0.3)	149.09223 (+0.1)		10.7	
2-methyl-2H-indazol-3-amine or 6-amino-2-methyl-2H-indazole	$C_8H_9N_3^{+b}$	147.07965 (0)				
trichlorophenol	$C_6H_3OCl_3^+$	195.92449 (−2.3)	196.92845 (+0.6)		6.1	
2,4,7,9-tetramethyl-5-decyn-4,7-diol (Surfynol 104)	$C_{14}H_{22}^+$ (M-2H_2O)	190.17201 (−0.7)	191.17554 (−0.1)		15.7	
	$C_{10}H_{15}O^+$	151.11225 (−0.3)	152.11581 (+0.7)	153.11934 (+5.6)	12.0	0.9
dimethadione	$C_5H_7NO_3^+$	129.04272 (+1.0)	130.04608 (+1.2)		6.1	
1,1'-(1,3 or 1,4-phenylene)bis-ethanone	$C_{10}H_{10}O_2^+$	162.06820 (+0.7)			11.8	
	$C_9H_7O_2^+$	147.04437 (−1.6)	148.04844 (+3.0)	149.04942 (−5.3)	10.6	
ditertbutylmethylphenol	$C_{15}H_{24}O^+$	220.18292 (+0.9)	221.18629 (+0.7)		16.8	
1- or 2-naphthol	$C_{10}H_8O^+$	144.05705 (−3.2)			11.2	

diTertbutylethylphenol	$C_{16}H_{26}O^+$	234.19815 (−0.9)	235.20157 (−0.9)	236.20551 (+2.5)	18.0	1.8
diethyltoluamide	$C_{12}H_{16}NO^{+c}$	190.12299 (−1.0)				
methylbenzotriazole	$C_7H_7N_3^+$	133.06406 (+0.5)	134.06748 (+0.9)		7.9	
hydroxymethylnaphthalenedione	$C_{11}H_8O_3^+$	188.04724 (−0.6)	189.05012 (−3.2)		13.2	
ethyl citrate	$C_9H_{15}O_5^{+d}$	203.09163 (−1.6)	204.09483 (−2.7)		11.1	
tris(chloroethyl)phosphate	$C_6H_{12}O_4PCl_2^+$ (M-Cl)	248.98467 (−1.4)	249.98771 (−3.0)	250.98216 (0)	67.2	
terbuthylazine	$C_9H_{16}N_5Cl^{+e}$	229.10949 (+0.3)	230.11308 (+1.3)	231.10665 (0)		
butylbenzenesulfonamide	$C_7H_8NO_2S^+$	170.02734 (−1.4)	171.03115 (+3.9)	172.02337 (−1.5)	9.0	4.6
tris(chloropropyl)phosphate	$C_8H_{16}O_4PCl_2^{+f}$ (M-Cl)	277.01656 (+0.8)	278.02002 (+0.9)	279.01371 (+0.8)	9.7	69.3
tris(dichloroisopropyl)phosphate	$C_8H_{13}O_4PCl_5^{+g}$ (M-Cl)	378.89928 (−0.3)	379.90263 (−0.6)	380.89635 (−0.4)	8.9	157.5
diphenylhydantoin (Dilantin)	$C_{15}H_{12}N_2O_2^+$	252.08992 (+0.2)	253.09296 (−1.1)		17.4	
	$C_{13}H_{10}N^+$	180.08106 (−1.5)				

[a] Experimental exact mass best agreed with $C_8H_{10}O^+$, but did not exclude $C_4H_{13}ONP^+$

[b] Experimental exact mass best agreed with $C_8H_9N_3^+$, but did not rule out $C_4H_{12}N_4P^+$

[c] Experimental exact mass best agreed with $C_{12}H_{16}NO^+$, but did not rule out $C_8H_{19}N_2OP^+$ and $C_6H_{17}N_5P^+$

[d] Experimental exact mass did not rule out $C_8H_9N_7^+$

[e] Experimental exact masses best agreed with $C_9H_{16}N_5Cl^+$, but did not rule out $C_5H_{19}N_6PCl^+$

[f] Experimental exact masses and relative abundances did not exclude four other compositions

[g] Experimental exact masses and relative abundances did not exclude four other compositions

138

concentrations. Because some of the compounds are pharmaceuticals, they have been manufactured for the specific purpose of producing biological effects in animals. Their known mechanisms of action, potential side effects, and effective doses in test animals might provide a basis from which to begin an assessment of their potential for causing adverse health effects in animals in the aquatic environment. Any such assessment will certainly present a great challenge, given the array of biological effects that some pharmaceuticals can produce in a single species and the vast differences in biological response and sensitivity sometimes displayed by different species to the same test chemical. In addition, the exposure received by organisms in an aquatic environment can be expected to be quite different from that received by animals deliberately dosed in a laboratory for product safety testing. In the latter case, animals typically receive great doses administered for a limited period of time. In contrast, animals exposed in the aquatic environment are likely receiving longer-term, lower-dose exposures by different routes (e.g., uptake across integument or gill rather than intraperitoneal injection or oral gavage). Until toxicological relevance is determined, we can only show that these compounds are able to enter the aquatic environment in trace concentrations with unknown impacts to the aquatic ecosystem.

References

(1) LaBounty, J. F.; Horn, M. J. *J. Lake Reservoir Manage.* **1997**, *13*, 95-108.
(2) Prentki, R. T.; Paulson, L. J. In *Aquatic Resources Management of the Colorado River Ecosystem*; Adams, V. D., Lamarra, V. A., Eds.; Ann Arbor Science: Ann Arbor, MI, 1983, pp 105-123.
(3) Paulson, L. J.; Baker, J. R. In *Proceedings of the Symposium on Surface Water Impoundments*; Stefan, H. G., Ed.; American Society of Civil Engineering: New York, 1981, pp 1647-1658.
(4) Bevans, H. E.; Goodbred, S. L.; Miesner, J. F.; Watkins, S. A.; Gross, T. S.; Denslow, N. D.; Schoeb, T. *Water-Resources Investigations Report 96-4266* **1996** U.S. Geological Survey.
(5) Talmage, S. S. *Environmental and Human Safety of Major Surfactants: Alcohol Ethoxylates and Alkylphenol Ethoxylates*; Lewis Publishers: Boca Raton, FL 1994.
(6) Routledge, E. J.; Sheahan, D.; Desbrow, C.; Brighty, G. C.; Waldock, M.; Sumpter, J. P. *Environ. Toxicol. Chem.* **1998**, *32*, 1559-1565.
(7) Giesy, J. P.; Snyder, E. M. In *Principles and Processes for Evaluating Endocrine Disruption*; SETAC Press, 1998, pp 155-237.

(8) Jobling, S.; Noylan, M.; Tyler, C. R.; Brighty, G.; Sumpter, J. P. *Environ. Sci. Technol.* **1998**, *32*, 2498-2506.

(9) Folmar, L. C.; Denslow, N. D.; Rao, V.; Chow, M.; Crain, D. A.; Enblom, J.; Marcino, J.; Guillette Jr., L. J. *Environ. Health Perspect.* **1996**, *104*, 1096-1101.

(10) Gamble, A.; Sherry, J.; Parrott, J.; Hodson, P.; Solomon, K. *poster* **1997**, *SETAC, San Francisco, California.*

(11) Snyder, S. A.; Keith, T. L.; Verbrugge, D. A.; Snyder, E. M.; Gross, T. S.; Kannan, K.; Giesy, J. P. *Environ. Sci. Technol.* **1999**, *33*, 2814-2820.

(12) Snyder, S. A.; Snyder, E.; Villeneuve, D.; Kurunthachalam, K.; Villalobos, A.; Blankenship, A.; Giesy, J. In *Analysis of Environmental Endocrine Disruptors*; Keith, L. H., Jones-Lepp, T. L., Needham, L. L., Eds.; American Chemical Society: Washington, DC, 2000, pp 73-95.

(13) Zacharewski, T. *Environ. Sci. Technol.* **1997**, *31*, 613-623.

(14) Grange, A. H.; Sovocool, G. W. *Rapid Commun. Mass Spectrom.* **1998**, *12*, 1161-1169.

(15) Grange, A. H.; Donnelly, J. R.; Sovocool, G. W.; Brumley, W. C. *Anal. Chem.* **1996**, *68*, 553-560.

(16) Grange, A. H.; Brumley, W. C. *J. Am. Soc. Mass Spectrom.* **1997**, *8*, 170-182.

(17) Snyder, S. A.; Villeneuve, D. L.; Snyder, E. M.; Giesy, J. P. *Environ. Sci. Technol.* **2000**, *unpublished.*

(18) Panter, G. H.; Thompson, R. S.; Sumpter, J. P. *Aquat. Toxicol.* **1998**, *42*, 243-253.

(19) Kramer, V. J.; Miles-Richardson, S.; Pierens, S. L.; Giesy, J. P. *Aquat. Toxicol.* **1998**, *40*, 335-360.

(20) Purdom, C. E.; Hardiman, P. A.; Bye, V. J.; Eno, N. C.; Tyler, C. R.; Sumpter, J. P. *Chemistry and Ecology* **1994**, *8*, 275-285.

(21) Halling-Sorensen, B.; Nielsen, S. N.; Lanzky, P. F.; Ingerslev, F.; Lutzhoft, H. C. H.; Jorgensen, S. E. *Chemosphere* **1998**, *36*, 357-393.

(22) Desbrow, C.; Routledge, E. J.; Brighty, G. C.; Sumpter, J. P.; Waldock, M. *Environ. Sci. Technol.* **1998**, *32*, 1549-1558.

(23) Daughton, C. G.; Ternes, T. A. *Environ. Health Perspect.* **1999**, *107 (suppl. 6)*, 907-938.

Personal Care Products

Chapter 8

Occurrence and Fate of Synthetic Musks in the Aquatic System of Urban Areas

Polycyclic and Nitro Musks as Environmental Pollutants in Surface Waters, Sediments, and Aquatic Biota

Th. Heberer[1], A. These[1], and U. A. Grosch[2]

[1]Institute of Food Chemistry, Technical University of Berlin, Sekr. TIB 4/3-1,
Gustav-Meyer-Allee 25, 13355 Berlin, Germany
[2]Berlin Fishery Board, Havelchaussee 149/151, 14055 Berlin, Germany

Worldwide, more than 5000 tons per year of synthetic musks (most of them polycyclic musks) are produced and used as fragrances in various scented consumer products. In sewage treatment, these compounds are not completely removed. Thus, residues of polycyclic musks are discharged via municipal sewage effluents into the receiving waters. In surface water samples of conurbations they are found at concentrations up to the µg/L-level. Due to their high lipophilicity, they accumulate in sewage sludges, aquatic sediments, and at the top of the aquatic food chain where they are found at concentrations up to more than 100 mg/kg lipid in fishes and mussels.

Nitro musks (Figure 1) and polycyclic musks (Figure 2) are used widely as fragrances in cosmetics, detergents, and other scented products. Although they are chemically quite different, both compound classes are summarized as "synthetic musks" because of their typical musk-like odor. Due to their low rates of biological and biochemical degradation and due to their high lipophilic character, synthetic musks are potential environmental contaminants (1,2).

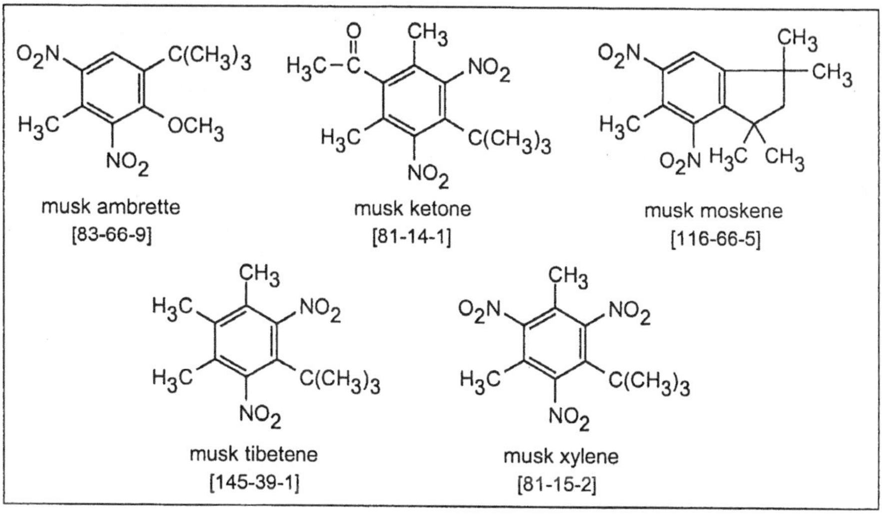

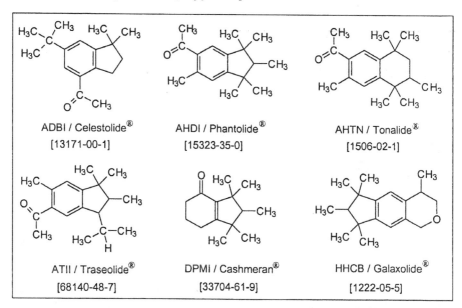

Figure 1. Structural formulae, trade names, and CAS Registry numbers (in parenthesis) of five important nitro musks.

Figure 2. Structural formulae, abbreviations, trade names, and CAS Registry numbers (in parenthesis) of six environmentally important polycyclic musks.

In 1981, Yamagishi et al. (*3,4*) reported the identification of musk xylene (1-tert-butyl-3,5-dimethyl-2,4,6-trinitrobenzene; log K_{ow} 4.9) and musk ketone (1-tert-butyl-3,5-dimethyl-2,6-dinitro-4-acetylbenzene; log K_{ow} 4.3) in fishes and in water samples from the river Tama and from Tokyo Bay. More than 10 years later, several reports by Rimkus and Wolf (*5-11*) gained much attention and caused concern about the environmental fate of the nitro musks.

In 1987, the worldwide production of nitro musks was estimated at 2500 tons per year (*12*). The general concern about the environmental fate and the toxicology of nitro musks resulted in a decrease in their use as fragrances whereas the importance of the polycyclic musks is increasing (*1,13*). In 1996, the global annual production of polycyclic musks was estimated at more than 5000 tons (*13*). Rebmann et al. (*13*) estimated the contribution of the polycyclic musks to the global production of synthetic musks at 85% compared to only 12% for nitro musks. Recently, several investigations have shown that polycyclic musks have a similar environmental behavior to the nitro musks.

Synthetic Musks in Sewage and Surface Water Samples

Synthetic musks are not completely removed by adsorption and/or degradation during biological sewage treatment. Thus, these compounds are constantly discharged into the receiving waters without any seasonal variation. In investigations of influents and effluents of two U.S. sewage treatment works (STWs), Simonich et al. (*14*) found removal rates of 81.4% and 94.8% following secondary treatment for musk ketone and musk xylene, respectively. In another study, Eschke et al. (*15*) reported, however, that only 50% of musk ketone and 82% of musk xylene were removed during sewage treatment. The differences in the removal rates were even larger for the two polycyclic musks 7-acetyl-1,1,3,4,4,6-hexamethyltetrahydronaphthaline (AHTN; Tonalide®, Fixolide®) and 1,3,4,6,7,8-hexahydro-4,6,6,7,8,8-hexamethylcyclopenta[ ]-2-benzopyran (HHCB; Galaxolide®, Abbalide®, Pearlide®). Simonich et al. (*14*) reported removal rates of 86.2 and 87.4% for AHTN and HHCB, respectively, whereas Eschke et al. (*16*) reported the removal of 60% for AHTN and only 34% for HHCB.

Nitro musk residues have been detected in sewage effluents up to the low µg/L-level (*15,17-19*). In river water samples (*4,15,19-23*), nitro musks were detected at concentrations at the low ng/L-level and in the North Sea (*21*) only traces of nitro musks were found. In recent investigations of influents and effluents of a STW, Gatermann et al. (*24*) showed that musk ketone and musk xylene are metabolized significantly into their mono-amino metabolites. These metabolites have been found at concentrations of approximately 300 ng/L in the sewage effluents, clearly exceeding (4-40 times) the concentrations of the parent compounds in these samples (*24*). The major amino metabolite was 2-amino musk ketone (250 ng/L), but 2-amino musk xylene (10 ng/L) and 4-amino musk xylene (34 ng/L) have also been identified in the effluents of this particular STW (*24*).

Polycyclic musks, especially HHCB and AHTN, have been found in sewage effluents at the μg/L-level (*14,16,17,19,25-27*). 4-Acetyl-1,1-dimethyl-6-tert-butylindene (ADBI; Celestolide®, Crysolide®) was also detected in sewage water samples, but only at lower concentrations (*19,25*). HHCB, AHTN, and ADBI have also been identified in surface water samples in Germany, Switzerland, The Netherlands, and Japan (*16,19,20,23,25-30*) and at trace-level concentrations (<1 ng/L) in the North Sea (*29*). In addition to the original compounds, Gatermann (cited in *37*) detected several impurities and by-products of the polycyclic musk compounds in wasterwater and river water samples. Recently, Franke et al. (*31*) identified 1,3,4,6,7,8-hexahydro-4,6,6,7,8, 8-hexamethylcyclo-penta[g]-2-benzopyran-1-one (Galaxolidone), an oxidation product of HHCB, in samples from the rivers Odra and Elbe. In laboratory experiments, the formation of this lactone has also been observed by Itrich et al. (*32*) in activated sewage sludge and in abiotic controls.

Polycyclic Musks in Surface Waters in Berlin, Germany

The highest concentrations of polycyclic musks (up to 20 μg/L) have been found in monitoring investigations of surface water samples from 30 locations in Berlin, Germany (*19*). In Berlin, municipal wastewater contaminations are of great environmental relevance due to the high contributions of STW effluents to the surface waters. This results from low surface water flows and large amounts of raw sewage produced by Berlin's population of around 3.5 million people. As may be observed in Figure 3, the concentration profiles for HHCB and AHTN measured in the Teltowkanal (a canal located in the southern districts of Berlin) show that peak concentrations are found at those sampling sites where sewage effluents have been discharged into the canal. The impact of sewage effluents on the surface water quality can be seen very clearly by the increase of synthetic musks concentrations. This indicator was found to be much more significant than any of the conventional chemical parameters, such as DOC or COD (*19*).

In the Wuhle, a small brook in Berlin that almost totally consists of sewage effluents, maximum concentrations of 12.5 μg/L for HHCB and 6.8 μg/L for AHTN were reported (*19*). These values are comparable to those measured by Simonich et al. (*14*) in sewage influents of an activated sludge wastewater treatment plant (13.7 μg/L HHCB and 10.7 μg/L AHTN) and may suggest poor removal of synthetic musks by the Berlin treatment plants. The different studies are, however, not directly comparable because several parameters have not been considered, e.g., the composition of the sewage waters (percentages of municipal or industrial sewage and rain drainage), dry weather and rainy periods, or the total drinking water consumption. Thus, the total drinking water usage per inhabitant in Germany is 127 liters per day compared with approximately 300 liters per day in the USA (33).

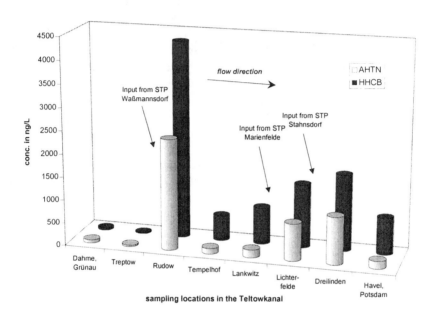

Figure 3. Concentration profiles measured for the polycyclic musks HHCB and AHTN in the Teltowkanal. Selected results from a surface water monitoring in Berlin, Germany. STP: sewage treatment plant. Adapted with permission from reference 19. Copyright 1999 John Wiley & Sons.

The investigations mentioned here give an impression about the contamination sources and the concentration levels of synthetic musks that are found in the receiving waters. The Berlin data probably show a worst-case scenario because Berlin has a high degree of urbanization, little producing industry, low natural surface water flow rates, and in this particular sampling series in September 1996, there was a dry weather period.

Synthetic Musks in Sewage Sludges

To date, there is only very limited data about the concentrations of musks in sewage sludges. In two Dutch studies, AHTN and HHCB were found at maximum concentrations of up to 34 and 63 mg/kg dry weight (d.w.), respectively (cited in 26). In a German study in Hesse (34), the highest concentrations of HHCB and AHTN (12-24 mg/kg d.w.) have been found in municipal STWs processing wastewater from private households. The concentrations were two or three orders of magnitude lower in sludges from industrial STWs (*34*). Musk ketone has only been found in non-digested sewage sludges; after digestion the concentrations were always below the detection limit (34). This could, of course, be explained by the formation of the metabolite 2-amino musk ketone (*24*). This conclusion was supported, recently, by a Swiss study (*35*) of synthetic musks and nitro musk amino metabolites in sewage sludges. Herren and Berset (*35*) detected the amino metabolites of musk xylene, musk ketone, and musk moskene at concentrations between 0.01 and 0.05 mg/kg d.w., exceeding the concentrations of the original compounds. Nevertheless, the concentrations of the nitro musks and their metabolites were found to be clearly below those of the polycyclic musks. Thus, HHCB and AHTN were found at concentrations between 0.7 and 12.1 mg/kg d.w., followed by ADBI, 6-acetyl-1,1,2,3,3,5-hexamethyldihydroindene (AHDI; Phantolide®) and 6,7-dihydro-1,1,2,3,3-pentamethyl-4(5H)indanone (DPMI; Cashmeran®) detected at concentrations below 1 mg/kg d.w. (*35*). In another recent study (*36*) of sewage sludges from three municipal STWs in Berlin (14 days collective samples), up to 11.4 mg/kg d.w. of HHCB and up to 5.1 mg/kg d.w. of AHTN were found. The concentrations measured for AHDI (up to 0.65 mg/kg d.w.), ADBI (up to 0.33 mg/kg d.w.) and ATII (up to 0.20 mg/kg d.w.) were significantly lower. DPMI was not detected in any of these samples (*36*).

Synthetic Musks in Suspended Particular Matter (SPM) and Aquatic Sediments

Winkler et al. (*23*) analyzed 31 SPM samples from the River Elbe and found HHCB (148-736 ng/g), AHTN (194-770 ng/g), ADBI (<4-43 ng/g), and musk ketone (4-22 ng/g). The mean concentrations of HHCB and AHTN were in the same order of magnitude as hexachlorobenzene and the PAHs and higher than those of PCBs and other organochlorine pesticides (*23*). Nevertheless, Winkler et al. (*23*) calculated that about 83% of AHTN, 92% of HHCB, and 95% of musk ketone is dissolved in the aqueous phase. In another study (*34*) with SPM samples from Hessian Rivers, the concentrations of HHCB and AHTN were found to be much higher. In particular, in three small rivers (Schwarzenbach, Rodau, Geräthsbach) with high portions of sewage effluents up to 13 μg/g d.w. of HHCB and AHTN have been found (*34*). Musk xylene and musk ketone were detected at maximum concentrations of 0.05 and 0.48 μg/g d.w., respectively. In a Dutch study by Breukel and Balk (cited in *26*) AHTN and HHCB were found in the rivers Rhine and Meuse at concentrations between 0.06-1.2 and 0.05-0.58 μg/g d.w., respectively. The data indicate that AHTN is bound stronger to SPM than HHCB (*23*). Thus, in many SPM samples, AHTN is found at concentrations similar to or even higher than those of HHCB.

There are few German data about the contamination of aquatic sediments. In 1996, sediment samples (n=8) from selected rivers in northern Germany were investigated by Lach and Steffen (cited in *37*). They found HHCB, AHTN, DPMI, musk xylene, and musk ketone at maximum concentrations of 54, 3.9, 0.7, 2.2, and 3.8 ng/g d.w., respectively. In Berlin, several aquatic sediment samples were collected from different sites with low and high percentages of municipal sewage (*36*). Up to 3.6 mg/kg d.w. of HHCB and up to 2.5 mg/kg d.w. of AHTN were found at contaminated sites. Again, the concentrations measured for AHDI (up to 0.14 mg/kg d.w.), ADBI (up to 0.05 mg/kg d.w.) and ATII (up to 0.03 mg/kg d.w.), were significantly lower. DPMI was not detected in the samples (*36*).

Synthetic Musks in Biota Samples

Many positive findings of nitro musks have been reported in samples of fish, other aquatic biota, human fat, and mothers milk (*5-11,15,20,38-41*). In 1994, positive findings of polycyclic musks in fishes and water samples from the river Ruhr at concentration levels higher than those of the nitro musks were reported by Eschke et al. (*16,25*). Meanwhile, in several investigations of fish samples from Germany, Denmark, Italy, the Czech Republic, and The Netherlands (*16,25,26,36,37,40-42*), HHCB and AHTN have been identified as the most important musk contaminants. Concentrations of up to 63.6 mg/kg lipid of HHCB and 57.9 mg/kg lipid of AHTN were found in eel samples collected from sewage ponds (*25*). ADBI and AHDI were also detected in a few samples but only at low concentrations. Polycyclic musks are found at high concentrations at the top of the aquatic food chain, and they are also found at considerable concentrations in human adipose tissue and mothers milk (*20,43,44*).

Although HHCB and AHTN are found at high concentrations in fish samples, the high log K_{ow} values of HHCB and AHTN (5.9 and 5.7) do not correlate with their bioconcentration factors (BCFs), which are significantly lower than expected (*37*). Rimkus (*37*) explained this phenomenon by a possible metabolism in fish. Franke et al. (*31*) studied the enantiomeric composition of HHCB (four enantiomers) and AHTN (two enantiomers) in fishes and mussels reared in a municipal sewage pond. In samples of crucian carp they observed pronounced deviations in the enantiomeric composition from racemic HHCB, whereas in samples of tench a deviation concerning the racemic AHTN was found. In samples of mussels, eel, and rudd no enantioselectivity was observed. However, within the limited set of biota no correlations between lipid levels, enrichment, and enantioselective biotransformation of HHCB and AHTN were seen (*31*). The highest lipid concentrations of HHCB (up to 155 mg/kg) and AHTN (up to 45 mg/kg) were found in mussels, tench, and crucian carp.

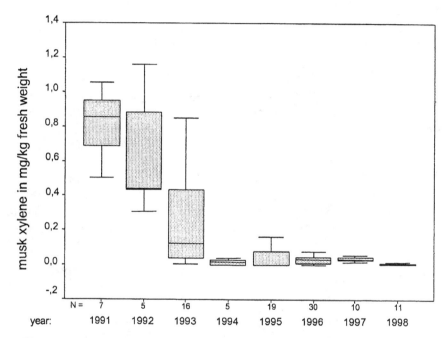

Figure 4. Box plot diagram showing the decreasing concentrations of musk xylene in eels caught between 1991 and 1998 in the Teltowkanal in Berlin, Germany. Reprinted from reference 36.

Synthetic Musks in Fishes from Rivers, Lakes and Canals in Berlin

The most comprehensive study on the occurrence of residues of synthetic musk compounds in fishes was commissioned by the Berlin Fishery Board between 1991 and 1998 (36).

More than 1500 fishes from different species such as eel, pike, pike perch, perch, bream, white bream, rudd, roach, tench, carp, and crucian carp caught in rivers, lakes, and canals in Berlin were analyzed for residues of nitro musks and polycyclic musks. Figure 4 shows a box plot diagram that demonstrates the decrease of musk xylene between 1991 and 1998 in eels caught from the Teltowkanal (36). This decrease is the result of the voluntary phasing out of the nitro musk use by the fragrance industry in Germany.

Between 1996 and 1998, 476 fishes from different surface waters inside and near to Berlin were analyzed for musk xylene, musk ketone, and also six polycyclic musks (36). Fish samples from contaminated surface waters (n=324) contained HHCB and AHTN at average fresh weight concentrations of 0.51 and 0.20 mg/kg (median 0.14 and 0.04 mg/kg), respectively. Maximum values of up to 384 mg/kg lipid of HHCB and 88 mg/kg lipid of AHTN were detected. ADBI, AHMI, ATII, musk xylene, and musk ketone were also detected in several samples, but at concentrations considerably lower than HHCB and AHTN. DPMI was always below the limit of quantitation. As one example, Table I shows the results measured for synthetic musks in eels caught from the Teltowkanal, a canal that is highly burdened with municipal sewage discharges.

Table I: Concentrations (in mg/kg fresh weight or lipid) of synthetic musks in eels from the Teltowkanal 1996-1998 (36).

	n	samples $<LOQ^a$	maximum value mg/kg	average mg/kg	median mg/kg	95% percentile mg/kg
HHCB[b]	51	0	4.80	1.48	1.44	3.7
AHTN[b]	51	0	2.30	0.70	0.57	1.40
AHDI[b]	21	0	0.21	0.07	0.05	0.20
ATII[b]	21	0	0.19	0.07	0.05	0.19
ADBI[b]	51	19	0.02	0.004	0.004	0.014
DPMI[b]	11	11	< LOQ	< LOQ	< LOQ	< LOQ
Musk ketone[b]	51	0	0.38	0.10	0.07	0.24
Musk xylene[b]	51	0	0.08	0.03	0.03	0.07
HHCB lipid[c]	51	0	18.5	6.39	6.17	14.7
AHTN lipid[c]	51	0	8.85	2.97	2.69	6.1
Fat %	51	-	36.0	23.9	25.2	33.8

[a] limit of quantitation (between 0.001 and 0.01 mg/kg fresh weight); [b] concentration in mg/kg fresh weight; [c] concentration in mg/kg lipid

Synthetic Musks in Northern American Biota Samples

A comparison of synthetic musk concentrations in samples of Canadian and European aquatic biota showed significant differences (*41*). In the Canadian biota samples, musk ketone (up to 17.7 mg/kg lipid) dominated, followed by HHCB (up to 3.0 mg/kg lipid). Musk xylene and AHTN were only found at relatively low concentrations in these samples. These results contrast with those measured in Europe where the concentrations of HHCB and AHTN exceeded those of the nitro musks by about one to three orders of magnitude (*16,25,36,37,41*). Gatermann et al. (*41*) explained these differences by different usage behaviors in Northern America and Western Europe and concluded that nitro musks, and especially musk ketone, still dominate in Northern America, whereas in Western Europe the polycyclic musks are prevalent as constituents of fragrances. Simonich et al. (*14*) opposed this assumption, because their influent data and the current industry volume (*45*) did not support this conclusion. They state, that as in Europe, the use of polycyclic musks (such as HHCB and AHTN) exceeds the use of nitro musks (musk xylene and musk ketone) in North America.

Acknowledgments

The authors thank Mr. B. Hatton for his help in preparing the manuscript.

References

Comprehensive reviews about production and use, occurrence, toxicology, environmental fate, and risk assessment of nitro musks and polycyclic musks have been written by Balk and Ford (*26,46-48*), Daughton and Ternes (*49*), Rimkus and Brunn (*1,2,37*), and Rebmann et al. (*13*). Schwartz et al. (*50*) calculated environmental concentrations of HHCB for the aquatic environment under consideration of various scenarios using environmental fate and distribution models laid down in the Technical Guidance Documents (TGD) and implemented in the European Union System for the Evaluation of Substances (EUSES).

1. Rimkus, G.; Brunn, H. "Synthetische Moschusduftstoffe - Anwendung, Anreicherung in der Umwelt und Toxikologie. Teil 1: Herstellung, Anwendung, Vorkommen in Lebensmitteln, Aufnahme durch den Menschen." *Ernährungs-Umschau* **1996**, *43*, 442-449.

152

2. Brunn, H.; Rimkus, G. "Synthetische Moschusduftstoffe - Anwendung, Anreicherung in der Umwelt und Toxikologie. Teil 2: Toxikologie der synthetischen Moschusduftstoffe und Schlußfolgerungen." *Ernährungs-Umschau* **1997**, *44*, 4-9.

3. Yamagishi, T.; Miyazaki, T.; Horri, S.; Kaneko, S. "Identification of musk xylene and musk ketone in freshwater fish collected from Tama River,Tokyo." *Bull. Environ. Contam. Toxicol.* **1981**, *26*, 656-662.

4. Yamagishi, T.; Miyazaki, T.; Horri, S.; Akijama, K. "Synthetic musk residues in biota and water from Tama River and Tokyo Bay (Japan)." *Arch. Environ. Contam. Toxicol.* **1983**, *12*, 83-89.

5. Rimkus, G.; Wolf, M. "Rückstände in Regenbogenforellen aus Teichwirtschaften." *Lebensmittelchemie* **1993**, *47*, 26-31.

6. Rimkus, G.; Wolf, M. "Nachweis von Nitromoschusverbindungen in Frauenmilch und Humanfett." *Dtsch. Lebensm. Rdsch.* **1993**, *89*, 103-107.

7. Rimkus, G.; Wolf, M. "Rückstände und Verunreinigungen in Fischen aus Aquakultur. 2.Mitteil. Nachweis von Moschus Xylol und Moschus Keton in Fischen." *Dtsch. Lebensm. Rdsch.* **1993**, *89*, 171-175.

8. Geyer, H.J.; Rimkus, G.; Wolf, M.; Attar, A.; Steinberg, C.; Kettrup, A. "Synthetische Nitromoschus-Duftstoffe und Bromocyclen. - Neue Umweltchemikalien in Fischen und Muscheln bzw. Muttermilch und Humanfett." *UWSF - Z. Umweltchem. Ökotox.* **1994**, *6*, 9-17.

9. Rimkus, G.; Rimkus, B.; Wolf, M. "Nitro musks in human adipose tissue and breast milk." *Chemosphere* **1994**, *28*, 421-432.

10. Linkerhägner, M.; Stan, H.J.; Rimkus, G. "Detection of nitro musks in human fat by capillary gas chromatography with atomic emission detection (AED) using programmed temperature vaporization (PTV)." *J. High Resolut. Chromatogr.* **1994**, *17*, 821-826.

11. Rimkus, G.; Wolf, M. "Nitro musk fragrances in biota from freshwater and marine environment." *Chemosphere* **1995**, *30*, 641-651.

12. Barbetta, L.; Trowbridge, T.; Eldib, I.A. "Musk aroma chemical industry." *Perfumer Flavorist* **1988**, *13*, 60-61.

13. Rebmann, A.; Wauschkuhn, C.; Waizenegger, W. "Bedeutung der Moschusduftstoffe im Wandel der Zeit." *Dtsch. Lebensm. Rdsch.* **1997**, *93*, 251-255.

14. Simonich, S.L.; Begley, W.M.; Debaere, G.; Eckhoff, W.S. "Trace analysis of fragrance materials in wastewater and treated wastewater." *Environ. Sci. Technol.* **2000**, *34*, 959-965.

15. Eschke, H.-D.; Traud, J.; Dibowski, H.-J. "Analytik und Befunde künstlicher Nitromoschussubstanzen in Oberflächen- und Abwässern sowie Fischen aus dem Einzugsgebiet der Ruhr." *Vom Wasser* **1994**, *83*, 373-383.

16. Eschke, H.-D.; Traud, D.; Dibowski, H.-J. "Untersuchungen zum Vorkommen polycyclischer Moschus-Duftstoffe in verschiedenen Umweltkompartimenten - Nachweis und Analytik mit GC/MS in Oberflächen-, Abwässern und Fischen (1. Mitteilung)." *Z. Umweltchem. Ökotox.* **1994**, *6*, 183-189.

17. Paxeus, N. "Organic pollutants in the effluents of large wastewater treatment plants in Sweden." *Wat. Res.* **1996**, *30*, 1115-1122.

18. Heberer, Th. *Identification and Quantification of Pesticide Residues and Environmental Contaminants in Ground and Surface Water Applying Capillary Gas Chromatography - Mass Spectrometry* (in German); Wissenschaft & Technik Verlag: Berlin, 1995; pp. 272-273.

19. Heberer, Th.; Gramer, S.; Stan, H.-J. "Occurrence and distribution of organic contaminants in the aquatic system in Berlin. Part III: Determination of synthetic musks in Berlin surface water applying solid-phase microextraction (SPME)." *Acta Hydrochim. Hydrobiol.* **1999**, *27*, 150-156.

20. Müller, S.; Schmid, P.; Schlatter, C. "Occurrence of nitro and non-nitro benzenoid musk compounds in human adipose tissue." *Chemosphere* **1996**, *33*, 17-28.

21. Gatermann, R.; Hühnerfuss, H.; Rimkus, G.; Wolf, M.; Franke, S. "The distribution of nitrobenzene and other nitroaromatic compounds in the North Sea." *Mar. Pollut. Bull.* **1995**, *30*, 221-227.

22. Hahn J. "Untersuchungen zum Vorkommen von Moschus-Duftstoffen in Umwelt, Lebensmitteln und Humanmilch." *Lebensmittelchemie* **1996**, *50*, 77.

23. Winkler, M.; Kopf, G.; Hauptvogel, C.; Neu, T. "Fate of artificial musk fragrances associated with suspended particulate matter (SPM) from the River Elbe (Germany) in comparison to other organic contaminants." *Chemosphere* **1998**, *37*, 1139-1156.

24. Gatermann, R.; Hühnerfuss, H.; Rimkus, G.; Attar, A.; Kettrup, A. "Occurrence of musk xylene and musk ketone metabolites in the aquatic environment." *Chemosphere* **1998**, *36*, 2535-2547.

25. Eschke, H.-D.; Dibowski, H.-J.; Traud, D. "Untersuchungen zum Vorkommen polycyclischer Moschus-Duftstoffe in verschiedenen Umweltkompartimenten - Befunde in Oberflächen-, Abwässern und Fischen sowie in Waschmitteln und Kosmetika (2. Mitteilung)." *Z. Umweltchem. Ökotox.* **1995**, *7*, 131-138.

26. Balk, F.; Ford, R.A. "Environmental risk assessment for the polycyclic musks AHTN and HHCB in the EU. I. Fate and exposure assessment." *Toxicol. Lett.* **1999**, *111*, 57-79.

27. Verbruggen, E.M.J.; van Loon, W.M.G.M.; Tonkes, M.; van Duijn, P.; Seinen, W.; Hermens, J.L.M. "Biomimetic extraction as a tool to identify chemicals with high bioconcentration potential: An illustration by two fragrances in sewage treatment plant effluents and surface waters." *Environ. Sci. Technol.* **1999**, *33*, 801-806.

28. Franke, S.; Hildebrandt, S.; Schwarzbauer, J.; Link, M.; Francke, W. "Organic compounds as contaminants of the Elbe river and its tributaries. Part II: GC/MS screening for contaminants of the Elbe water." *Fresenius J. Anal. Chem.* **1995**, *353*, 39-49.

29. Bester, K.; Hühnerfuss, H.; Lange, W.; Rimkus, G.G.; Theobald, N. "Results of non target screening of lipophilic organic pollutants in the German bight II: Polycyclic musk fragrances." *Chemosphere* **1998**, *32*, 1857-1863.

30. Yun, S.-J.; Teraguchi, T. ; Zhu, X.-M., Iwashima, K. "Identification and analysis of dissolved organic compounds in the Tamagawa River" (in Japanese). *J. Environ. Chem.* **1994**, *4*, 325-333.

31. Franke, S. Meyer C.; Heinzel N.; Gatermann R.; Huhnerfuss H.; Rimkus G.; Konig W.A.; Francke W. "Enantiomeric composition of the polycyclic musks HHCB and AHTN in different aquatic species." *Chirality* **1999**, *11*, 795-801.

32. Itrich, N.R.; Simonich, S.L.; Federle, T.W. "Biotransformation of the Polycyclic Musk, HHCB, During Sewage Treatment." Poster at the SETAC 19th Annual Meeting, Nov. 14th-15th 1998, Charlotte, NC, USA. Cited in (31).

33. Engelhard, T. "Trinkwasser: Schluck für Schluck ein Kunstprodukt." *Geo Magazin*, February 2000, 42-52.

34. Fooken, C.; Gihr, R.; Häckl, M.; Seel, P. "Orientierende Messungen gefährlicher Stoffe - Landesweite Untersuchungen auf organische Spurenverunreinigungen in hessischen Fließgewässern, Abwässern und Klärschlämmen, 1991–1996." In (ed.: Hessische Landesanstalt für Umwelt) Umweltplanung, Arbeits- und Umweltschutz, Heft 233, 1997, Wiesbaden, Germany, pp. 89-96.

35. Herren, D.; Berset, J.D. "Nitro musks, nitro musk amino metabolites and polycyclic musks in sewage sludges. quantitative determination by HRGC-ion-trap-MS/MS and mass spectral characterization of the amino metabolites." *Chemosphere* **2000**, *40*, 565-574.

36. Heberer, Th., Fromme, H.; Jürgensen, S. "Synthetic musks in the aquatic system of Berlin as an example for urban ecosystems." In (Rimkus G.G. ed.) *Synthetic Musk Fragrances in the Environment*; Springer-Verlag: Heidelberg, in preparation.

37. Rimkus, G.G. "Polycyclic musk fragrances in the aquatic environment." *Toxicol. Lett.* **1999**, *111*, 37-56.

38. Hahn J. "Untersuchungen zum Vorkommen von Moschus-Xylol in Fischen." *Dtsch. Lebensm. Rdsch.* **1993**, *89*, 175-177.

39. Liebl, B.; Ehrenstorfer, S. "Nitro musks in human milk." *Chemosphere* **1993**, *27*, 2253-2260.

40. Hajslova, J.; Gregor, P.; Chladkova, V.; Alterova, K. "Musk compounds in fish from Elbe River." *Organohalogen Compd.* **1998**, *39*, 253-256.

41. Gatermann, R.; Hellou, J.; Hühnerfuss, H.; Rimkus, G.; Zitko, V. "Polycyclic and nitro musks in the environment: A comparison between Canadian and European aquatic biota." *Chemosphere* **1999**, *38*, 3431-3441.
42. Draisci, R.; Marchiafava, C.; Ferretti, E.; Palleschi, L.; Catellani, G.; Anastasio, A. "Evaluation of musk contamination of freshwater fish in Italy by accelerated solvent extraction and gas chromatography with mass spectrometric detection." *J. Chromatogr. A* **1998**, *814*, 187-197.
43. Rimkus, G.G.; Wolf, M. "Polycyclic musk fragrances in human adipose tissue and human milk." *Chemosphere* **1996**, *33*, 2033-2043.
44. Eschke, H.D.; Dibowski, H.J.; Traud, J. "Nachweis und Quantifizierung von polycyclischen Moschus-Duftstoffen mittels Ion-Trap GC/MS/MS in Humanfett und Muttermilch." *Dtsch. Lebensm. Rdsch.* **1995**, *91*, 375-379.
45. Research Institute for Fragrance Materials (RIFM), NJ, USA; unpublished data cited in (14).
46. Ford, R.A. "The safety of nitromusks in fragrances – A review." *Dtsch. Lebensm. Rdsch.* **1998**, *94*, 192-200.
47. Ford, R.A. "The human safety of the polycyclic musks AHTN and HHCB in fragrances – A review." *Dtsch. Lebensm. Rdsch.* **1998**, *94*, 268-275.
48. Balk, F.; Ford, R.A. "Environmental risk assessment for the polycyclic musks AHTN and HHCB in the EU. II. Effect assessment and risk characterisation." *Toxicol. Lett.* **1999**, *111*, 81-94.
49. Daughton, C.G.; Ternes, T.A. "Pharmaceuticals and personal care products in the environment: agents of subtle change?" *Environ. Health Perspect.* **1999**, *107* (suppl. 6), 907-938.
50. Schwartz, S.; Berding, V.; Matthies, M. "Aquatic fate assessment of the polycyclic musk fragrance HHCB Scenario and variability analysis in accordance with the EU risk assessment guidelines." *Chemosphere* **2000**, *41*, 671-679.

Chapter 9

Ecotoxicology of Musks

D. R. Dietrich and Y.-J. Chou

Environmental Toxicology, University of Konstanz, Fach X918,
D-78457 Konstanz, Germany

Due to the fact that both nitro and polycyclic musks and their metabolites are found in the aquatic environment and appear to accumulate in some of the species, the past and most recent research has focused mainly on possible ecotoxicological effects of musks in aquatic rather than terrestrial species. The compilation of the newest available data for aquatic interactions demonstrates in general that neither parent compounds nor the metabolites of nitro and polycyclic musks pose any significant hazard for the aquatic ecosystem. The observation that amphibians appear more susceptible to endocrine modulating effects of xenobiotics than other species mandates that the interactions of the nitro musk metabolites with the *Xenopus laevis* estrogen receptor, as presented in this review, are investigated in more detail. Such an investigation appears warranted despite the fact that all observed adverse interactions of nitro and polycyclic musks occur at concentrations several orders of magnitude higher than those detected in the environment.

Introduction

The yearly global production of nitro and polycyclic musk fragrances has been estimated to be approximately 2000 (for the year 1988) and 5600 tons, respectively (*1-3*). The use of musks as fragrances and fragrance fixatives in a wide array of personal care products (e.g., washing detergents, detergents in general, perfumes, lotions, soaps and shampoos, cosmetics, etc.) stipulates that most of these compounds will appear in municipal sewage treatment plants (STP). The removal of nitromusks (NMs) and polycyclic musks (PCMs) during municipal sewage treatment processes has been estimated at approximately 60-80% and 40-60%, respectively. The higher retention of NMs in the STP are explained by the presence of the aromatic ring and thus higher affinity for particles, a rather low water solubility, and a moderately high lipophilicity [K_{ow}: 4.9 and 4.3 for musk xylene (MX) and musk ketone (MK), respectively]. In contrast, PCMs have a high water solubility, despite their inherently high lipophilicity (K_{ow}: 5.7 and 5.9 for AHTN and HHCB, respectively; see next section for abbreviations) and biological stability (*4*). In view of the lipophilicity of NMs and PCMs and their broad form of application, it is not surprising to find these compounds as contaminants in the aquatic environment. Indeed, the concentrations detected in environmental samples range from ng/L to µg/L in effluent and surface waters and from µg/kg lipid to mg/kg lipid in aquatic organisms (*3, 5-8*). Furthermore, most recent analyses point to NM and PCM metabolites as being of greater environmental concern, due to greater metabolic stability and environmental persistence and thus higher concentrations present in biological samples, e.g., in fish flesh, than the respective parent compounds (*1, 2, 9-11*).

All available analytical data, while showing the capability of musk fragrances to bioconcentrate in various aquatic species, do not demonstrate any capacity of these compounds for biomagnification in the aquatic ecosystem. The capacity for "bioconcentration/bioaccumulation" must be differentiated in that for musk compounds this appears more likely to be a function of momentary exposure of the species in question, rather than that of a lifetime up-concentration from a chronically contaminated environment. Indeed, age class analyses of fish taken from the Elbe river demonstrated no significant differences in tissue levels of NMs and PCMs from younger and older fish of the same species (*12*). The concentrations of musk fragrances in the aquatic environment, including species, e.g., fish, are highly related to the distance from the STP (*11*). In consequence and contrary to the situation with PCBs, the potential for toxicological effects resulting from musk exposure stems largely from the actual concentrations the species are exposed to via the ambient water *in situ* (*13*). In view of this the following paragraphs represent primarily a compilation of data for the acute, subacute, and "potential" for subchronic-chronic toxicity of musk fragrances in "target" species (algae, daphnia, fish, and amphibians) and not with imaginable but highly unlikely indirect effects in other species.

Acute and Subacute Toxicity

The acute toxicity and potential environmental effects of NMs and PCMs were summarized in several publications either using the EU-Technical Guidance Documents as a basis for environmental risk assessment (*4, 14-15*), test procedures in conformity with OECD guideline 201 and 202 for testing of chemicals (*16-17*), or test procedures identical or analogous to ASTM guideline E 1439-91 (*18*). The latter publications include studies with algae (*P. subcapitata*), *Daphnia magna*, bluegill sunfish (*L. macrochirus*), rainbow trout (*O. mykiss*), zebrafish (*D. rerio*), fathead minnow (*P. promelas*), and the South African clawed frog (*X. laevis*). The most prominent results are compiled in Table I.

The main focus of the latter studies was on musk xylene (MX), musk ketone (MK) and the three polycyclic musks AHTN (7-acetyl-1,1,3,4,4,6-hexamethyltetraline), HHCB (1,3,4,6,7,8-hexahydro-4,6,6,7,8,8-hexamethyl-cyclopenta-(g)-2-benzopyran) and ADBI (4-acetyl-1,1-dimethyl-6-*tert*-butylindane). Additional data can be found for the three amino-metabolites of MX and MK (*18*) as well as for musk moskene, tibetene, and ambrette (*17*). Toxicity of either NMs and PCMs was observed at rather high concentrations of these respective compounds (Table I), i.e., in many cases at or exceeding the inherent water solubilities (Table II).

The mechanism(s) involved in the acute toxicity of the NMs and PCMs is presently unknown. However, a generalized narcosis, as previously demonstrated for various other organic compounds in fish and amphibians (*19*), may be suggested in view of the high concentrations necessary to induce acute mortality (*17, 18, 20*) and the erratic behavior noted with daphnia (*17*). The latter findings are contrasted by the report of Behechti et al., (*16*) who found acute toxicity of low concentrations of the amino-metabolites of MX, especially of the 4-amino-MX in *D. magna* (EC_{50} = 250 ng/L; 95% CI 230-280 ng/L). Whether the findings of Behechti et al. are generally applicable to other aquatic organisms is presently unknown. It is, however, a fact that these toxic concentrations lie approximately 1-2 orders of magnitude above those found in STP effluents and 3 orders of magnitude above those found in surface waters (*1, 2*).

In contrast, more specific effects are noted when embryos of *X. laevis* and *D. rerio* are exposed to PCMs but not NMs (*18, 20*). Both *D. rerio* and *X. laevis* embryos present with a significant increase in malformations (Fig. 1 and 2a). Surprisingly, while all three PCMs (ADBI, AHTN, HHCB) induced malformations in zebrafish embryos, malformations are observed only in ADBI treated *X. laevis* embryos (Fig. 2a.). While ventro-dorsal curvature of the tail was the most prominent and characteristic malformation for PCM exposure in both species, the concentrations necessary to induce malformations in *D. rerio* were approx. one order of magnitude lower than those necessary to produce the same effects in the amphibian embryos. Of the PCMs tested, AHTN demonstrated the greatest degree of teratogenicity, with the steepest dose-

Table I: Compilation of acute and subacute toxicity values obtained with nitro and polycyclic musks in various species

Species	Endpoint	MX [mg/L]	MK [mg/L]	MM [mg/L]	AHTN [mg/L]	HHCB [mg/L]	ADBI [mg/L]	Ref.
Algae	$EC_{50\ growth}$	NE^a	0.244	-	>0.797	>0.854	-	4,
	$EC_{50\ biomass}$	NE^a	0.118	-	0.468	0.723	-	14-15
Daphnia magna	24hr EC_{50}	NE^a	NE^a	NE^a	-	-	-	4,
	21d IC_{50}	0.680	0.338-0.675	NE^a	0.341	0.293	-	14-15,
	21d EC_{50} rep.	-	0.169-0.338	NE^a	0.244	0.282	-	17
O. mykiss	96hr LC_{50}	>1000	-	-	-	-	-	4,
	21d LC_{50}	-	>0.50	-	-	-	-	14
Lepomis macrochirus	96hr LC_{50}	1.20	-	-	-	-	-	4,
	21d LC_{50}	-	-	-	0.314	0.452	-	14-15
Danio rerio	14d LC_{50-} adult fish	0.4	-	-	-	-	-	14, 18-
	96hr LC_{50-} embryo	>0.4	>0.4	>0.4	>0.67	>0.67	>1.0	19
	96hr EC_{50-} embryo-hatching	>0.4	>0.4	>0.4	>0.67	>0.67	>1.0	
	96hr EC_{50-} embryo-teratog.	>0.4	>0.4	>0.4	0.18	0.39	0.69	
	96hr EC_{50-} embryo-growth	>0.4	>0.4	>0.4	>1.0	>1.0	>1.0	
P. promelas	32d LC_{50-} embryo-adult	-	-	-	0.100	>0.140	-	15
Xenopus laevis	96hr LC_{50-} embryo	>0.4	>0.4	>0.4	> 2.0	>2.0	>4.0	18-19
	96hr EC_{50-} embryo-teraogen	>0.4	>0.4	>0.4	>4.0	>4.0	>4.0	
	96hr EC_{50-} embryo-growth	>0.4	>0.4	>0.4	>1.0	>2.0	>4.0	

^{a}: NE, No effect found at concentrations exceeding compound solubility in H_2O (Table II)

Table II: Calculated water solubilities of nitro and polycyclic musks

	MK [mg/L]	MX [mg/L]	MM [mg/L]	AHTN [mg/L]	HHCB [mg/L]
H_2O sol.	1.9^a; 0.46^b	0.49^a; 0.15^b	0.046^b	1.25^c	1.75^c

^{a}: (14); ^{b}: (17); ^{c}: (11)

160

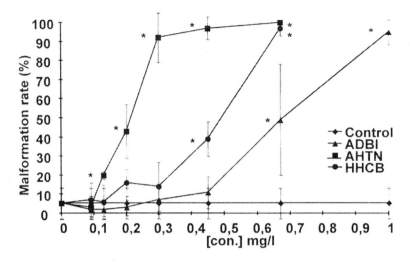

*Figure 1. Malformation in early-life-stage Danio rerio following 96 hours of exposure to polycyclic musks (n=3). (ANOVA and Dunnett's T test. * p<0.05).(Reproduced with permission from (20))*

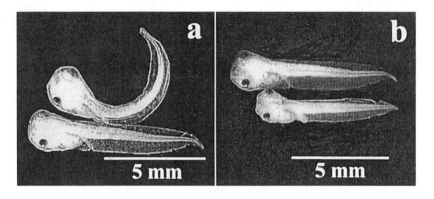

Figure 2. Malformation and growth inhibition in early-life-stage xenopus (Xenopus laevis) exposed to polycyclic musks for 96 hours. a) ADBI treatment (4 mg/L) (top) and control (bottom): occurrence of ventro-dorsal curvature. b) Control (top) and AHTN treatment (2 mg/L) (bottom): stunted growth).(Reproduced with permission from (20))

response curve (Fig. 1), while ADBI was teratogenic at high concentrations only. AHTN-induced malformations appear to be specific for cyprinid embryos, as tail-loss was noted in *P. promelas* embryos exposed to 0.067 or 0.14 mg/L AHTN, while no malformations were observed in *X. laevis* embryos exposed to AHTN or HHCB (20) or in *P. promelas* exposed to HHCB (15). Of the three PCMs tested in a semi-static embryotoxicity test with *X. laevis*, AHTN and HHCB demonstrated a significant and dose-dependent effect on growth at concentrations below those which were acutely toxic to the embryos (Fig. 2b, Table I). No effects on growth were observed in zebrafish embryos, as the doses necessary to induce a significant growth inhibition exceeded those inducing acute toxicity (Table I). Similar effects were noted in *P. promelas* exposed to 0.140 mg/L HHCB but not for AHTN (*15*).

The comparison of the NM and PCM concentrations found in environmental samples (*1-3, 9-10, 21-22*) with those concentrations inducing acute and subacute toxicity in various aquatic species, as discussed above, strongly suggests that NMs and PCMs do not pose an acute risk for the aquatic ecosystem. This conclusion is also supported by the instrumentalized risk assessment processes for NMs and PCMs using the EU-Technical Guidance Documents (*4, 14-15*), which predict no effects of these musk fragrances in the aquatic environment.

Subchronic-Chronic Toxicity

At present, only limited data are available for assessing the risk to the aquatic environment, i.e., the populations of aquatic species exposed subchronically or chronically to low concentrations of parent compounds and metabolites of NMs and PCMs. In general, there are three potential adverse interactions of xenobiotics with the health and sustainability of a population that are of primary importance: (i) an extremely high incidence of pathological changes, e.g., tumors (*23*) resulting from genotoxic or a tumor promoting activity; (ii) suppression of the immune system and thus higher susceptibility of the population to pathogens (*24*); and (iii) endocrine modulation affecting the reproductive success of the population.

Neither the parent compounds nor the metabolites of NMs and PCMs have been demonstrated to possess carcinogenic activity, with the exception of a species-specific promotion of liver tumors at high concentrations of MX observed in mice (*25*). This process was shown to be not of genotoxic (*26-27*), but rather of an epigenetic nature, i.e., driven by the induction of microsomal enzymes, particularly those of the CYP2B family (28), and the pattern of induction was consistent with that observed for phenobarbital, the classical CYP2B inducer and mouse liver carcinogen (*29-30*).

No information is as yet available regarding the potential interaction of NMs and PCMs on immune parameters of aquatic species. However, the present expectation is that no immune-suppressive activity is to be expected in aquatic species as no evidence was found suggesting immune suppressive activity of these compounds in mammalian species exposed subchronically or chronically to high concentrations of these compounds (*25, 31-32*).

Although the present database on potential endocrine modulating activity of NMs and PCMs is still rather scant, the compilation of mammalian data and data from *in vitro* assays with cells and tissue homogenates from aquatic species suffices for a primary assessment, at least of the potential (anti)estrogenic activity of these compounds. Neither subchronic or chronic administration of NMs, PCMs or mixtures of NMs and PCMs (25, 31-32) suggests any form of (anti)estrogenic activity in rodent species. The basis for this assessment was organ weight and histopathological examination of the uterus, seminal vesicles, mammary gland, testes, ovaries, and vaginas. These findings are corroborated by a study of Seinen et al. (33) who exposed juvenile mice to high dietary levels of AHTN and HHCB and found no evidence for an increase in uterine weight. On the other hand, the same scientists reported a very weak estrogenic activity of both compounds using ERα- and ERβ-dependent gene transcription assays with human embryonal kidney 293 cells. The reported estrogenic activity was approximately six to eight orders of magnitude lower than the endogenous ligand estradiol (E_2). The latter findings demonstrated that only extremely high concentrations of AHTN and HHCB have measurable estrogenic potency and that the current levels of wildlife and human exposure to these compounds are too low to induce any estrogenic effects in the exposed species. The interaction of the PCMs with the hepatic estrogen receptor(s) of rainbow trout, carp, or the amphibian *X. laevis* was also shown in an *in vitro* competitive binding assay (Fig. 3). In comparison to the endogenous ligand E_2, approximately four orders of magnitude higher concentrations of AHTN were necessary to elicit the same degree of ligand competition (IC_{50}) in the *X. laevis* receptor binding assay. Very weak binding of AHTN and HHCB were found in the rainbow trout receptor binding assay (34), corroborating the findings by Seinen et al. (33). Neither AHTN nor HHCB, but ADBI bound to the carp estrogen receptor (34), corroborating earlier findings by Smeets at al. (35), who investigated AHTN and HHCB induced synthesis of vitellogenin in carp hepatocytes *in vitro*. Neither of the two compounds was capable of inducing vitellogenin in this system suggesting that these compounds do not interact with the fish estrogen receptor(s) to the degree or with the high concentrations necessary for estrogen dependent gene transcription. Although metabolites of AHTN and HHCB, as found in environmental samples (3, 9-10), were not analyzed for (anti)estrogenic activity, it can safely be assumed that these metabolites were also formed during incubation of the primary carp hepatocytes used as the screening method for estrogenic activity. If indeed these metabolites had any form of estrogenic activity the lack of vitellogenin induction in the carp hepatocyte system suggests that the metabolites were not formed in adequate concentrations to have an estrogenic effect. Overall it can be concluded that the current environmental PCM levels are too low to induce estrogenic effects in aquatic species.

In contrast to the PCMs neither of the two nitro musk parent compounds (MX and MK) had any competitive binding activity to either the rainbow trout or the Xenopus estrogen receptor(s). However, amino-metabolites of MX and MK, formed during the sewage treatment process, were able to bind to the estrogen receptors of rainbow trout (Fig. 4) and *X. laevis* (Fig. 5). The concentrations of the 2-amino-MX metabolite necessary to displace 50% of the

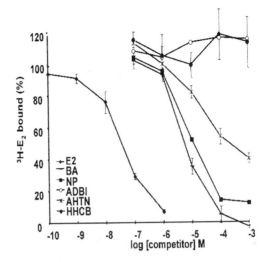

Figure 3. Competitive binding of estradiol (E_2), bisphenol A (BA), nonylphenol (NP), and polycyclic musks to Xenopus ER. The incubation concentrations were 10^{-10}-10^{-6} M for E_2, 10^{-7}-10^{-3} M for BA, NP, and PCMs. IC_{50} values (n=3) were 24.0 ± 0.5 nM (E_2), 3.7 ± 0.1 μM (BA), 24.0 ± 0.6 μM (NP) and 257 ± 6 μM (AHTN). [Reproduced with permission from (34)]

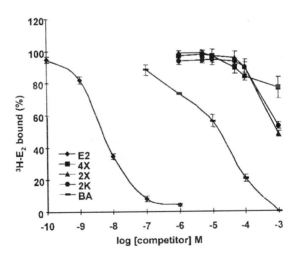

Figure 4. Competitive binding of estradiol (E_2), bisphenol-A (BA), 4-NH_2-MX (4X), 2-NH_2-MX (2X) and 2-NH_2-MK (2K) to the rainbow trout ER. The incubation concentrations were 10^{-10}-10^{-7} M for E_2, 10^{-7}-10^{-3} M for BA, and 10^{-6}-10^{-3} M for amino metabolites. IC_{50}s were 5.3 ± 1.2 nM for E_2, 8.8 ± 1.8 μM for BA and 1.2 ± 1.1 mM for 2X. [Reproduced with permission from (38). Copyright 1999 Elsevier Science Ireland Ltd.]

164

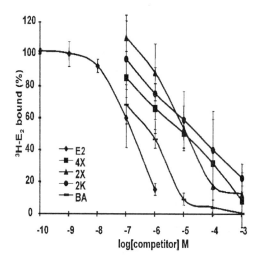

Figure 5. Competitive binding of estradiol (E₂), bisphenol-A (BA), 4-NH₂-MX (4X), 2-NH₂-MX (2X), and 2-NH₂-MK (2K) to the Xenopus ER. The incubation concentrations were 10^{-10}-10^{-7} M for E_2, 10^{-7}-10^{-3} M for BA, 4X, 2X, and (2K). IC₅₀s were 187 ± 76 nM for E2, 441 ± 247 nM for BA, 30.8 ± 28.5 μM for 4X, 12.9 ± 10.3 μM for 2X and 70.1 ± 88.3 μM for 2K. [Reproduced with permission from (38). Copyright 1999 Elsevier Science Ireland Ltd.]

endogenous ligand at the rainbow trout estrogen receptor(s) was approximately six orders of magnitude greater than that of the endogenous ligand (E₂) itself, again demonstrating that unrealistically high concentrations of these metabolites were needed to elicit any estrogenic activity in rainbow trout (Fig. 4).

Surprisingly the binding curves derived from the *X. laevis* estrogen receptor binding assay, demonstrated that all three known amino-metabolites of MX and MK were able to compete with the endogenous ligand. The concentrations necessary for competition were only 2-3 orders of magnitude higher than those of E₂ (Fig. 5). Furthermore, the concentrations of 2-amino-MX necessary for E₂ competition at the *X. laevis* estrogen receptor(s) were nearly 3 three orders of magnitude lower than those needed for competing at the rainbow trout estrogen receptor(s). The latter suggests that there are some species-specific susceptibilities with regard to potential estrogenic activities of nitro musk metabolites. Indeed, the findings in the *X. laevis* system (Fig. 5) are unique in that these *in vitro* findings were indicative for the endocrine modulating effects observed for bisphenol A (BA) *in vivo* (36). Chronic exposure of *X. laevis* embryos to low concentrations of BA induced a feminization of male embryos (37). Although the above *in vitro* systems may be indicative that some of the NM metabolites and PCMs may have the potential for endocrine modulation in aquatic species, the mere interaction of a xenobiotic with the estrogen receptor(s) of a given aquatic species does not imply that this interaction will also lead to all of the specific associated downstream events. Indeed, an investigation of the estrogenic activity of complex STP effluents using several *in vitro* assay systems demonstrated that while an interaction with the rainbow trout ER(s) was observed, no simultaneous inductions of ER and vitellogenin mRNA in primary rainbow trout hepatocyte cultures were detectable (*39-40*).

The concurrent chemical analysis of these STP effluents revealed the presence of ethoxylates and plant steroids in ng-µg/L quantities, thus strongly suggesting that, with the exception mentioned above, high concentrations of these estrogenic xenobiotics are necessary to elicit a demonstrable endocrine modulating effect at the individual or population level.

Conclusions

Although the present data base for ecotoxicological effects of NMs and PCMs and of their respective metabolites is still too small for a concluding risk assessment, there is little evidence that would suggest that these compounds, despite their overt presence in environmental samples, generally would have an adverse impact on the aquatic ecosystem. The concentrations of musk fragrances in the aquatic environment are highly related to the distance to the STP (*11*). Indeed, as indicated also via the comparison between the tissue levels of various ages of fish exposed to NMs and PCMs, no biomagnification within the same species (age classes) or various trophic levels appears to occur (*12*). In consequence and contrary to the situation with PCBs, the potential for toxicological effects resulting from musk exposure stems largely from the actual concentrations the species are exposed to via the ambient water *in situ* (*13*) and this risk appears to be negligible when using the presently available database for risk estimation. However, as pointed out above, amphibians appear to be more susceptible to endocrine modulating compounds than most of the species investigated so far (*37*). In light of this, the interaction of the MX and MK metabolites with the estrogen receptor of *X. laevis* (*38*) must be taken more seriously and should encourage others to investigate the mechanisms of this interaction, the potential effects, and risks associated with these amino metabolites for amphibians in more detail.

Acknowledgments

We would like to thank Drs. Werner Kloas and Ilka Lutz, presently at the Institut für Gewässer und Binnenfischerei, for their assistance with the *X. laevis* estrogen receptor assay and Dr. Evelyn O'Brien for critically reading the manuscript. We would also like to acknowledge the Arthur und Aenne Feindt Foundation (Hamburg, Germany) for financial support of this project and the Landesgraduiertenförderung Baden-Württemberg (Germany) for the Ph.D. stipend in support of Y-J. Chou.

166

References

1. Gatermann, R.; Hühnerfuss, H.; Rimkus, G.; Attar, A.; Kettrup, A. *Chemosphere* **1998**, *36*, 2535-2547.
2. Rimkus, G. G.; Hühnerfuss, H.; Gatermann, R. *Toxicol. Lett.* **1999**, *111*, 5-15.
3. Rimkus, G. G. *Toxicol. Lett.* **1999**, *111*, 28-37.
4. Van de Plassche, E. J.; Balk, F. *Environmental risk assessment of the polycyclic musks AHTN and HHCB according to the EU-TGD*; National Institute of Public Health and the Environment: Bilthoven, The Netherlands, 1997.
5. Eschke, H.-D.; Traud, J.; Dibowski, H.-J. *Vom Wasser* **1994**, *83*, 373-383.
6. Eschke, H.-D.; Traud, J.; Dibowski, H.-J. *UWSF-Z. Umweltchem. Ökotox.* **1994**, *6*, 183-189.
7. Eschke, H.-D.; Dibowski, H.-J.; Traud, J. *UWSF-Z. Umweltchem. Ökotox.* **1995**, *7*, 131-138.
8. Rimkus, G.; Wolf, M. *Chemosphere* **1995**, *30*, 641-651.
9. Biselli, S.; Gatermann, R.; Kallenborn, R.; Rimkus, G. G.; Hühnerfuss, H. *Analyses of a transformation product of the polycyclic musk compound Galaxolide*; 10th Annual Meeting of SETAC Europe, Brighton, UK, 2000; 1, pp 156-157.
10. Gatermann, R.; Rimkus, G.; Hecker, M.; Biselli, S.; Hühnerfuss, H. *Bioaccumulation of synthetic musks in different aqautic species*; 9th Annual Meeting of SETAC Europe, SETAC Bruxelles: Leipzig, 1999; 1, pp Abstract.
11. Balk, F.; Ford, R. A. *Toxicol. Lett.* **1999**, *111*, 57-79.
12. Hajslova, J.; Gregor, P.; Chladkova, V.; Alterova, K. *Organohal. Comp.* **1998**, *39*, 253-256.
13. Rimkus, G. G.; Butte, W.; Geyer, H. *Chemosphere* **1997**, *35*, 1497-1507.
14. Tas, J. W.; Balk, F.; Ford, R. A.; van de Plassche, E. J. *Chemosphere* **1997**, *35*, 2973-3002.
15. Balk, F.; Ford, R. A. *Toxicol. Lett.* **1999**, *111*, 81-94.
16. Behechti, A.; Schramm, K.-W.; Attar, A.; Niederfellner, J.; Kettrup, A. *Wat. Res.* **1998**, *32*, 1704-1707.
17. Schramm, K.-W.; Kaune, A.; Beck, B.; Thumm, W.; Behechti, A.; Kettrup, A.; Nickolova, P. *Wat. Res.* **1996**, *30*, 2247-2250.
18. Chou, Y.-J.; Dietrich, D. R. *Toxicol. Lett.* **1999**, *111*, 17-25.
19. McCarty, L. S.; Mackay, D.; Smith, A. D.; Ozburn, G. W.; Dixon, D. G. *Environ. Toxicol. Chem.* **1992**, *11*, 917-930.
20. Chou, Y.-J.; Prietz, A.; Dietrich, D. R. *Toxicol. Lett.* **2000**, *in preparation*.
21. Gatermann, R.; Hellou, J.; Hühnerfuss, H.; Rimkus, G.; Zitko, V. *Chemosphere* **1999**, *38*, 3431-3441.
22. Bester, K.; Hühnerfuss, H.; Lange, W.; Rimkus, G. G.; Theobald, N. *Wat. Res.* **1998**, *32*, 1857-1863.

23. Gardner, G. R.; Yewich, P. P.; Harshbarger, J. C.; Malcolm, A. R. *Environ. Health Perspect.* **1991**, *90*, 53-66.

24. Prietz, A.; Fleischhauer, V.; Hitzfeld, B. C.; Dietrich, D. R. *Effects of stream water on immune parameters of brown trout (Salmo trutta f.)*; 10th Annual Meeting of SETAC Europe, SETAC Europe: Brighton, UK, 2000; 1, pp 110.

25. Maekawa, A.; Matsushima, Y.; Onodera, M.; Shibutani, M.; Ogasawara, H.; Kodama, Y.; Kurokawa, Y.; Hayashi, Y. *Fundam. Chem. Toxicol.* **1990**, *28*, 581-586.

26. Api, A. M.; Ford, R. A.; San, R. H. C. *Fundam. Chem. Toxicol.* **1995**, *33*, 1039-1045.

27. Api, A. M.; Pfitzer, E. A.; San, R. H. C. *Fundam. Chem. Toxicol.* **1996**, *34*, 633-638.

28. Lehman-McKeeman, L. D.; Caudill, D.; Vasallo, J. D.; Pearce, R. E.; Madan, A.; Parkinson, A. *Toxicol. Lett.* **1999**, *111*, 105-115.

29. Lehman-McKeeman, L. D.; Caudill, D.; Young, J. A.; Dierckman, T. A. *Biochem. Biophys. Res. Commun.* **1995**, *206*, 975-980.

30. Lehman-McKeeman, L. D.; Johnson, D. R.; Caudill, D. *Toxicol. Appl. Pharmacol.* **1997**, *142*, 169-177.

31. Api, A. M.; Ford, R. A. *Toxicol. Lett.* **1999**, *111*, 143-149.

32. Fukuyama, M. Y.; Easterday, O. D.; Serafino, P. A.; Renskers, K. J.; North-Root, H.; Schrankel, K. R. *Toxicol. Lett.* **1999**, *111*, 175-187.

33. Seinen, W.; Lemmen, J. G.; Pieters, R. H. H.; Verbruggen, E. M. J.; van der Burg, B. *Toxicol. Lett.* **1999**, *111*, 161-168.

34. Chou, Y.-J.; Dietrich, D. R. *Aquat. Toxicol.* **2000**, *in preparation.*

35. Smeets, J. M.; Rankouhi, T. R.; Nichols, K. M.; Komen, H.; Kaminski, N. E.; Giesy, J. P.; van den Berg, M. *Toxicol. Appl. Pharmacol.* **1999**, *157*, 68-76.

36. Lutz, I.; Kloas, W. *Sci. Total Environ.* **1999**, *225*, 49-57.

37. Kloas, W.; Lutz, I.; Einspanier, R. *Sci. Total Environ.* **1999**, *225*, 59-68.

38. Chou, Y.-J.; Dietrich, D. R. *Toxicol. Lett.* **1999**, *111*, 27-36.

39. Pfluger, P.; Wasserrab, B.; Knörzer, B.; Dietrich, D. R. *Complex chemical mixtures in two German rivers: embryotoxicity and GC-MS analysis.*; 10th Annual Meeting of SETAC Europe, SETAC Europe: Brighton, UK, 2000; 1, pp 135-136.

40. Wasserrrab, B.; Pfluger, P.; Knörzer, B.; Dietrich, D. R. *Detection of endocrine modulating activity in water samples using primary trout hepatocytes and liver homogenates.*; 10th Annual Meeting of SETAC Europe, SETAC Europe: Brighton, UK, 2000; 1, pp 148.

Chapter 10

Environmental Risks of Musk Fragrance Ingredients

Froukje Balk[1], Han Blok[1], and Daniel Salvito[2]

[1]Haskoning Consulting Engineers and Architects, P.O. Box 151,
6500 AD Nijmegen, The Netherlands
[2]Research Institute of Fragrance Materials, Two University Plaza,
Suite 406, Hackensack, NJ 07601

Synthetic musk fragrance ingredients are used in many personal care products and in household cleaning products. Representatives are the polycyclic musks (AHTN and HHCB) and the nitromusks (musk ketone and musk xylene). The environmental fate and behavior of these substances are reviewed. Comparison of environmental concentrations with toxicity data shows that the risks of these substances for aquatic and terrestrial organisms, as well as for the aquatic food chain, are low.

Introduction

Personal care products are marketed for direct use by the consumer and have intended end uses primarily on the human body, according to the definition of Daughton and Ternes (1). This category includes cosmetics, toiletries, and fragrances. The odor of these products is a highly important element and often a key factor for the consumer. Fragrance ingredients are used not only in personal care products but also in many consumer and household products such as detergents and cleaning products. In fact, their use in the categories of soaps, fabric softeners, detergents, and cleaners accounts for more than half of the total use, and less than 40% is used in personal care products. The concentration of fragrances varies strongly per product type or category. In colognes and eau de toilette, the concentration may be between 5 and 8%, whereas in perfume extracts it may be up to 20%. Approximate concentrations in cosmetics like skin and hair care products and bath/shower products are 0.5%, and in products where they are intended to mask product odors, it is less than 0.1%. For detergents, the concentrations range from 0.1% to 5% (*2,3*).

Synthetic musks are important ingredients in fragrances because of their typical musky scent and their fixative properties. Synthetic musks comprise a series of chemicals that emulate the odor of the musk deer and musk rats, but have different chemical structures: nitromusks (NMs) and polycyclic musks (PCMs). The most important are the polycyclic musks with 6-acetyl-1,1,2,4,4,7-hexamethyltetraline (AHTN) and 1,3,4,6,7,8-hexahydro-4,6,6,7,8,8-hexamethylcyclopenta-γ-2-benzopyran (HHCB) as the major products. Together they represent about 95% of the EU market and 90% of the US market for all polycyclic musks. Other members of this group are ADBI, AHMI, and AITI. Chemical names, structures, and characteristics are given in Table I. The nitromusks make less than 8% of the synthetic musks volume in the EU and less than 6% of the US volume. The main representatives of this group are musk ketone and musk xylene.

Fragrance ingredients have attracted attention in the past decade following the detection of certain NMs and the PCMs in samples of surface water, fish and human adipose tissue, and milk in Germany (*4,5,6,7,8,9*). These findings prompted more environmental sampling and analysis in other countries and initiated a series of discussions on the environmental safety of these substances. In the OSPAR (Oslo and Paris Commissions for the Prevention of Marine Pollution) Action Plan of 1998, musk compounds were included in the category of 'diffuse sources and groups of substances' to be considered for action. In the Netherlands, on behalf of the Ministry of the Environment (VROM) an environmental risk assessment was carried out on the NMs (musk ketone and musk xylene) by the Dutch National Institute of Public Health and the Environment RIVM (*10,11*). At the same time, human health effects of the NMs and PCMs were evaluated by the EU Scientific Committee on Cosmetics. In the meantime, the NMs were also included on the EU's Third Priority List related to the EU Existing Chemicals Regulation (EEC 793/93). With The Netherlands as the rapporteur, RIVM will proceed with a risk assessment for the environment, for the consumer, and for occupational health. Simultaneously with these events, the Research Institute of Fragrance Materials (RIFM) initiated a risk assessment for the two most important PCMs. The reports have been adopted by the Dutch authorities (3) and published (12,13). Both AHTN and HHCB are on the EU's Fourth Priority List for the Existing Chemicals Regulation. During the process, more physico-chemical and ecotoxicity data are being generated to refine the risk assessment results. The environmental risks of musk ketone and musk xylene and for AHTN and HHCB will be reviewed here.

Emission

Use Volume

The use volumes for Europe have been surveyed by RIFM (Research Institute of Fragrance Materials) and by IFRA (International Fragrance Association) since 1992. There has clearly been a general downward trend in the use of the four major musks, as in 1998 the use volume for each was reduced roughly by a factor of 2 or more as compared with 1992 (Table I).

Table I: Use volume [metric tons] of Nitromusks and Polycyclic Musks in Europe

	Europe		
	1992	*1995*	*1998*
Nitromusks			
Musk ketone	124	61	40
Musk xylene	174	110	86
Moskene	17	5	
Musk tibetene	3	0.8	
Polycyclic musks			
AHTN	885	585	385
HHCB	2400	1482	1473
ADBI		34	18
AHMI		3	19
AITI		40	2

Release to the environment

Due to their application in detergents, household cleaning products, and bath and hair care products, most of the production volume of the NMs and the PCMs is released to the sewer after use. The share of the fragrance ingredients used in these categories is estimated at 75% of the total use volume in the EU

(Fig. 1). For practical reasons, it is assumed for the environmental risk assessment that the remaining 25% (personal care, toiletries, perfumes) will be washed off after use and also be released to the sewer as well.

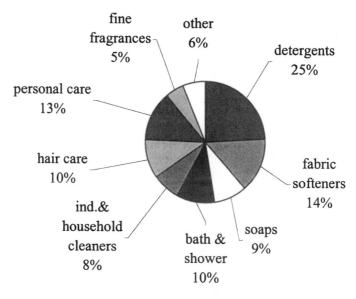

Figure 1. The use of fragrance oil in the European Union.

Environmental fate and behavior

Identification and physico-chemical characteristics

Nitromusks are nitrated aromatics and polycyclic musks are substituted indanes and tetralins. The characteristics of representatives of both groups of synthetic musks is given in Table II. Common physico-chemical characteristics are their hydrophobic behavior (log Kow > 4.5) and poor solubility in water, at a level of 0.1 to 2 mg/L. Based on the lipophilicity of the substances, they are expected to adsorb strongly to organic materials such as sewage sludge, sediments, and lipid tissue.

Table II. Characteristics of musk fragrance ingredients.

Identification and characteristics [a]		*Structure*

Nitromusks

Musk ketone
CAS: 81-14-1
3,5-dinitro-2,6-dimethyl-4-
t-butyl-acetofenone

$C_{14}H_{18}N_2O_5$
MW 294.3
S 1.9 mg/L [b]
log Kow 4.3
Vp 0.04×10^{-3} Pa [b]

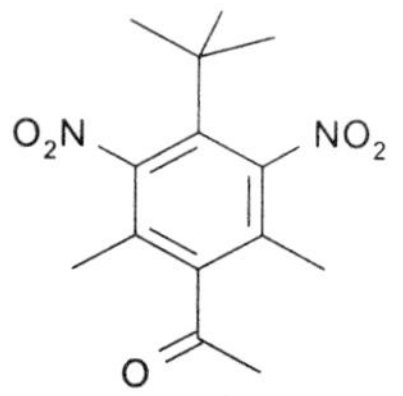

Musk xylene
CAS: 81-15-2
2,4,6-trinitro-1,3-dimethyl-
5-t-butylbenzene

$C_{12}H_{15}N_3O_6$
MW 297.3
S 0.49 mg/L [b]
log Kow 4.9
Vp 0.03×10^{-3} Pa [b]

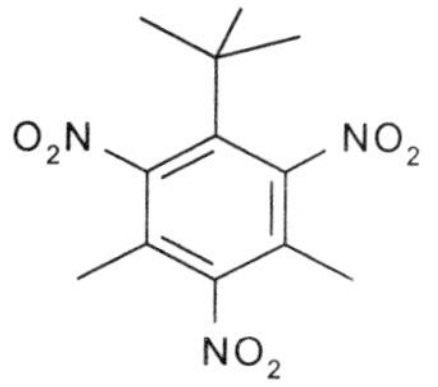

Moskene
CAS: 116-66-5
1,1,3,3,5-pentamethyl-4,6-
dinitroindane

$C_{14}H_{18}N_2O_4$
MW 278.3
S 0.17 mg/L [b]
log Kow 5.39 [b]
Vp 0.00023 Pa [b]

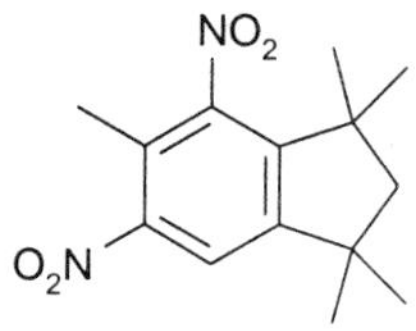

Musk tibetene
CAS: 145-39-1
1-ter-butyl-3,4,5-trimethyl-
2,6- dinitrobenzene

$C_{13}H_{18}N_2O_4$
MW 266.3
S 0.29 mg/L [b]
log Kow 5.18 [b]
Vp 0.00058 Pa [b]

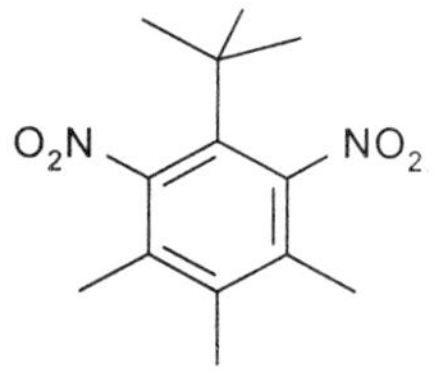

Polycyclic musks

AHTN
CAS: 21145-77-7
and 1506-02-1
Tonalid®, Fixolide®
7-acetyl-1,1,3,4,4,6-
hexamethyl-1,2,3,4-
tetrahydronaphtalene

$C_{18}H_{26}O$
MW 258.4
S 1.25 mg/L
log Kow 5.7
Vp 0.0608 Pa

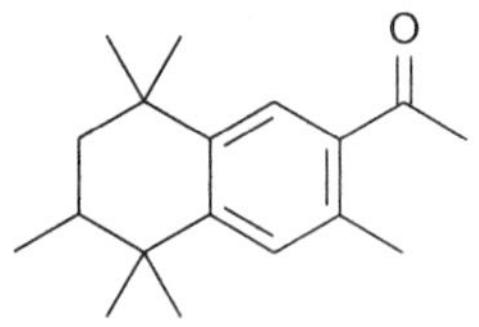

Table II. *Continued*

HHCB
CAS: 1222-05-5
Galaxolide 50®, Abbalide® $C_{18}H_{26}O$
1,3,4,6,7,8-hexahydro-   MW 258.4
4,6,6,7,8,8-hexa-   S 1.75 mg/L
methylcyclopenta[gamma]-   log Kow 5.9
2-benzopyran   Vp 0.0727 Pa

ABDI $C_{17}H_{24}O$
CAS: 13171-00-1   MW 244.4
Celestolide®, Crysolide®   S 0.22 mg/L [b]
4-acetyl-6-tert-butyl-1,1-   log Kow 5.93 [b]
dimethylindan   Vp 0.0192 Pa [b]

AITI $C_{18}H_{26}O$
CAS: 68140-48-7   MW 258.4
Traseolide®   S 0.09 mg/L [b]
5-acetyl-1,1,2,6-   log Kow 6.31 [b]
tetramethyl-3-isopropyl-   Vp 0.0091 Pa [b]
indan

AHMI $C_{17}H_{24}O$
CAS: 15323-35-0   MW 244.4
Phantolid®   S 0.25 mg/L [b]
5-acetyl-1,1,2,3,3,6-   log Kow 5.85 [b]
hexamethylindan   Vp 0.0196 Pa [b]

[a] MW: molecular weight; S: water solubility, Vp: vapor pressure

[b] estimated value (*14*); log Kow tends to be overestimated for these substances

Bioaccumulation

The synthetic musks have attracted attention because they were detected in fish samples. With log Kow between 4.3 and 5.9, a high bioaccumulation potential is expected. However, most experiments with fish show bioaccumulation factors (fresh weight), between 500 and 1500. Bioconcentration factors (BCFs) are summarized in Table III. Field-derived BCFs have been included for comparison. These values tend to be lower than the ones determined in the laboratory. However, it is always difficult to determine the exposure conditions for fish caught in surface waters (varying water concentrations, exposure to both dissolved and sorbed contaminants, presence of other substances, history, etc.).

The test design of the bioaccumulation experiments with musk ketone, AHTN, and HHCB included the measurement of metabolites. For these three substances, it could be established that they are transformed to more polar metabolites in a relatively short time and that these metabolites are rapidly excreted by the fish. Elimination half-lives were estimated at 2.5 days for musk ketone, 1 to 2 days for AHTN, and 2 to 3 days for HHCB. This indicates that the accumulation of these substances is highly reversible (11,12).

Degradability

The biodegradability of musk ketone, musk xylene, AHTN, and HHCB has been investigated in standard tests. These tests follow the oxygen consumption or carbon dioxide production over time. Under the conditions of these standard tests (OECD TG 301 or 302 for ready or inherent biodegradation), the mineralization of these substances is slow. In the two-phase closed bottle test (ISO 10708), AHTN and AHMI showed a low level of mineralization (12-21%) (12,13). The PCMs, however, are not persistent and they are biotransformed. AHTN, HHCB, and AHMI were shown to degrade in certain soil samples and in cultures inoculated with soil microorganisms to various more polar metabolites within a few days to a week. In another study with an air-borne fungus, later on amended with a soil slurry, HHCB was transformed to its lactone and to more polar structures and some CO_2. In a microcosm study with four soil types, an average of 14% HHCB remained after one year, showing that HHCB was disappearing with a half-life of approximately 4 months (12). When added to activated sludge, HHCB was transformed to its lactone (log Kow 4.0) and HHCB-hydroxy acid (log Kow 0.5). After 24 h, the overall metabolite pattern

Table III.   **Bioaccumulation of musk fragrance ingredients.**

Substance	log K_{ow}	BCF [method]	Source
Nitromusks			
musk ketone	4.3	Rainbow trout $(8+21)$d-BCF_{ww} = 1380 [^{14}C]	(*11*) (*15*)
		Xenopus 11d-BCF_{ww} = 3 [GC]	(*16*)
		environmental samples BCF = 1100	
musk xylene	4.9	Bluegill sunfish $(16+12)$d-BCF_{ww} = 1600 [^{14}C]	(*11*) (*17*)
		Rainbow trout $(190+250)$d-BCF_{ww} = 4400 [GC/MS]	(*15*) (*15*)
		Zebrafish 4d-BCF_{ww} = 432 [^{14}C] [a]	
		Xenopus 4d-BCF_{ww} = 47 [^{14}C] [a] 11d-BCF_{ww} = 12 [GC]	(*18*)
		Rainbow trout 21d-BCF_{ww} 10 to 60 [GC/MS]	(*16*)
		environmental samples BCF = 4100	
moskene	5.4	Xenopus 11d-BCF_{ww} = 12 [GC]	(*15*)
Polycyclic musks			
AHTN	5.7	Zebrafish $(14+26)$d-BCF_{ww} = 600 [GC/MS]	(*19*)
		Bluegill sunfish $(28+28)$d-BCF_{ww} = 597 [^{14}C, LC/HPLC]	(*12*)
		environmental samples: Eel BCF_{ww} = 200 to 650 non-eel BCF_{ww} = 50 to 145	(*12*)
		environmental samples: Eel BCF_{ww} = 410 [b] Rudd BCF_{ww} = 460 [b] Tench BCF_{ww} = 280 [b] Mussel BCF_{ww} = 560 [b]	(*20*)

Continued on next page.

Table III. *Continued*

HHCB	5.9	Zebrafish (14+26)d-BCF$_{ww}$ = 624 [GC/MS]	(*19*)
		Bluegill sunfish (28+28)d-BCF$_{ww}$ = 1584	(*12*)
		[^{14}C, TLC/HPLC]	
		environmental samples:	(*12*)
		Eel BCF$_{ww}$ = 150 to 600	
		non-eel BCF$_{ww}$ = 49 to 188	
		environmental samples:	(*20*)
		Eel BCF$_{ww}$ = 290 [b]	
		Rudd BCF$_{ww}$ = 230 [b]	
		Tench BCF$_{ww}$ = 620 [b]	
		Mussel BCF$_{ww}$ = 500 [b]	

[a] The different values may be explained by the transformation into metabolites that are included in detection by ^{14}C and not by GC.
[b] Recalculated from BCF$_{Lipid}$ with fraction lipids (data for a sewage pond in SH, Germany)

increased in polarity as time progressed. Analysis by reversed-phase HPLC showed three peaks in the extracts with log Kow < 0.1, 2.1, and 3.1. The half-life of HHCB under these conditions was 21 h (*21*). A tentative mass balance calculation for the sludge digestion process, assuming no biodegradation during the sewage treatment process, showed that approximately 40 to 45% of AHTN and HHCB is removed during sludge digestion, probably by primary degradation (*12*).

For the NMs, musk ketone and musk xylene, it was shown that during sewage treatment a major metabolism pathway is the reduction of the nitro-groups to form 2-amino-MK, 4-amino-MX, and 2-amino-MX (*22*). These metabolites are more polar, e.g., log Kow of 4-amino-MX is 3.8 (*23*).

The photodegradation of four nitromusks and AHTN was examined by irradiating an aqueous solution with artificial UV light (mercury-high pressure burner OMNILAB TQ 150) at room temperature. The measured half-lives under these conditions are in the range of 2.0 to 8.2 min for the nitromusks and 1.25 min for AHTN. This degradation was also attained under sunlight but at a lower rate. At termination of the experiment with AHTN (after 5 hours), no metabolites could be detected even after 1000-fold concentration (*24,25*). The atmospheric oxidation was estimated as the rate in the reaction with hydroxyl radicals. Assuming a concentration of 1.5×10^6 OH/cm^3, a light period of 12 h per day and 25°C, the atmospheric half-lives for the nitromusks vary between 5 and 13 days. Under those conditions, the estimated half-lives for the polycyclic musks vary between 3 and 16 hours (*14*).

Environmental pathways

The major portion of the musk fragrances is used in private households in detergents, cleaning products, and personal care products that end up in domestic wastewater after use. During sewage treatment, a fraction of the load may be degraded, and the remainder partitions between the water phase and the organic matter of the activated sludge. Due to the lipophilic character of these substances, a high portion is sorbed to sludge. Removal percentages vary between 50 and more than 90%, depending on the treatment plants and sampling strategy (*12*). After discharge, the effluent of the treatment plants is diluted in surface water where the remainder of the substances further partitions between the suspended matter or sediment in the river, accumulates in fish and other aquatic organisms, and may be further biodegraded. Predatory birds or mammals may take up the material accumulated in fish.

In some states, sewage sludge is disposed to land where it is possible that musks will end up in the terrestrial compartment. However, for AHTN, HHCB, and AHMI, it was shown that these substances may be degraded in soil. For the nitromusks these experiments have not been conducted. If present in soil, the musks may be available for terrestrial organisms (insects, worms, and other invertebrates) and be taken up by their predators (insectivores, small mammals, and birds).

In the risk assessments carried out for musk ketone and musk xylene, as well as for AHTN and HHCB in Europe, the aquatic and terrestrial compartments were considered (*11,12,13*). For these risk assessments, calculations have been performed using the European Union System for the Evaluation of Substances (EUSES) (*26*). EUSES conservatively estimates the environmental fate and concentrations of the substances in a region modeling the highly populated western part of Europe (200 × 200 km^2 and 20 × 10^6 inhabitants) (*27*). For AHTN and HHCB, the estimated concentrations in sludge were substituted with actual measured concentrations.

In the meantime, more data from environmental samples have become available allowing a better estimation of the daily input per inhabitant of some musk ingredients to the sewage treatment plant (STP) and the distribution and degradation during the sewage treatment process and in the soil. This results in a flow diagram where the environmental pathways predicted by EUSES have been refined using empirical data. Figure 2 shows the environmental fate of HHCB as an example.

Example: Environmental fate of HHCB

In the EU risk assessment, it was originally assumed that the total use volume of HHCB would be discharged to the sewer. This implied a daily discharge of 11.1 mg per person (inhabitant equivalent or I.E.). Based on the standard scenario of EUSES, the predicted concentrations in influent, effluent,

178

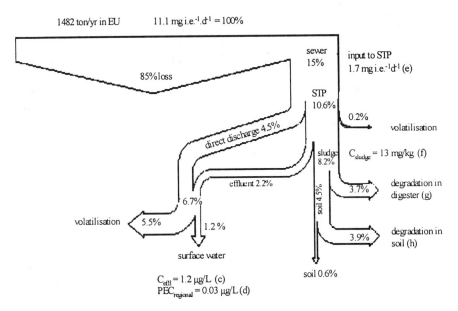

1482 ton/yr in EU   11.1 mg i.e.⁻¹.d⁻¹ = 100%

85% loss

sewer
15%

input to STP
1.7 mg i.e.⁻¹d⁻¹ (e)

STP
10.6%

0.2%

volatilisation

direct discharge 4.5%

sludge
8.2%

C_{sludge} = 13 mg/kg (f)

effluent 2.2%

soil 4.5%

3.7%

degradation in
digester (g)

6.7%

volatilisation   5.5%   1.2%

3.9%

degradation in
soil (h)

surface water

soil 0.6%

C_{eftl} = 1.2 µg/L (c)
$PEC_{regional}$ = 0.03 µg/L (d)

*Figure 2.   Fate of HHCB in the 'regional environment' of   EUSES.
See text for further explanation.*

and sludge, however, were higher than the measured concentrations by a factor of 10 to 20, which implies that the fraction discharged to the sewer may be considerably lower than 1. Therefore the mass flow of HHCB has been further analyzed based on the partition processes between the compartments in EUSES and refinements derived from measurements.

Effluent and sludge concentrations were measured in various European countries. The median effluent concentrations in these data sets varied between 1.4 and 2.3 µg/L (see also Fig. 3). Median concentrations in sludge were approximately 14 mg/kg in primary sludge, 4 to 28 mg/kg in activated sludge, and approximately 20 mg/kg in digested sludge (more details in next section). According to the tentative mass balance in the STP, the removal of HHCB during sludge digestion was estimated at 45% (*12*) [g in Fig. 2]. The input to the STP per I.E. is derived from the amount of HHCB on primary and activated sludge (1.35 mg/d) and from the effluent in the STP (0.34 mg/day), totalling 1.7 mg/d per i.e. [e in Fig. 2]. This is 15% of the estimated discharge per I.E. of 11.1 mg/day. In the EUSES model, sludge is used as an amendment on agricultural land. In mesocosm experiments where HHCB was applied to four different soil types, an average of 14% HHCB remained in the soil after 1 year (*12*) [h in Fig. 2].

EUSES assumes that 70% of the wastewater is treated in an STP and that 30% is directly discharged to surface water. Based on the above data and calculations and on the partition coefficients between the compartments, the generalized mass flow for HHCB as presented in Figure 2 predicts a concentration in sludge of 13 mg/kg [f in Fig. 2], an effluent concentration of 1.2 µg/L and a 'steady state' concentration in surface water of 0.03 µg/L in remote regions [c and d in Fig. 2]. Actually, the concentrations in a region will decrease from 1.2 µg/L in almost undiluted effluent directly after discharge, to 0.03 µg/L. The frequency of measured surface water concentrations in Figure 4 shows that measured HHCB concentrations fall in this range.

The modeling of volatilization in the regional model of EUSES is based on the physico-chemical parameters of HHCB and the relative volumes of the environmental compartments. The loss of material by volatilization proves to be an important process, which may be explained by the large air volume as compared with the volumes of the other compartments [b in Figure 2]. In the same way, volatilization could be the major loss process during use of the household products or cosmetics. After all, they can only serve their function as a fragrance when they are in the air. Therefore, the fraction in the top of the scheme in Figure 2 [a] that is lost between the 'use per inhabitant' and 'detected in the STP' may be explained to a large extent by volatilization during/after use.

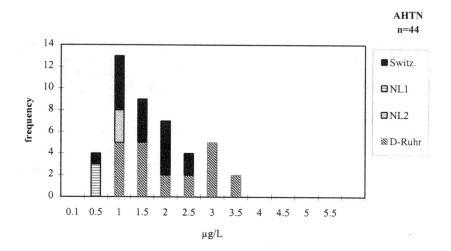

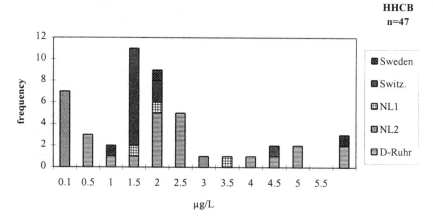

Figure 3. Concentrations of AHTN and HHCB in STP effluents from Sweden (31), Switzerland (33), Netherlands NL1 (29), NL2 (30), D-Ruhr (6).

Environmental concentrations

Musks have been detected in surface waters and sediment, biota, and in wastewater treatment plants. Often the results of these studies are summarized as a minimum and a maximum with occasionally a mean or median and a 90th-percentile value. Thus it is often difficult to gain an insight in the relevance of the values. They may be part of a large or a limited sampling activity, and the maximum may be quite extreme. For a proper assessment of the environmental risk, the probability of occurrence of the environmental concentrations is of importance. Therefore a frequency distribution is more useful than a list of extremes.

Concentrations in sewage treatment plants

Concentrations were measured in effluent samples of sewage treatment plants on the Ruhr (*6*), in Hamburg and Schleswig-Holstein in Germany (*22,28*), in The Netherlands (*29,30*), in Sweden (*31*), and in Switzerland (*32,33*) between 1993 and 1998. The concentration of musk ketone in effluents seems to vary around 0.2 µg/L, ranging from below the detection limits, to exceptional values around 1 µg/L. Figure 3 presents the frequency distributions of the available data for AHTN and HHCB. The median and 90th-percentile are 1.3 and 2.9 µg/L for AHTN and 1.6 and 4.5 µg/L for HHCB. The generic effluent concentration of 1.2 µg/L predicted above (see Fig. 2) for HHCB corresponds to the 36th-percentile in the distribution frequency of the measured data. This frequency distribution may be used as a reference for comparison with other effluent sample concentrations.

Concentrations measured in sludge in Germany (*34*), The Netherlands (*35,29*), and Switzerland (*36*) range roughly between 1 and 30 mg/kg for both AHTN and HHCB and between the detection limit to 0.06 mg/kg for musk ketone and musk xylene (*12,37*).

Surface water concentrations

The concentrations found in surface water are highly variable, with higher concentrations clearly related to samples in close proximity to STP discharge points. The concentrations reported from a large number of studies are summarized in Figure 4. For HHCB, the median of the concentrations encountered in the Western European surface waters (n=209, excluding the North Sea samples) is 0.07 µg/L, and the 90th-percentile is 0.5 µg/L. Some extremely high concentrations were observed in surface waters in Berlin in

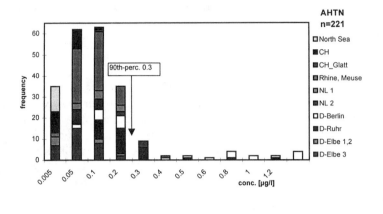

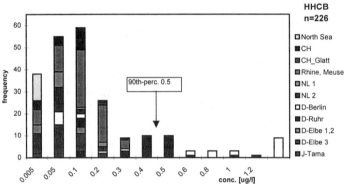

Figure 4. Concentrations of AHTN and HHCB in surface water.
North Sea: German Bight (38), CH: Swiss rivers and lakes (33), CH: Glatt (39),
Rhine and Meuse Dutch border (40), Netherlands surface waters NL1 (29),
NL2 (30), Germany: D-Berlin, water system Berlin (41), D-Ruhr(6), D-Elbe1
(38), D-Elbe2 (42), D-Elbe3 (43), Japan: Tama river (44).
(Adapted with permission from reference 12, Copyright 1999 Elsevier Science).

which a high proportion of the flow consists of sewage treatment plant effluents. Of all the samples, 27% are $\leq$ 0.03 µg/L, which corresponds to the generic concentration in remote areas as predicted above (see Fig. 2). Likewise, the concentrations of AHTN and musk ketone in over 200 surface water samples are presented in Figures 4 and 5. The median and 90th-percentiles for Western Europe are 0.07 and 0.3 µg/L for AHTN and <0.005 µg/L and 0.04 µg/L for musk ketone.

Biota

Several authors have published concentrations measured in fish from surface waters as well as from aquaculture and effluent ponds. For an environmental risk assessment, however, the relevance of the measurements is

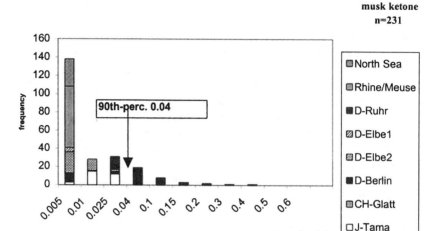

*Figure 5. Concentrations of musk ketone in surface water. North Sea (45),
Rhine and Meuse at the Dutch borders (40), D-Ruhr (4), Elbe-1 (38), Elbe-2
(43), D-Berlin (41), CH: Switzerland, Glatt (39), J-Tama: Japan (16).*

determined by the defined risk groups, i.e., aquatic organisms in the natural
environment and their predators. In that respect, data from fishponds or effluent
treatment ponds are not relevant. Data for fish and shellfish from surface water
(rivers, lakes) in Germany, The Netherlands, and Italy (*4,5,11,12,28,46,47,20,
50*) have been combined into frequency distributions for the various substances.
Overall median and 90th-percentile concentrations are 0.01 and 0.10 mg/kg
(fresh weight) for AHTN as well as for HHCB. For musk ketone, the
concentrations in fish are lower by an order of magnitude. Comparatively, in
North America, it was just the other way around: concentrations of musk ketone
in Canada were at the level of AHTN and HHCB in Europe, and an order of
magnitude above the local Canadian AHTN and HHCB levels. Musk ketone
levels were two orders of magnitude above musk xylene levels in Canada (*48*).

Effects

The aquatic toxicity of musk ketone, musk xylene, and AHTN and HHCB
is reviewed in Tas et al. (*11*) and Balk and Ford (*13*). NOECs (No Observed
Effect Concentrations) are available for the effects on the growth of algae, the

reproduction of *Daphnia magna,* and the growth and/or embryo development of various fish species. The toxicity data are summarized in Figure 6. The figure includes the Predicted No Effect Concentration for the aquatic ecosystem, $PNEC_{water}$, derived as described in the EU Technical Guidance Document for Risk Assessment [EU-TGD] (*27*); the lowest NOEC is divided by an assessment factor of 10 (50 for musk xylene).

For musk ketone, musk xylene, and moskene, neither mortality, growth inhibition, nor malformation was found to occur in zebrafish and Xenopus embryos after 96 h at 400 µg/L (nominal). After 11 days, however, viability of the Xenopus larvae seemed to be reduced by 15 to 20% (*15*). No binding of musk ketone, musk xylene, or moskene to the estrogen receptor of rainbow trout and Xenopus was observed (*23*). These data do not influence the level of the PNEC for these nitromusks. For high concentrations of AHTN (>67 µg/L), loss of caudal fin was observed. At the next higher concentration (140 µg/L), larvae survival was 18% (*13*). The toxicity of the amino metabolites of musk xylene to *D. magna* was shown to be at the same level as for musk xylene (*49*). Other aquatic toxicity data are not available.

The terrestrial toxicity of musk ketone, musk xylene, AHTN, and HHCB has also been reported (*11, 13*). For musk ketone, AHTN, and HHCB, tests were carried out with earthworms and springtails, whereas for musk xylene only an earthworm test was available. The PNEC for musk xylene was derived by equilibrium partitioning from the $PNEC_{water}$, and for the other substances an assessment factor of 50 was applied to the lowest test result after standardization for organic carbon according to the EU-TGD (*27*). The $PNEC_{soil}$ is 0.64 mg/kg dw for musk ketone, 0.23 mg/kg dw for musk xylene, and 0.32 mg/kg dw for AHTN and HHCB. These data are summarized in Table IV.

The toxicity for predators in the natural environment is derived from data available for rats. The NOAELs (No Observed Adverse Effect Level) are first converted to a dose in the daily food by multiplication with a factor of 10 for musk ketone and musk xylene, and 20 for AHTN and HHCB, respectively. For the derivation of the $PNEC_{predator}$, the converted NOAEL is divided by an assessment factor of 30 according to the EU-TGD (*27*). The $PNEC_{predator}$ is 25 mg/kg food (ww) for musk ketone, 8 mg/kg ww for musk xylene, 10 mg/kg ww for AHTN, and 100 mg/kg ww for HHCB (*11,13*).

Risk assessment

In an environmental risk assessment, risk is expressed as the quotient of the exposure and the no-effect concentrations for the relevant organisms (Exposure

Concentration/PNEC). When the risk quotient is below 1, the environmental concentrations are not expected to cause an adverse effect. For an adequate risk evaluation, the exposure concentration should be presented in a probabilistic way and not by the extremes. For surface water, the 90th-percentiles of the measured concentrations were used, as these were derived from large data sets. Likewise, concentrations measured in fish were used to assess the risk to predatory birds and mammals. For the risk assessment for soil organisms, the measured concentrations in sludge were used to predict the concentration in agricultural soil. The exposure concentrations and no-effect concentrations used to calculate the risk quotients are given in Table V. The table shows that for musk ketone and musk xylene, all risk quotients are below 0.01, whereas for AHTN and HHCB, the risk quotients range from below 0.01 to 0.10.

Table V. Environmental concentrations (EC) in various compartments, Predicted No Effect Concentration (PNEC), and risk quotient (EC/PNEC) for musk fragrance ingredients.

	Musk ketone	Musk xylene	AHTN	HHCB
Aquatic organisms				
surface water concentration 90th-percentile [µg/L]	0.04 [a]	0.01 [b]	0.30 [c]	0.50 [c]
PNEC [µg/L] [d]	6.3 [b]	1.1 [b]	3.5 [c]	6.8 [c]
Risk quotient	*0.006*	*0.009*	*0.09*	*0.07*
Fish-eating predators concentration in fish,				
max. or 90th-percentile [mg/kg ww]	0.066 [b]	0.095 [b]	0.10 [ce]	0.10 [ce]
PNEC [mg/kg ww]	25 [b]	8 [b]	10 [c]	100 [c]
Risk quotient	*0.003*	*0.012*	*0.01*	*0.001*
Soil predicted conc. in soil based on measured sludge concentrations [mg/kg ww]	0.001 [f]	0.001 [f]	0.029 [c]	0.032 [c]
PNEC [mg/kg ww]	0.64 [b]	0.23 [b]	0.32 [c]	0.32 [c]
Risk quotient	*0.002*	*0.004*	*0.09*	*0.10*

[a]data in Figure 5; [b]data from (*11*); [c]data from (*12,13*); [d]PNEC in Figure 6; [e]data from (*7,12,50,20*); [f]modified calculations from (*11*) using sludge data from (*51*)

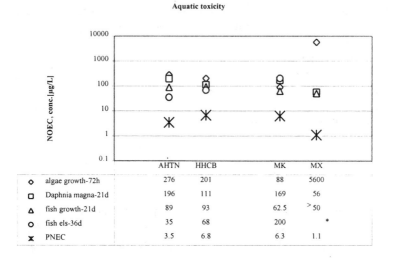

Figure 6. Aquatic toxicity data and the Predicted No Effect Concentration in the aquatic compartment PNEC (13). * fish growth test duration for MX was 14 days

Table IV. Terrestrial and mammalian toxicity.

	Musk ketone[a]	Musk xylene[a]	AHTN[b]	HHCB[b]
Earthworm				
8 week NOEC reprod.	32	> 50	105	45
14 d NOEC growth [mg/kg soil]				
Springtail				
4 week NOEC reprod. [mg/kg soil]	100		45	45
Rat				
90 d NOAEL [mg/kg bw/d]	75	24	15	150

[a] data from *11*; [b] data from *13*

Discussion and Conclusion

In the present context, the musk fragrance ingredients are included in the Pharmaceutical and Personal Care Products group (PPCPs). In comparison with other members of the PPCPs, concentrations observed in effluents and surface water of the musk with the highest use volume, namely HHCB, are lower than for many of the other PPCPs detected in these compartments (*1*). As a subset of the PPCPs, however, the musks are *not* intended to be biologically active and the aquatic toxicity of many pharmaceuticals is higher by orders of magnitude (*52*). This is not meant to trivialize the environmental relevance of the musk ingredients, but rather, to put the issue in a broader perspective.

Musk fragrance ingredients have attracted the attention of the EU, OSPAR, and various national authorities because they have been detected in environmental and human samples. As a consequence, the substances are included in lists with recognized persistent and highly bioaccumulating groups of substances such as dioxins and PCBs. The available data, however, show that this is not justified. AHTN and HHCB are rapidly metabolized and excreted by fish and the measured bioaccumulation is considerably lower than that predicted on the basis of log Kow. Furthermore, the substances are biodegradable and biotransformable, as well as photodegradable.

The risk assessments presented here are the result of an iterative process where initially the exposure was based on 'worst case' model predictions, and the effect assessment was based on a limited data set. Triggered by risk quotients above 1, empirical data were generated resulting in a more realistic evaluation of the environmental risks. In the present phase, the risk assessments still include various conservative assumptions, in particular for the estimation of the exposure and toxicity levels for soil organisms. Current environmental risk quotients of the NMs (musk ketone and musk xylene) and the PCMs (AHTN and HHCB) are at or below 0.1 for aquatic organisms, fish-eating predators, and terrestrial organisms. Keeping in mind the conservative approach taken, these data are reassuring that the environmental risks of these substances are low.

References

1. Daughton, C.G.; Ternes, T.A. *Environ. Health Perspect.* **1999**, *107(suppl. 6)*, 907-938.
2. De Groot, A.C.; Frosch, P.J. *Contact Dermatitis* **1997**, *36*, 57-86.
3. Van de Plassche, E.J.; Balk, F. *Environmental risk assessment of polycyclic musks AHTN asnd HHCB according to the EU-TGD;* RIVM report no. 601

503 008; National Institute of Public Health and the Environment RIVM: Bilthoven, NL, 1997.

4. Eschke, H.-D.; Traud, J.; Dibowski, H.-J. *Vom Wasser* **1994**, *83*, 373-383.
5. Eschke, H.-D.; Traud, J.; Dibowski, H.-J. *UWSF. Z. Umweltchem. Ökotox.* **1994**, *6*, 183-189.
6. Eschke H.-D.; Dibowski, H.J.; Traud, J. *UWSF.Z. Umweltchem. Ökotox.* **1995**, *7*, 131-138.
7. Eschke H.-D.; Dibowski, H.J.; Traud, J. *Dt. Lebensm. Rdsch.* **1995**, *91*, 375-379.
8. Rimkus, G.; Wolf, M. *Chemosphere* **1995**, *30*, 641-651.
9. Rimkus, G.; Wolf, M. *Chemosphere* **1996**, *33*, 2033-2043.
10. Tas, J.W.; Van de Plassche, E.J. *Initial environmental risk assessment of musk ketone and musk xylene in the Netherlands in accordance with the EU-TGD;* RIVM report 601503 002; National Institute of Public Health and the Environment RIVM: Bilthoven, NL, **1996**.
11. Tas, J.W.; Balk, F.; Ford, R.A.; Van de Plassche, E.J. *Chemosphere* **1997**, *35*: 2973-3002.
12. Balk, F.; R.A. Ford *Toxicol. Lett.* **1999**, *111*, 57-79.
13. Balk, F.; R.A. Ford *Toxicol. Lett.* **1999**, *111*, 81-94.
14. *SRC's Estimation Software;* Environmental Science Center, Syracuse Research Corporation, NY, **1996**.
15. Chou, Y.-J.; Dietrich, D.R. *Toxicol. Lett.* **1999**, *111*, 17-25.
16. Yamagishi, T.; Miyazaki, S.; Horii, S.; Akiyama, K. *Arch. Environ. Contam. Toxicol.* **1983**, *12*, 83-89.
17. Rimkus, G.G.; Butte, W.; Geyer, H.J. *Chemosphere* **1997**, *35*, 1497-1507.
18. Boleas, S.; Fernandez, C.; Tarazona, J.V. *Bull. Environ. Contam. Toxicol.* **1996**, *57*, 217-222.
19. Butte, W.; Ewald, F. *Kinetics of accumulation and clearance of the polycyclic musk compounds Galaxolide (HHCB) and Tonalide (AHTN);* Poster University Oldenburg, Germany, **1999**.
20. Rimkus, G. *Toxicol. Lett.* **1999**, *111*, 37-56.
21. Itrich, N.R.; Simonich, S.L.; Federle, T.W. *Biotransformation of the polycyclic musk, HHCB, during sewage treatment;* Environmental Science Department, Procter and Gamble, Poster SETAC 19th Annual Meeting November **1998**, Charlotte, NC USA..
22. Gatermann, R.; Hühnerfuss, H.; Rimkus, G.; Attar, A.; Kettrup, A. *Chemosphere* **1998**, *36*, 2535-2547.
23. Chou, Y.-J.; D.R. Dietrich *Toxicol. Lett.* **1999**, *111*, 27-36.
24. Butte, W.; Schmidt, S; Schmidt, A. *Photochemical degradation of nitrated musk compounds;* Carl von Ossietzky University Oldenburg, Germany. Poster presented at 'Analytika', April 1996.

25. Willenborg, R.; Butte, W. *Photochemischer Abbau polycyclischer Moschusverbindungen;* Carl von Ossietzky University Oldenburg, Germany. Poster presented at 'Tag der Chemie', November 1997.

26. *EUSES, the European Union System for the Evaluation of Substances;* RIVM, NL, Ed.; European Chemicals Bureau (EC/JRC): Ispra, Italy, 1996.

27. *Technical Guidance Document in support of Directive 96/67/EEC on risk assessment of new notified substances and Regulation (EC) No. 1488/94 on risk assessment of existing substances;* Office for Official Publications of the EC: Luxembourg, Lux., 1997; Part II.

28. Rimkus G.; Gatermann, R.; Hühnerfuss, H. *Toxicol. Lett.* **1999**, *111*, 5-15.

29. Verbruggen, E.M.J.; Van Loon, W.M.G.M.; Tonkes, M.; Van Duijn, P; Seinen, W.; Hermens, J.L.M. *Environ. Sci. Technol.* **1999**, *33*, 801-806.

30. Rijs, G.B.J.; A.J. Schäfer *Musken;* RIZA Report 99.006 (in Dutch). Institute for Inland Water Management and Waste Water Treatment RIZA, Lelystad, NL, 1999.

31. Paxéus, N. *Wat. Res.* **1996**, *30*, 1115-1122.

32. SAEFL *Occurrence of nitromusk compounds in the aquatic environment in Switzerland.* Swiss Agency for the Environment, Forests and Landscape SAEFL, Berne, CH, 1995.

33. SAEFL *Occurrence of polycyclic musk compounds in the aquatic environment in Switzerland.* Swiss Agency for the Environment, Forests and Landscape SAEFL, Berne, CH, 1998.

34. Sauer, J.; Antusch, E.; Ripp, Ch. *Vom Wasser* **1997**, *88*, 49-69.

35. Blok J. *Measurement of musk fragrances in sludges of sewage treatment plants in The Netherlands.* Report to RIFM, BKH Consulting Engineers, Delft, NL, 1997.

36. Herren, D.; Berset, J.D. *Chemosphere* **2000**, *40*, 565-574.

37. *Background Document on Musk Xylene and other Musks, Series Point and Diffuse Sources No. 101*, OSPAR Commission (2000), Copenhagen, Denmark. ISBN 094956553.

38. Bester, K.; Hühnerfuss, H.; Lange, W.; Rimkus, G.G.; Theobald, N. *Wat. Res.* **1998**, *32*, 1857-1863.

39. Müller, S.; Schmid, P.; Schlatter, C. *Chemosphere* **1996**, *33*, 17-28.

40. Breukel, R.M.A.; Balk, F. *Musken in Rijn en Maas;* RIZA Werkdocument 96.197x. National Institute for Inland Water management and Waste Water Treatment RIZA, Lelystad, NL.

41. Heberer, Th.; Gramer, S.; Stan, H.-J. *Acta Hydrochim. Hydrobiolog.* **1999**, *27*, 150-156.

42. Lagois, U. *Wasser – Abwasser* **1996**, *137*, 154-155.

43. Winkler, M.; Kopf, G.; Hauptvogel, C.; Neu, T. *Chemosphere* **1998**, *37*, 1139-1156.

190

44 Yun, S.-J.; Teraguchi, T.; Zhu, X.-M.; Iwashima, K. *J. Environ. Chem.* **1994**, *4*, 325-333 (in Japanese).

45. Gatermann, R.; Hühnerfuss, H.; Rimkus,G.; Wolf, M.; Franke, S. *Marine Pollut. Bull.* **1995**, *30*, 221-227.

46. Geyer, H.J.; Rimkus, G.; Wolf, M; Attar, A.; Steinberg, C.; Kettrup, A. *UWSF - Z. Umweltchem. Oekotox.* **1994**, *6*, 9-17.

47. Gatermann, R.; Rimkus, G.; Hecker, M.; Biselli, S.; Hühnerfuss, H. *Bioaccumulation of synthetic musks in different aquatic species.* Poster presented at SETAC Europe 9th Annual Meeting, 25-29 May 1999, Leipzig, Germany.

48. Gatermann, R.; Hellou, J.; Hühnerfuss, H.; Rimkus, G.; Zitko, V. *Chemosphere* **1999**, *38*, 3431-3441.

49. Giddings, J.M.; Salvito, D.T.; Putt, A.E. *Wat. Res.* **2000**, *34*, 3686-3689.

50. Draisci, R.; Marchiafava, C.; Ferretti, E.; Palleschi, L.; Catellani, G.; Anastasio, A. *J. Chromatogr. A* **1998**, *814*, 187-197.

51. *Background Document on Musk Xylene and other Musks, Series Point and Diffuse Sources No. 101,* OSPAR Commission (2000), Copenhagen, Denmark. ISBN 094956553.

52. Halling-Sørensen, B.; Nors Nielsen, S.; Lanzky, P.F.; Ingerslev, F.; Holten Lützhøft, H.C.; Jørgensen, S.E. *Chemosphere* **1998**, *36*, 357-393.

Waste Treatment

Chapter 11

Drugs in Municipal Sewage Effluents: Screening and Biodegradation Studies

Edda Möhle and Jörg W. Metzger[*]

Institute of Sanitary Engineering, Water Quality and Solid Waste Management,
Department of Hydrochemistry and Hydrobiology, University of Stuttgart,
Bandtäle 2, 70569 Stuttgart, Büsnau, Germany

A gas chromatography/mass spectrometry screening method was developed for the identification of low molecular weight, polar, and semipolar substances in the complex matrix of wastewater. This method allowed for the identification of 14 drugs and metabolites in nine municipal sewage plants in the region of Stuttgart, Germany. To get information about the degree of elimination (adsorption or degradation) of different drugs in a municipal sewage plant, tests under aerobic conditions were performed. A batch reactor containing individual drugs or a mixture of several different drugs in environmentally relevant concentrations in contact with a suspension of activated sludge was coupled on-line with high performance liquid chromatography/tandem mass spectrometry. The concentrations of the drugs and their metabolites in the batch reactor were detected after electrospray-ionization by a triple quadrupole-mass spectrometer. Sampling, filtration, and analysis were carried out automatically for the first 10 hours in intervals of 30 minutes and afterwards in longer intervals. In this way, concentration-time curves were obtained. The concentrations of most of the examined drugs determined after 15 minutes were much lower than the initial concentrations. This 'fast elimination' was explained primarily with the adsorption of the drugs on the activated sludge. For most of the compounds examined, only this effect was observed. In contrast, for some substances an additional subsequent slow decrease of the concentrations was observed within several hours or even days until concentrations less than 1% of the initial concentrations were reached. This effect was assumed to be mainly caused by biodegradation and/or transformation. To get information about the stability of glucuronic acid conjugates of drugs in contact with sludge, acetamidophenyl glucuronide was examined. It could be shown that the conjugate was cleaved and free acetamidophenol was formed, which was then degraded.

Introduction

Due to the presence of persistent organic pollutants, the effluents of wastewater treatment plants still contain varying amounts of organic carbon even after biological treatment. The organic carbon of the effluent is determined by means of summary parameters including dissolved organic carbon (DOC) or chemical oxygen demand (COD). The individual compounds contributing to the DOC or the COD are generally known only in part. The determination of all single substances in wastewater is impractical due to their great number, their low concentrations and the lack of knowledge about their identities. The residual DOC in effluents is mainly responsible for the pollution of surface waters with environmental substances. The ecology of the surface waters can be disturbed by single substances, and also the recovery of drinking water with sufficient quality from surface waters is difficult in the presence of environmentally relevant compounds. Therefore it is important to screen the effluent for hardly biodegradable substances. These include known products, which have attracted scientific and public interest for years, such as plant-protection agents. Furthermore, it is necessary to screen the effluent for the possible presence of chemicals that so far have not been identified in wastewater. The intention of this study was to develop a gas chromatography-mass spectrometry (GC-MS) method, allowing for the identification of hardly biodegradable substances, within a particular polarity range, in wastewater.

GC-MS was chosen for analysis. For electron impact-mass spectrometry (EI-MS), large libraries of mass spectra exist, which allow a straightforward identification of unknown substances. However, the application of both GC and MS are restricted to less polar substances. For more polar substances, derivatization is required. To identify unknown substances via their mass spectra, the measurements had to be performed in the scan mode rather than in the more sensitive single ion monitoring (SIM) mode.

The screening method was developed with model substances having log P_{OW} (octanol/water partition coefficients) between –0.38 and 2.46. Most of the substances that can be detected in the effluent at the low ppb level are polar substances with a limited tendency to adsorb to the activated sludge. To increase the sensitivity, an enrichment procedure was applied, which has been described previously (1).

A selection of drugs, which had been detected in the different secondary effluents of the wastewater treatment plants with this method, were examined with regard to their elimination behavior in contact with activated sludge under aerobic conditions. For these investigations, high-performance liquid chromatography coupled with tandem mass spectrometry (HPLC-MS-MS) was coupled on-line to a batch reactor filled with activated sludge of a municipal sewage plant. An advantage of LC-MS-MS in comparison with GC-MS is that the analysis is possible directly from aqueous solution. Additionally, the method is time-saving, because it does not require off-line solid phase extraction (SPE). The detection limits of the LC-MS-MS method in the SIM mode were comparable to the detection limits of GC-MS in the scan mode.

Experimental

Materials

The solvents ethyl acetate and diethyl ether were obtained from Merck (Darmstadt, Germany), both analytical grade. 2-(2-Ethoxyethoxy)-ethanol and N-methyl-N-nitroso-p-toluenesulfonamide were purchased from Aldrich (Steinheim, Germany), purity > 99 %. Drugs of analytical purity were obtained from Synopharm (Barsbüttel, Germany), Sigma (Deisenhofen, Germany), and Aldrich (Steinheim, Germany).

Stock solutions of all compounds for the GC-MS analysis were prepared in ethyl acetate. Solid phase extraction was carried out with a vacuum manifold using glass cartridges with a volume of 3 mL filled with 200 mg LiChrolut EN, Merck (Darmstadt, Germany).

For the elimination studies, the stock solutions of the water-soluble substances were prepared in water, those of the other substances in ethyl acetate.

The biodegradation/elimination studies were carried out with activated sludge from the aeration tank of the wastewater treatment plant in Stuttgart-Büsnau, Germany. The dry solids were approximately 4 g/L and the sludge volume index (SVI) was approximately 100 mL. If necessary, the activated sludge was diluted 1:10 or 1:5 with the supernatant fluid of the activated sludge.

Sample Preparation

The samples (2 x 1 L) were adjusted to two different pH values, pH 12 after addition of EDTA (3 g) to prevent the precipitation of calcium carbonate, and pH 2. The samples were filtered through fiberglass filters. After conditioning of the solid phase material, the sample solution was added. Conditioning of the SPE cartridges was performed with ethyl acetate (2 x 2 mL) and finally with deionized water (2 mL). The water samples (1 L) were then passed through the cartridges at a flow rate of approximately 10 mL/min. After drying the cartridges for 30 min by flushing with nitrogen, the analytes were eluted with ethyl acetate (2 x 2 mL). The eluates were concentrated to 1 mL under a gentle stream of nitrogen. The eluates from the 'pH 12 samples' were injected directly into the gas chromatograph. Those from the 'pH 2 samples' were derivatized with diazomethane.

Derivatization

The residues of the 'pH 2 samples' were derivatized with diazomethane. A mixture of diethyl ether and 2-(2-ethoxy-ethoxy)-ethanol (5 mL; 1:1) and KOH (5 mL, 40 %) was placed in a special apparatus for methylation. After addition of N-methyl-N-nitroso-p-toluenesulfonamide (ca. 1 mg), diazomethane was formed and transferred into the sample until a yellow color was obtained. The excess diazomethane was evaporated overnight. Ethyl acetate was added to the residue to give a final sample volume of 1 mL. One μL of this solution was injected into the GC-MS.

GC-MS

A mass spectrometer MSD 5972 (Hewlett Packard) combined with a gas chromatograph 6890 (Hewlett Packard) fitted with a 25-m x 250-μm x 0.25-μm DB-5ms capillary column was used for measurement.

One μL of sample was injected using hot splitless injection. The injection port temperature was 280°C. The oven temperature was held at 60°C for 1 min following injection, then programmed at 40°C/min to 80°C, then at 3°C/min to 100°C, which was held for 10 min, followed by 30°C/min to 250°C, which was held for 5 min. The flow rate was set to 1 mL/min. Transfer line temperature was 280°C. The mass-range in the scan mode was 47 - 350 amu.

Batch reactor coupled on line with LC-MS-MS

The batch reactor, a 2-L Woulff flask filled with 2 L of activated sludge, was coupled on-line to the LC-MS-MS instrument (Fig. 1). Single drugs or a mixture of several drugs at environmentally relevant concentrations, i.e., at the ppb-level, were added into a batch reactor filled with activated sludge. The concentrations of the drugs in the reactor before addition of the drugs was generally < 10% of the added concentration. The change of the concentration with time was measured on-line using LC-MS-MS. Changes in concentration of oxygen and nutrient and external carbon sources during the experiment can principally affect the biodegradation of the drugs. For simplification of the experimental set-up, and since the experiments lasted only a relatively short period of time (max. 72 h), neither an external organic carbon source nor further nutrients was added.

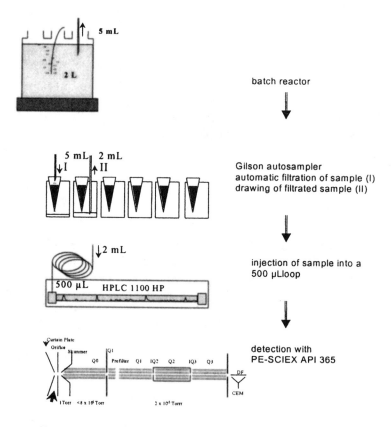

Figure 1: Experimental apparatus for the studies of the elimination of drugs by sewage sludge with HPLC-MS-MS

During the testing period, the sludge was stirred and aerated. Ten min prior to sampling, stirring and aeration were stopped to allow the sludge to settle. Samples (5-mL) were taken by an autosampler (Gilson 232 XL, Abimed, Langenfeld, Germany) and filtered automatically through glass wool. After filtration, the filtrate (2 mL) was injected with the autosampler into the 500-μL HPLC sample loop. The eluents were 5mM aqueous ammonium acetate (solvent A) and 5% acetonitrile/5mM aqueous ammonium acetate 99:1 (solvent B). The eluent composition at start time was 95% A and 5% B. Under these conditions the sample was enriched at one end of the HPLC column. For HPLC, a liquid chromatograph HP 1100 (Hewlett-Packard, Waldbronn, Germany) fitted with a microbore column LUNA, 100 mm x 2 mm, 3-μm particle size (Phenomenex, Hösbach, Germany) was used. Gradient elution was used to separate the substances. The conditions of the elution are listed in Table I. The flow rate was set to 0.3 mL/min.

Table I: HPLC-gradient (solvent A: 5mM aqueous ammonium acetate; solvent B: acetonitrile/5mM aqueous ammonium acetate 99:1)

time [min]	solvent B [%]
0.20	5
9.00	80
10.00	100
12.00	100

Mass spectrometric measurements were performed using an API 365 mass spectrometer (PE-Sciex, Thornhill, Canada) combined with a turbo ion spray interface or a heated nebulizer (APCI) interface. The turbo ion spray interface was used mainly as an ion-source. The voltage on the electrospray needle was +4800 V and the turbo spray temperature 400 °C.

The substances were mainly detected in the positive ion mode and MRM (multiple reaction monitoring) with nitrogen as collision gas. The cleavage of the glucuronide conjugates was measured in the negative mode using APCI as interface.

Sampling

Samples (2.5 L) were taken from nine different wastewater treatment plants in the vicinity of Stuttgart, Germany, in March, June, September, and December 1997. The capacity of the sewage treatment plants expressed in the total number of inhabitants and population equivalents varied from 20,000 (Eglosheim) to

1,000,000 (Mühlhausen). The samples of the effluent were taken automatically with autosamplers over a period of 24 h as combined samples.

Results and Discussion

Screening-method

By comparison of the experimental mass spectrum of each sample with the spectrum in the NIST-libraries, several drugs and metabolites of drugs (totalling 14) in municipal sewage effluents could be identified.

These drugs belong to different chemical groups and have different medicinal applications (2). The antiepileptics carbamazepine, primidone, and pheneturide could be identified. Out of the group of analgesics, we found diclofenac, propyphenazone, and a metabolite of dipyrone, acetylamino-antipyrine. Furthermore, hydrocodone and dihydrocodeine, two antitussives, were detected. Pentoxifylline is a drug for widening blood vessels. Crotamiton is applied in the form of an ointment against itching. Amantadin is used against influenza. Lupanine, which could be detected with over 90 percent congruence with the NIST-spectrum, is a metabolite of sparteine, a medicine for heart and circulatory system disorders. The reference substances for pheneturide and lupanine have not been available, so these results still have to be confirmed.

The consumption per year of single drugs is very high. The consumption can be calculated using the prescriptions per annum (3) multiplied with the size of a daily dose (4). The calculated consumption in Germany for diclofenac is 27–55 tonnes, depending on the size of the daily dose. In addition, diclofenac can be used as a salve, so the real consumption is significantly higher. The consumption of carbamazepine amounts to 17-138 tonnes. Although, only approx. 1–2% of carbamazepine is excreted in unmetabolized form, it is found in the aquatic environment.

In Table II, the results of a *qualitative* analysis are shown. Some of the drugs, such as carbamazepine or diclofenac, were detected in all samples investigated. Other substances, such as lupanine could be detected only in some cases.

To eliminate matrix effects, quantification was carried out by the method of standard additions. For eight (listed in Table III) out of the 12 drugs identified in municipal sewage treatment plants (listed in Table II), standards were available. Therefore, only for these was a quantification possible. Because of the different chemical structures of the drugs, quantification using a single internal standard was difficult. The standard solutions were added to the sample prior to solid phase extraction.

Table II: Identification of several drugs in the effluents of nine municipal sewage treatment plants

	Mühlhausen				Möhringen				Ditzingen				Fellbach				Hegnach			
	March	June	Sept	Dec	March	June	Sept	Dec	March	June	Sept	Dec	March	June	Sept	Dec	March	June	Sept	Dec
Amantadine	d				d		d	d	d		d	d		d	d	d	d		d	
Carbamazepine	d	d	d	d	d	d	d	d	d	d	d	d	d	d	d	d	d	d	d	
Crotamiton			d				d						d	d		d		d		
Diclofenac	d	d	d	d	d	d	d	d	d	d	d	d	d	d	d	d	d	d	d	
Dihydrocodeine	d	d	d	d	d	d	d	d	d	d	d	d		d	d	d	d	d	d	
Hydrocodone	d	d	d	d		d	d	d	d	d	d	d		d	d	d	d	d	d	
Lupanine			d								d				d					
Metabolit of dipyrone	d	d	d	d	d	d	d	d	d	d	d	d			d			d		
Pentoxifylline				d							d	d			d	d	d	d		
Pheneturide		d	d		d	d	d	d	d					d	d			d		
Primidone	d	d	d	d	d	d	d	d	d	d	d	d			d		d	d	d	
Propyphenazone		d			d	d	d			d				d		d	d	d		

	Waiblingen				Eglosheim				Hoheneck				Poppenweiler			
	March	June	Sept	Dec	March	June	Sept	Dec	March	June	Sept	Dec	March	June	Sept	Dec
Amantadine	d	d	d					d			d	d			d	d
Carbamazepine	d	d	d	d		d	d	d		d	d	d		d	d	d
Crotamiton			d													
Diclofenac	d	d	d	d		d	d	d		d	d	d		d	d	d
Dihydrocodeine	d	d	d	d		d	d	d		d	d	d			d	d
Hydrocodone	d	d	d	d		d	d	d		d	d	d		d	d	d
Lupanine								d				d				
Metabolit of dipyrone	d	d	d	d		d	d	d		d	d	d			d	d
Pentoxifylline						d		d				d				
Pheneturide	d	d		d		d	d			d	d	d		d	d	
Primidone		d	d	d		d	d	d		d	d	d		d	d	d
Propyphenazone							d				d				d	

d	Identification of substances by mass spectrum	▨	no sampling

With the standard additions method, the results presented in Table III were obtained. The table shows the ions, which were selected for quantification, the number of samples, which were quantified by the standard additions method, the median values, and the limits of detection and quantification. The maximum concentrations determined were between 130 ng/L for crotamiton and 6,220 ng/L for diclofenac.

Two drugs of special interest that were detected in several effluents were the antitussives, dihydrocodeine and hydrocodone. The consumption of these two drugs is significantly lower in comparison with the drugs mentioned above.

Table III: Quantification using the standard additions method

	LOD[a] in ng/L	LOQ[b] in ng/L	N[c]	N > LOQ*	Median in μg/L	c_max in μg/L	Ion m/z
Carbamazepine	50	150	8	8	1.31	1.76	193
Crotamiton	40	120	3	1		0.13	120
Diclofenac	70	210	9	9	1.04	6.22	214
Dihydrocodeine	100	300	8	8	1.47	4.06	301
Hydrocodone	100	300	8	7	0.72	1.94	299
Pentoxifylline	70	210	8	1		0.23	221
Primidone	80	250	8	7	0.56	0.67	146
Propyphenazone	20	60	4	4	0.27	0.42	215

[a] LOD: Limit of detection, [b] LOQ: Limit of quantitation; [c] N: Number of samples

The daily dose averages only between 7–30 mg/d. Besides the medicinal use, dihydrocodeine is also used as a substitute for addicting narcotics (5). The daily dose for this use amounts from 1,200 up to 4,500 mg/d. Until February 1998, doctors in Germany were allowed to prescribe the substitute without state control. Since February 1998, the prescription underlies stricter regulations, so that perhaps the detectable concentrations of dihydrocodeine in the effluent will be lower in the future.

In the case of these two drugs, samples from the influents were investigated because we suspected hydrocodone to be formed in the sewage treatment plant by oxidation of dehydrocodeine. In fact, higher concentrations of hydrocodone in the effluents than in the influents were observed indicating formation of hydrocodone only during wastewater treatment. In the influent, dihydrocodeine was detected at a concentration of 5.04 μg/L and in the effluent at 3.59 μg/L whereas hydrocodone was found at a concentration of 0.45 μg/L in the influent and 1.94 μg/L in the effluent. An experiment that additionally supported the interpretation that hydrocodone is formed from dihydrocodeine is described below.

Biodegradation/elimination studies

The compounds listed in Table III were examined with regard to their elimination behavior, except for hydrocodone, which was excluded since it was assumed to be formed from dihydrocodeine during the wastewater treatment process. Also examined were caffeine, which is known to have a good biodegradability, as well as acetamidophenol, which could not be found in effluents despite the relatively high number of prescribed daily doses.

Elimination by adsorption

Table IV shows the results of a batch experiment with 10 substances, which were simultaneously investigated. The decrease in concentrations versus time is shown. The chosen initial concentrations of the single substances differ because of different detection limits and the fast elimination kinetics of some compounds.

The concentrations for most of the drugs measured after 15 min were much lower than their initial concentrations. It can be assumed that this initial fast elimination process is primarily caused by adsorption of the drugs to the activated sludge. Most of the compounds, such as acetamidoantipyrine, crotamiton, diclofenac, primidone, and propyphenazone, show exclusively this kind of concentration-time dependence. The concentrations of these five drugs after this initial drop in concentration kept relatively constant. The variation coefficient (relative scatter in data to the mean) of these concentrations determined in 30 measurements within 2 days lay between 7 and 15%. Considering the fact that sampling and filtration as well as analysis were carried out automatically, these coefficients of variation are acceptable.

Elimination by biodegradation and/or transformation

In the case of caffeine, acetamidophenol, and pentoxifylline, an additional subsequent slow decrease of concentrations was observed within several hours or even days down to concentrations less than 1% of the initial concentrations.

Figure 2 shows the decrease in concentration of acetamidophenol, caffeine, and pentoxifylline within 20 hours. The initial concentrations were approx. 25 µg/L of pentoxifylline and 50 µg/L of acetamidophenol and caffeine. With lower concentrations, detection would have been impossible after 2 hours due to the fast elimination kinetics of these substances. For this experiment, the activated sludge was diluted 1:10 with the supernatant fluid of the activated sludge. We assume that this second effect of elimination is mainly the result of biodegradation and/or transformation.

Acetamidophenol is the chemical name of the analgesic paracetamol. The daily dose per person is 2,000 to 3,000 mg/d. By multiplying the daily dose with the number of prescriptions per annum in Germany, a mass of consumption of 182–272 tonnes per annum can be calculated (not considering the consumption of drugs containing acetamidophenol in combined preparations, which amount to 80 million daily doses). The good biodegradability of acetamidophenol is

202

perhaps the reason that the substance could not be detected in the secondary effluent.

Table IV: Batch reactor with a mixture of 10 compounds

	c μg/L] t = 0	c [μg/L] t = 15 min	Adsorption [%]	c [μg/L] t = 1:15 h	c [μg/L] t = 4:15 h	c [μg/L] t=10:30 h	V_k [%]
Acetamidoantipyrine	4.8	1.0	78	0.8	0.9	0.9	9
Acetamidophenol	51.7	31.8	39	22.5	9.7	n.n.	
Carbamazepine	5.0	3.1	37	3.1	3.2	3.6	7
Caffeine	52.7	31.7	40	19.1	0.4	n.n.	
Crotamiton	5.0	4.4	11	4.0	4.4	4.4	5
Diclofenac	48.0	46.2	4	38.7	39.3	49.0	13
Dihydrocodeine	4.9	1.4	73	1.3	1.1	1.2	
Pentoxifylline	24.2	17.7	27	13.7	10.8	3.0	
Primidone	19.9	15.8	21	14.8	16.0	16.0	7
Propyphenazone	5.0	2.8	45	2.2	2.2	3.3	15

Concentration at t = 0 min, 15 min, 1.25 h, 4.25 h, 10.5 h after addition; 'adsorption': decrease of the concentration during t = 0 min to t = 15 min; variation coefficient V_k refers to 30 measured values in 50 h.

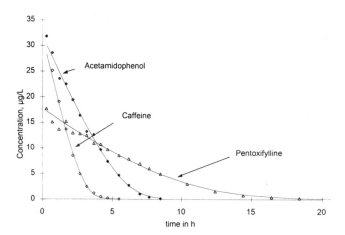

Figure 2: Concentration-time curves of pentoxifylline, acetamidophenol, and caffeine (initial concentrations 24.2 μg/L, 51.7 μg/L, 52.7 μg/L, respectively; activative sludge diluted 1:10, 0.4 g/L dry solids)

Caffeine and pentoxifylline are chemically related. The substitution of the methyl group for a 5–oxo-hexyl group at the N-atom leads to a different degradation process. The rate of biodegradation for caffeine (calculated by using the linear range of the concentration-time curve) is 9x faster than that of pentoxifylline.

Cleavage of conjugates

Many drugs will be metabolized in the human body as glucuronides or sulfates. With the described batch-reactor, it is possible to follow the cleavage of conjugates on-line, such as with acetamidophenol glucuronide (Fig. 3), resulting in the release of free acetamidophenol, which is then degraded. Because of the improved detection limit, these substances were measured in the negative ion mode with the heated nebulizer interface (APCI). The activated sludge was diluted 1:10. The initial concentration was 1.34 mg/L.

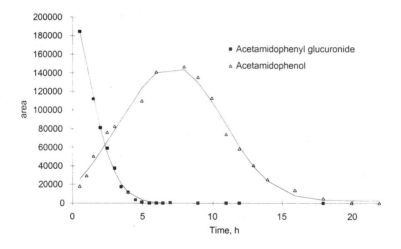

Figure 3: Cleavage of acetamidophenyl glucuronide

Detection of primary metabolites

As mentioned before, hydrocodone was assumed to be a metabolite of dihydrocodeine. With the LC-MS-MS method this could be proven. After addition of 10 µg/L dihydrocodeine into the batch reactor, dihydrocodeine and hydrocodone were measured using MRM. The dihydrocodeine concentration decreased at a similar rate as hydrocodone was formed (activated sludge diluted

1:5, 0.8 g dry solids). Thus the presence of hydrocodone in the secondary effluent is not caused exclusively by its medicinal use.

Furthermore, we examined the metabolism of carbamazepine. Only 1–2% of the carbamazepine is excreted in unmetabolized form (2). Carbamazepine is one of the drugs that was found in every secondary effluent and a drug which was also identified in surface waters (6). In the human body, the two main metabolites formed are carbamazepine epoxide and carbamazepine–diol (7). After addition of carbamazepine to the reactor, the initial concentration decreased within the first 15 min and then reached a constant level. No metabolites could be identified using the MRM-method. For carbamazepine no further degradation was observed.

Conclusions

Screening with GC-MS allowed the individual identification of a small part of the dissolved organic carbon in the effluent of wastewater treatment plants. By measurement in the scan mode, new environmentally relevant compounds could be detected. In this study, 14 drugs and metabolites of drugs could be identified and some of them subsequently quantified using the standard additions method.

With the described batch reactor coupled on-line with LC-MS-MS, it was possible to follow elimination (adsorption, biodegradation) of the detected drugs in environmentally relevant concentrations under conditions similar to those found in biological wastewater treatment plants. Experiments are possible with single substances as well as with mixtures of substances. The advantages of low initial concentrations are the almost real conditions and short lag-phases. Short lag-phases mean short test durations. High initial concentrations can simulate better degradation rates than can be obtained with low concentrations.

In common biodegradation studies, such as the modifications of the Zahn-Wellens-test (8–10), the elimination will be measured with summary parameters, e.g., DOC. The detection limit of this method is significantly higher in comparison with the described LC-MS-MS method. The measurement of DOC always involves the inoculum or the activated sludge besides the substance of interest. Secondly, the method provides a hint on elimination caused by adsorption versus elimination based on biodegradation and/or transformation. An unambiguous differentiation, however, requires further experiments (e.g., after inactivation of the sludge or using radiolabeled compounds). If the transformation pathway of the substances is known, possible metabolites can also be measured in the same run.

In summary, despite the simplicity of the batch reactor and the experiment itself, the test conditions are closer to the reality of wastewater treatment plants than conventional biodegradation tests.

Acknowledgments

The authors thank the 'Federal Ministry of Education and Research (BMBF)' for financial support and also Prof. Eichelbaum (Dr. Margarete Fischer-Bosch Institute for Clinical Pharmacology, Robert-Bosch-Hospital, Stuttgart) for a gift of carbamazepine-diol.

References

1. Möhle, E.; Horvath, S.; Merz, W.; Metzger, J.W. *Vom Wasser* **1999**, *92*, 207-223.
2. Mutschler, E.: Arzneimittelwirkungen, *Lehrbuch der Pharmakologie und Toxikologie*, 6. Aufl., Wissenschaftliche Verlagsgesellschaft GmbH, Stuttgart 1991.
3. Schwabe U.: *Arzneiverordnungsreport '97*, Aktuelle Daten, Kosten, Trends und Kommentare, Gustav Fischer Verlag, Stuttgart/Jena 1997.
4. *Rote Liste 1996*, Arzneimittelverzeichnis des BPI, Hrsg.: Bundesverband der pharmazeutischen Industrie e.V., ECV Editor Cantor Verlag für Medizin und Naturwissenschaften GmbH 1996.
5. Roider, G.; Drasch, G.; von Meyer, L.; Eisenmenger, W. *Pharm. Z.* **1996**, 16, 1369-1377.
6. Sacher, F.; Lochow, E.; Bethmann, D.; Brauch, H.-J. *Vom Wasser* **1998**, 90, 233-243.
7. Saris, L.A.; Brekelmans, G.J.F.; van der Linden, G.J.; Rademaker, R.V.; Edelbroek, P.M. *J. Chromatogr. B* **1997**, 691, 409-415.
8. Schönberger, H. *Z. Wasser-Abwasser-Forsch.* **1991**, 24, 118-128.
9. Pagga, U. *Korrespondenz Abwasser* **1995**, 42, 263-270.
10. *OECD Guideline* 302 B for testing of chemicals vom 17.07.1992. Zahn-Wellens-Test.

Chapter 12

Concerns about Pharmaceuticals in Water Reuse, Groundwater Recharge, and Animal Waste

Jörg E. Drewes[1] and Laurence S. Shore[2]

[1]National Center for Sustainable Water Supply, Department of Civil and Environmental Engineering, Arizona State University, Tempe, AZ 85287-5306
[2]Department of Hormone Research, Kimron Veterinary Institute, Bet Dagan, Israel 50250

Residuals of pharmaceuticals and/or personal care products can be found in sewage effluent, sewage sludge, and animal waste. They can enter surface water via direct sewage discharges into streams and lakes, and via surface runoff from fields that have received applications of animal manure or sewage sludge. They can reach groundwater via infiltration of rain or irrigation water in manured or sludge-amended fields and subsequent downward movement of the water to underlying aquifers. They can also enter groundwater below waste-affected streams or lakes that have higher water levels than the local groundwater table. Where sewage effluent or sewage-contaminated water is used for irrigation, most of the irrigation water goes back to the atmosphere via evaporation from the soil and transpiration from the plants. Thus, the salts, nitrate, pharmaceuticals, and other chemicals that were in the effluent are leached out of the root zone with a small fraction of the irrigation water with which they were applied. This concentrated percolation water then moves down to underlying aquifers where it can contaminate the groundwater. Another pathway for pharmaceutical residues to move to groundwater is via artificial recharge of groundwater with surface water that is contaminated with sewage effluents or animal wastes. Artificial recharge of groundwater is a preferred practice in water reuse, especially for indirect potable reuse of sewage effluent because the effluent receives additional treatment (soil-aquifer treatment or geopurification) as it moves through the underground formations. This additional treatment can lead to substantial removal of pharmaceuticals. Artificial recharge also breaks up the direct connection between the sewage treatment plant and the water supply system, which enhances the aesthetics and public acceptance of potable water reuse. In addition to the various pathways in which surface water containing pharmaceuticals and other residual chemicals of concern can join groundwater, this chapter also discusses fate and occurrences of pharmaceuticals in sewage treatment plants and in the underground environment, as well as their significance in terms of health effects and the need for more research.

Introduction

Increasing populations and finite water supplies inevitably will lead to water shortages and the need for best management of water resources. The present global population of about 6 billion people is projected to double by the end of this century. Most of the population increases will take place in the Third World countries where major water and sanitation problems already exist (*1* and references therein). Industrialized countries with high living standards are expected to grow at a much slower pace and reach essentially constant or even declining populations, except where immigration rates are high as in the United States. More and more rural people are projected to migrate to the cities, giving rise to multimillion-people mega cities with mega water needs, mega sewage flows, and mega water problems. Future water availabilities are clouded by uncertainties in climate (global change) with some parts of the world possibly receiving more precipitation, others less, and, in general, more weather extremes (floods and droughts). Water management increasingly must be integrated, meaning that all aspects of the projects must be considered (*1*). Such holistic approaches include demand management practices like water conservation, recycling and reuse of water, and transfer of water to uses with higher economic returns (i.e., from agricultural irrigation to municipal and industrial uses). It also includes water quality management, public health, environmental and ecological aspects, socio-cultural aspects, storage in times of water surplus for use in times of water shortage, conjunctive use of surface water and groundwater, regional approaches, weather modification, public involvement, conflict resolution, sustainability, and economics.

New dams are increasingly difficult to build because of environmental issues, public opposition, costs, and lack of good sites (*1*). Thus, additional storage of water increasingly must be underground, in aquifers via artificial recharge. Since all but the shallowest aquifers have essentially zero evaporation losses, long-term underground storage (years to decades) or water banking is feasible. In contrast, evaporation rates from surface reservoirs in warm climates range from about 1.5 to 2.5 m/yr, which makes long-term storage less effective. Artificial recharge also is desirable in water reuse, because the effluent undergoes "geopurification" by soil-aquifer treatment (SAT). Artificial recharge also enhances the aesthetics and public acceptance of water reuse since it breaks up the pipe-to-pipe connection of direct reuse and the water comes out of a well rather than out of a sewage treatment plant.

The three main thrusts of integrated water management most likely will be more storage, more water conservation and water reuse, and more transfer of water from agricultural to municipal use. Since the latter is mostly non-consumptive, most water for municipal use comes back as sewage effluent, which can be used for agricultural irrigation, cooling water and other industrial uses, groundwater recharge, municipal water supplies, and other purposes. While such water reuse is touted as an excellent practice because recycling often is the best form of natural resources management, it also gives all the contaminants in the effluent a chance to permeate plants, soils, and aquifers, in addition to the surface waters or water courses where most sewage effluents are now discharged. Quality issues for chemical constituents in potable and non-potable reuse originate from their resistance to both sorption and biodegradation during wastewater treatment and subsequent soil-aquifer treatment. This behavior might introduce compounds associated with a broad spectrum of health concerns into receiving aquifers. A potential concern is bioactive chemicals including active ingredients in pharmaceuticals and personal care products (PPCPs) together with their metabolites and transformation products, when sewage contaminants are released into the environment (2, 3). Concentrations observed so far typically are in the nanogram to low microgram per liter range. As populations increase, the soil-plant-groundwater route for the spread of contaminants in sewage and other wastewater will become increasingly significant.

Animal wastes, especially manures from farm animals that receive regular doses of hormones and/or antibiotics, may contribute to the spread of pharmaceuticals into soils and plants and from there into surface water via direct surface runoff as these manures are spread on land for fertilizer and soil conditioning benefits. The pharmaceuticals can also move to groundwater via infiltration and subsequent deep percolation of rain or irrigation water below manured fields (4 - 7). Thus, there are a number of pathways for chemicals in surface water and soil to move to aquifers, and groundwater is not immune to contamination by PPCPs. However, when it comes to human health and other biological effects, there are a lot more questions than answers and much more research is needed.

Irrigation with Sewage Effluent

Standards or guidelines for irrigation with sewage effluent are pathogen driven. The most stringent requirements apply to unrestricted irrigation, which includes sprinkler irrigation of food crops consumed raw where the water contacts edible portions, and also urban irrigation of parks, playgrounds, golf courses, landscaping, school grounds, and residential yards. For this type of irrigation, California requires sewage treatment so that the median total coliform level does not exceed 2.2 per 100 mL and the virus inactivation is at least 5 log cycles (8). The prescribed treatment consists of primary and secondary treatment to produce an oxidized effluent that is then filtered and disinfected. If chlorine is used, the chlorine dose must be at least 5 mg/L with a minimum contact time of 90 min. This could also create disinfection by-products (DBPs) like tri-halomethanes, halo-acetic acid, and the "new" by-product N-nitrosodimethylamine (NDMA), which is an extremely carcinogenic compound formed by the reaction of chlorine with dimethylamine (DMA). The California Department of Health Services has set an NDMA drinking water action level of 20 ng per liter (9). However, adequate dose-response relations for humans are not available.

The California standards for irrigation with sewage effluent are followed in the U.S. and in most other industrialized countries. For countries with less developed economies, high-technology treatments often are too expensive and too difficult to properly operate and maintain. For those conditions, the World Health Organization (WHO) has developed less stringent standards that can be achieved by in-series lagooning. Detention times in the system then should be long enough (a month, for example) to produce an effluent that has a fecal coliform count of less than 1,000/100 mL and not more than one helminth egg per liter (11). In essence, these systems use time as disinfectant, which does not create any disinfection byproducts. Other quality parameters were not considered. The WHO guidelines were based on epidemiological studies of disease outbreaks due to consumption of raw vegetables irrigated with sewage effluent of various treatments and qualities. While the risk of disease is low with WHO guidelines, it is not as low as with the essentially zero-risk California standards. Filtration of the lagoon effluent through simple intermittent sand filters constructed as sand filled, plastic lined basins or pits with a gravel drainage layer at the bottom (12) can give additional quality improvements such as removal of suspended solids, giardia, cryptosporidium, and helminthic eggs as typically achieved with conventional in-plant sand filters (9).

Sewage irrigation of crops that are not brought raw into the kitchen and are cooked or baked before human consumption can be done with less treatment, for example primary and secondary treatment and chlorination. While fiber and seed crops could technically be irrigated with undisinfected secondary effluent or even primary effluent, primary and secondary treatment and disinfection are preferred to protect the health of farm workers and others coming into contact with the effluent (9).

From an agronomic standpoint, the effluent should also meet the normal chemical requirements for irrigation water, such as salt content, pH, sodium adsorption ratio, nitrogen, and heavy metals and other toxic trace elements (13). Most sewage effluents meet these criteria, except where sewer systems receive major industrial waste discharges. Concentrations of household and industrial chemicals in sewage effluent can be high in areas where the in-house water use is low. In some countries, the in-house water use may only be one-tenth of that in the U.S., which significantly increases the concentrations of sewage contaminants in the effluent. At these concentrations, total nitrogen, which normally is considered an asset of sewage irrigation (free fertilizer) could actually have adverse effects on crops (14, 15) and on underlying groundwater where it could cause nitrate levels to rise above the maximum contaminant level (MCL) of 10 mg/L as nitrate-nitrogen for drinking.

The environmental and agronomic effects of irrigation with treated sewage water from sedimentation and oxidation ponds are not always straightforward and other chemicals can also be important. For example, treated sewage water can contain relatively large amounts of steroidal estrogen (up to about 300 ng/L) which can cause legumes to increase their endogenous phytoestrogen (e.g., coumestrol) levels. The elevated phytoestrogen levels in the legumes can then cause infertility when the legumes are fed to sheep and cattle (16, 17). Also, estrogen concentrations of as little as 10 ng/L can significantly increase growth of alfalfa (18) and decrease root growth in mung beans (19).

Unfortunately, the effects of sewage irrigation on underlying groundwater are seldom if ever considered. Even though sewage irrigation looks good upon first considerations, a closer in-depth look reveals potentially serious threats to the groundwater (20). Based on salt balance considerations, irrigation water amounts D_i given in dry climates must always be in excess of the amounts that go back to the atmosphere as evapotranspiration (ET, evaporation from soil plus transpiration from plants) so that excess water can move through the root zone and down to underlying groundwater. Salts and other chemicals that have entered the soil with the irrigation water are then leached out of the root zone with this leaching water, which is also called drainage or deep percolation water, D_p. Thus D_p is equal to D_i - ET (ignoring rainfall), and the salt concentrations in the deep percolation water are D_i/D_p times larger than those in the irrigation water. The ratio ET/D_i is called the irrigation efficiency E. An efficiency of 100% is only sustainable when distilled water is used for irrigation, as typified by rain. For normal irrigation waters, salt contents often are in the range of 100 to 1000 mg/L, so that irrigation efficiencies must be well below 100% to produce enough leaching water for maintaining a salt balance in the root zone. This is essential for sustainable irrigated agriculture without build-up of salts in the root zone. Good irrigation efficiencies typically are about 80% and they can be achieved with good quality irrigation water (low salt content), good irrigation systems, and good irrigation management. Thus, if a crop needs an ET of 1.2 m of water for its growing season and 1.5 m of irrigation water are applied, the irrigation efficiency is 80% and the amount of drainage water is 0.3 m. The concentrations of all conservative chemicals (those that resist transformation/ degradation) in the drainage water then are 1.5/0.3 (5 times) higher (less in areas with significant rainfall) than those in the irrigation water. This applies to all refractory chemicals not taken up by the plant or adsorbed by the soil. Part of the nitrogen is absorbed by plants and some can also be removed by denitrification. However, this may not be sufficient and nitrate levels in the drainage water could still be too high for drinking (21), especially in areas where the in-house water use is low and, hence, the sewage is concentrated. Phosphates and heavy metals can be immobilized in the root zone, especially in calcareous soils or other soils with a high pH (22). Most biodegradable organic carbon also is removed in the root zone. In addition to salts, nitrate, and pesticides (21), the deep percolation water thus can also contain PPCPs that were in the sewage effluent, disinfection byproducts (DBPs), and humic and fulvic acids.

Some PPCPs have already been detected in groundwater near sewage contaminated streams (23, see also section "Riverbank Filtration"). The DBPs consisted of those already in the drinking water as it was chlorinated in the water treatment plant, plus those formed by chlorination in the sewage treatment plant. Humic and fulvic acids are the major components of dissolved organic carbon (DOC) in drinking water. They are also generated as soluble microbial products (SMPs) during wastewater treatment due to the decomposition of organic material, and they are formed on and in the soil as dead plant material and other biomass decomposes (24, 25). Organic matter such as humic substances in groundwater can form a new suite of DBPs when affected groundwater is pumped up and chlorinated for drinking. Most DBPs are biodegradable in the underground environment (26), but others may be more refractory.

Much more needs to be known about the fate of PPCPs in sewage-irrigated crops, in the root zone, in the unsaturated zone below the root zone, and in the aquifer. If groundwater levels are high and artificial drainage is needed, the drains or pumps could discharge the drainage water into surface water and introduce PPCPs and other chemicals that can adversely affect aquatic life and other users of that water. The concentration of refractory PPCPs in the drainage water could then be higher than in the sewage effluent used for irrigation, due to evaporation effects. With sprinkler irrigation, the effluent water wets the leaves, and the chemicals therein could enter the plant directly through the leaves, which could later be consumed by humans and animals. Because roots can also absorb chemicals, effluent irrigation systems that do not wet the foliage of the plants could still allow PPCPs and other organic chemicals to enter the plants. This can change the physiological processes in the plant, which in turn can have adverse effects on humans and animals consuming those plants, like the infertility in cattle due to consumption of sewage-irrigated alfalfa (17).

Deep percolation water below irrigated areas or other fields usually accumulates in the upper portion of the aquifer, as a result of vertical stacking. Since wells typically pump water from some lower depth interval, the effects of sewage irrigation on well water quality initially will be subtle and slow. When the salt content of the well water eventually becomes too high for drinking, membrane filtration like nanofiltration (NF) or reverse osmosis (RO) becomes necessary. Fortunately, this will also remove humic and fulvic acids, DBPs, and most of the PPCPs.

Adverse effects of sewage irrigation on groundwater will be especially serious in water-short areas like the Middle East, North Africa, and India where populations are increasing rapidly, irrigation is necessary, sewage effluent will become a major source of irrigation water, in-house water use is low, and groundwater is a major source of drinking water. These trends are on a collision course and integrated management principles must be applied to reach optimum solutions (1). Use of sewage effluent for irrigation may look like a good practice on the surface and in the short term, but, if in the long term, it makes underlying groundwater unsuitable for potable use, it would come at a heavy price.

Often, the root zone and soil mantle are considered effective removers and attenuators of pollutants (22). However, water tends to move downward along preferential paths like root holes, worm holes, cracks, and other macropores and also through paths of least resistance where soil permeabilities are higher than for the rest of the soil due to the innate heterogeneity of natural soils. Such preferential flow can give PPCPs and other contaminants a shortcut to the aquifer (21, 27). Preferential flow often is the reason that contaminant transport in the soil occurs much faster than predicted by models, and why contaminants show up at places much faster and much farther away than expected.

Transport and fate of chemicals in water moving downward through root zones basically are governed by four primary factors: (1) soil texture as it controls fluxes and specific surface areas for biofilm development and adsorption, (2) organic carbon in the soil as it enhances biodegradation through co-metabolism and secondary utilization (28), (3) macropores as they create preferential flow, and (4) evapotranspiration as it concentrates the chemicals in smaller amounts of drainage water in irrigated areas with low rainfall.

Artificial Recharge of Groundwater

Another pathway for surface water and its contaminants to reach groundwater is artificial recharge of aquifers. This is achieved mostly with surface infiltration systems (basins, channels), but sometimes also with infiltration trenches or shafts in the unsaturated or vadose zone, or with wells reaching into the aquifer for direct injection of the water (1, 12). Artificial recharge is used to store water in times of water surplus for use in times of water shortages, to reduce depletion of groundwater and associated land subsidence, to create hydraulic barriers against sea water intrusion or pollution plumes, and to clean up and store sewage effluent for reuse, mostly for indirect potable purposes and unrestricted irrigation. For recharge basins or channels, infiltration rates for year-round operation tend to be in the 30 to 300 m per year range, depending on soil type, water quality, and climate. Since evaporation rates are in the range of 1 to 2 m per year, concentration effects are insignificant so that concentrations of salt and other persistent chemicals such as non-biodegradable and non-adsorbable compounds in the water moving downward to the aquifer will be essentially the same as those in the recharge water itself.

How artificial recharge affects groundwater quality and the quality of water pumped from wells depends on the source of the recharge water, which may vary from pristine streams and lakes to streams receiving sewage effluent and other wastewater discharges, effluent-dominated streams, and 100% effluent straight from the sewage treatment plant. Recharge with conventionally treated sewage effluent (primary and secondary treatment followed by mild chlorination) and controlled recovery of the water from wells is increasingly done to enhance the quality of the effluent by soil-aquifer treatment (*12*) so that it can be used for essentially all purposes, including unrestricted irrigation, discharge into protected surface waters, and, with further treatment and/or dilution, potable use. For the latter, California has developed draft guidelines that specify: (1) the required treatment of the effluent before recharge, (2) the minimum depth to groundwater below the infiltration basins, (3) minimum underground retention times, (4) minimum distance of underground travel or horizontal separation, and (5) mixing ratios with native groundwater where it is pumped up again (Table I; *29*).

These guidelines are based on achieving absence of pathogens and a total organic carbon content (TOC) of sewage organics not exceeding 1 mg/L. One argument for this low number was that the TOC or dissolved organic carbon (DOC) still includes many unknown compounds (*29*), so that a conservative approach was needed. Current water reuse regulations have no MCLs for specific organic compounds.

The DOC concentration of secondary sewage effluents depends on what is in the raw influent and on the treatment applied. Wastewater treatment processes employing long in-plant detention times such as activated sludge treatment and nitrification-denitrification lead to a better effluent quality with respect to DOC concentration than do high-rate processes such as trickling filter treatment, which are less efficient in terms of DOC removal (*30*). Thus, DOC concentrations in treated effluents typically vary between 5 and 20 mg/L. However subsequent soil-aquifer treatment can efficiently remove DOC concentrations to levels of 2 to 5 mg/L with underground retention times of less than 30 days, depending on soil characteristics, dominant redox conditions, and hydraulic properties of the soil and aquifer (*12, 24, 30*). Based on a conservative value of 5 mg/L for the residual DOC, there should then be enough mixing with native groundwater of zero sewage TOC to produce an effluent derived TOC of less than 1 mg/L in the well water.

If the effluent also undergoes organics removal in the treatment plant by, for example, activated carbon filtration or reverse osmosis, the well water can then consist of a 50-50 mix of effluent water and native groundwater. Based on findings from river bank filtration systems and groundwater recharge sites using treated effluents, soil-aquifer treatment can lead to a preferred removal of assimilable organic carbon (AOC) and biodegradable organic carbon (BDOC) (*24, 30, 31*). Thus, the residual total organic carbon after SAT could be relatively richer in refractory PPCPs than the water going into the ground.

In a groundwater recharge system in Phoenix, Arizona, where secondary sewage effluent (activated sludge treatment) was used for soil-aquifer treatment, the lipid regulator clofibric acid used in cholesterol lowering drugs was found at concentrations of 0.3 to 1.6 µg/L in water from a 30-m deep piezometer extending about 10 m into an unconfined aquifer (*32, 33*). Biodegradable PPCPs likely will also be refractory in the underground environment because their concentrations tend to be so low that they are well below threshold values for microbiological degradation. On the other hand,

Table I. Proposed California guidelines for potable use of groundwater from aquifers recharged with sewage effluent (adapted from 29).

Contaminant Type	Type of Recharge	
	Surface Spreading	Subsurface Injection
Pathogenic Microorganisms		
Secondary Treatment	SS ≤ 30 mg/L	
Filtration	≤ 2 NTU	
Disinfection	4-log virus inactivation, ≤ 2.2 total coliform per 100 mL	
Retention Time Underground	6 months	12 months
Horizontal Separation	150 m	600 m
Regulated Contaminants	Meet all drinking water MCLs	
Unregulated Contaminants		
Secondary Treatment	BOD ≤ 30 mg/L, TOC ≤ 16 mg/L	
Reverse Osmosis	≤1 mg/L TOC of wastewater origin at drinking water well	100% treatment to TOC ≤ 1mg/L/RWC
Spreading Criteria for SAT 50% TOC Removal Credit	Depth to groundwater at initial percolation rate of: < 0.5 cm/min = 3 m < 0.7 cm/min = 6 m	NA
Mound Monitoring Option	Demonstrate feasibility of the mound compliance point	NA
Reclaimed Water Contribution in Well Water (RWC)	≤ 50 %	

SOURCE: Adapted with permission from reference 29. Copyright Marcel Dekker.

where the contaminated water used for recharge contains relatively high concentrations of biodegradable organic compounds, as reflected by relatively high BOD values, PPCPs may be microbiologically degraded through co-metabolism and secondary utilization (27). This could also enhance removal of PPCPs below the leach fields of septic tank systems where basically a primary effluent with a high BOD gives incidental groundwater recharge and eventually leads to water reuse.

Health effects studies on populations in parts of Los Angeles where the drinking water was affected by groundwater of wastewater origin as described in Table I indicated no adverse health effects (34, 35). Of course subtle and subclinical effects that could have been caused by residual PPCPs in the well water at very low concentrations probably would have escaped detection in these surveys, which looked only at morbidity and mortality data. For example, Sharpe and Skakkebaek (36) implicated environmental estrogens in the drastic reduction of sperm counts observed among Western men, while Colborn et al (16) suggested that environmental estrogen also may lead to increased rates of infertility, breast cancer, testicular cancer, and other hormone diseases and disorders. Widespread reproduction disorders already have been found in a variety of wildlife (16, 37), especially in birds, alligators, and turtles. Shore et al. (17) indicated that some of the reproductive disorders may be due to ingestion of phytoestrogens or exposure to numerous weakly estrogenic xenobiotics like DDT and methoxychlor, in addition to estradiol and estrone already present in the environment. More research on the various health effects is urgently needed, including dose-response relations so that acceptable maximum contaminant levels can be established.

PPCPs in Sewage Effluent

The main concern about PPCPs in groundwater recharged with sewage effluent relates to PPCPs that are not removed during wastewater treatment or which are not attenuated during travel through the subsurface and underground. Hormonally active agents (HAAs) have been determined to have reduced mobility in sediments and sludges due to their more hydrophobic character (6). The much higher concentrations of these chemicals in the adsorption layers relative to those in solution could even enhance biodegradation, thus making removal by adsorption more of a renewable process. Although a large portion of HAAs is removed via sorptive processes, the remaining low concentrations in the percolating water may be capable of exerting abnormal physiologic effects like endocrine disruption in aquatic biota (2).

In contrast, the polar and nonvolatile nature of many other drugs such as antibiotics, blood lipid regulators, antiepileptics, diagnostic contrast media and others increases not only their potential migration through soil and aquifers, but also prevents their adsorption from the liquid phase to solid phases. Refractory and polar compounds undoubtedly will move to and through the aquifer and show up in the water pumped from wells. Therefore, the specific processes used in wastewater treatment facilities play a key role in the introduction of pharmaceuticals into surface water and groundwater. In addition, direct discharge of untreated sewage to surface water and groundwater could be a major source of those PPCPs that are otherwise easily removed by conventional wastewater treatment processes (2).

Ternes (38) investigated the occurrence of 32 drug residues in German municipal wastewater treatment plants. In general, more than 60% of the drug residues that were detected at ppb or ppt levels in the influent were removed during treatment in the activated sludge process, particularly fenofibrate (lipid regulator metabolite), acetaminophen (paracetamol), and the metabolites of acetylsalicylic acid which were not detected in the effluent. Carbamazepine (antiepileptic drug), clofibric acid (lipid regulator), phenazone (antiphlogistic) and dimethylaminophenazone (antiphlogistic) showed low removal rates during this study. Although microbial degradation and sorption onto sludge flocs can potentially remove PPCPs, several reasons work against a sufficient removal during wastewater treatment (2). First, the concentrations of most pharmaceuticals are so low that the lower limits of enzyme affinities may not be met. Second, average detention times in wastewater treatment plants often are too short to guarantee a complete removal due to biodegradation processes. Third, many new forms of drugs and, to a lesser extent, completely new drugs are introduced to the market each year, each presenting an unknown factor in the biodegradation process. Studies focusing on the fate of PPCPs during travel through the subsurface are rare. In bank filtration systems affected by sewage effluents, researchers found several polar, non-volatile pharmaceuticals in German drinking water at concentration in the lower nanograms-per-liter range, with a maximum of 70 ng/L for clofibric acid (39). However, in the majority of samples analyzed, no drugs were observed. Health effects of these concentrations, which are minuscule in relation to prescribed doses, are not known.

Wenzel et al. (40) also studied the presence of estrogenic PCPPs in German sewage plant effluents, sewage sludge, compost and seepage water from compost deposits, sediments, and slurry. The estrogens examined were both synthetic (ethinylestradiol, mestranol) and natural (17-alpha-estradiol, estrone). The detection frequency in the effluent, was estrone > estradiol > ethinylestradiol > mestranol (13/9/4/3 of 41) and in sewage sludge estradiol > estrone > mestranol (10/7/3 of 38). In three sediment samples and in one seepage water sample, ethinylestradiol was detectable but it was not detectable in sludge. This difference in detection frequency could also show that the retention of PCPPs in sewage plant products and the detection methods differ in their efficiency by the nature of the medium. Furthermore, the retention of a PCPP by sludge is not necessarily predictable from its known physiochemical characteristics. For example, Stuer-Lauridsen et al. (41) have shown that the ability of furosemide, ibuprofen, oxytetracycline, and ciprofloxacin to bind to sludge is several orders higher than would predicted by the partitioning coefficients Kow or pKa of the compounds.

The steroidal sex hormones, estrone, estradiol, and testosterone, are a class of PPCPs of particular interest because they are naturally excreted into the environment from human and animal sources as well as extensively used as pharmaceuticals (a population of 6 million utilizes about 100 kg/yr of estradiol). In contrast to most PCPPs, the steroidal estrogens and the synthetic estrogens are physiologically active in the ng/L range that is frequently observed in surface waters. Indeed the best-documented widespread environmental effect of changes in gonadal histology of fish in UK estuaries, has been shown to be due to steroidal estrogen in sewage treatment plant effluents (*42*). Raw sewage contains large amounts of estrogen (40 to 300 ng/L) and testosterone (16 to 700 ng/L), which is highly variable depending on the source and dilution (*43*). Both steroidal estrogen (estradiol and estrone) and testosterone are readily detected in the effluent after conventional secondary and tertiary treatments, whereas filtration of effluents through sand almost completely removes both hormones (Table II).

High levels of steroidal estrogens have also been shown to occur in sewage treatment plant effluent for both UK (1.4 to 76 ng/L estrone; 2.7 to 48 ng/L estradiol) (*42*) and Germany (mean 17.5 ng/L estrone; 20.5 ng/L estradiol; 21.3 mestranol; 8.9 ethinyl estradiol) (*40*).

Riverbank Filtration

Increasingly stringent drinking water quality standards are prompting water supply companies to obtain their raw water from wells relatively close to a river (typically 50 to 200 m) rather than from the river itself. Pumping the wells then "pulls" river water through the aquifer, so that the water is filtered through the aquifer before it reaches the wells. This is a form of induced recharge of groundwater. The main objective of this riverbank filtration is to remove suspended solids, biodegradable organic compounds, DBP precursors as expressed by the DBP formation potentials, viruses, and other microorganisms (bacteria, protozoa like giardia and cryptosporidium, and other parasites). Additional benefits include dampening of peak concentrations of pollutants in the river due to accidental or unusual discharges of pollutants, moderating temperature peaks, and mixing with native groundwater. Significant removals of these contaminants, including TOC and DBP-precursors, have been obtained (*32, 44*). Zullei-Seibert (*23*) found low concentrations (below 4 μg/L) of lipid reducers, anti-rheumatics, and hormones in groundwater from artificial recharge and bank filtration systems. In one case, sewage effluent itself contained 29 pharmaceuticals, stream water 22 pharmaceuticals, and groundwater 5 pharmaceuticals. The pharmaceuticals in groundwater were diethylstilbestrol, estradiol, ethinylestradiol, clofibric acid, and fenofibrate. Thus, while SAT may effectively reduce PPCP concentrations, they can still be present in the groundwater.

Table II. Concentrations (ng/L) of hormones in sewage effluents after various treatments (*43*).

	Constructed wetlands water treatment plant in Maryland	
	testosterone (ng/L)	estrogen (ng/L)
raw sewage	669	53
sand filtration	5.6	7.3
wetland influent	6.0	2.0
wetland effluent	2.1	1.6
peat effluent	1.6	0.8
	Activated sludge plant in Tel Aviv	
	testosterone (ng/L)	estrogen (ng/L)
Raw sewage	208	48
Digestion tank		
top (aerobic)	118	39
bottom (anaerobic)	173	64
Secondary effluent	50	38
After sand percolation for three months	<3	<3

Animal Waste

So far, the main environmental concerns about the waste from farm animals like cows, pigs, chickens, and turkeys, have been nitrate contamination of groundwater and eutrophication of streams and lakes due to infiltration and surface runoff, respectively, from farms and manured fields (*45*). Fish ponds for aquaculture may also be a source of contamination of surface runoff and groundwater. Thus, the focus has been on nitrogen and phosphorus, as demonstrated by the chapters in the book edited by Steele (*45*). However in the introductory chapter by Sims (*46*), the issue of pharmaceuticals was brought up as follows:

"Nevertheless, questions are now arising more frequently about the potential environmental effects not only of trace elements, but of antibiotics, pesticides, and hormones in animal wastes. Limited research is available that characterizes the form of these elements and compounds in animal wastes and even less on any subsequent transformations that might occur in waste-amended soils that could affect their bioavailability and transport. Because of this, basing land application of animal wastes on criteria similar to that used for other solid wastes is premature. A prudent course of action at this time would be to identify and characterize the concentrations of any potential "pollutants" present in animal wastes and, based on existing scientific literature with other solid wastes, prioritize any research or monitoring efforts on the environmental fate of these waste constituents."

Pathogens like E. coli strains and cryptosporidium in animal manure that can cause adverse health effects in people consuming crops from manured fields also are of concern. Animal manure can be a source of natural steroids, estrogen, and testosterone (Table III). The use of animal manure for fertilization of fields and the concentration of large animal husbandry units on small areas (confined or concentrated animal feeding operations, CAFOs) increases the impact of the manure on watersheds. Shore et al. (4) state that "Although other manures are used for fertilizer, chicken manure has much higher concentrations of hormones as birds have much higher blood concentrations of hormones than mammals." Some data are presented in Table III. Chicken manure contains up to 533 ng estrogen/g dry wt and 670 ng testosterone/g. This is a sufficient quantity to cause hormonal effects in cattle that were fed chicken manure silage (47). Eluants from chicken manure piles were found to contain 630 ng/L of testosterone and 730 ng/L of estrogen (Shore, not published). The edge of field loss of 17 beta-estradiol from manured fields ranged from 1-3 ug/L (48). This is well above the level of 20-40 ng/L which can affect legumes (17) and fish gonads (49).

Testosterone and estrogen in animal manure apparently do not leave the environment by the same routes. Testosterone is leached by aqueous solution from the soil more readily than estrogen, probably due to the phenolic group of estrogen binding more firmly to organic matter. Estrogen has been shown to bind firmly to sandy soils (100 ng/kg) and can remain bound for several months after exposure. However, estrogen may not be readily biodegradable in soil and no known soil bacteria can metabolize it. Both estrogen and testosterone were found in streams throughout a watershed with heavy agricultural use and concentrations increased as the streams passed through areas with animal manures (4). However, testosterone but not estrogen was found in groundwater from drilled sampling wells on a farm that had used chicken manure for over 5 years (testosterone concentration 1.0±0.2 ng/L, estrogen concentration <0.10±0.02 ng/L) (50). This would indicate that estrogen reaches the watershed by surface runoff whereas

Table III: Hormones Content in Animal Manures

Source	Testosterone (ng/g d.w.)	Estrogen (ng/g d.w.)	Ref.
Immature broilers			
Females	133 ± 13	65 ± 7	(43)
Males	133 ± 12	14 ± 4	(43)
Laying Hens	254 ± 22	533 ± 40	(43)
Roosters	670 ± 95	93 ± 13	(43)
Chicken litter		133 µg/kg	(48)
Milk cows (slurry)	max: 640 µg/kg estrone		(40)
	max: 1229 µg/kg estradiol		
Pigs (slurry)	max: 84 µg/kg estrone		
	max: 64 µg/kg estradiol		(40)

testosterone reaches the watershed by both groundwater and surface water routes. From these observations, Shore et al. (*50*) concluded that "groundwater is not a major route for hormone transport." This was further confirmed by studies of estrogen and testosterone in surface water and groundwater in the Chesapeake Bay Watershed where there were sewage treatment plants and fields fertilized with chicken manure (*50*). Maximum concentrations of estrogen in runoff water from manured fields were 22 ng/L and of testosterone 215 ng/L, while the concentrations in groundwater ranged from 0.05 to 0.14 ng/L for estrogen and 0.50 to 1.5 ng/L for testosterone (*51*).

Two classes of pharmaceuticals used in animal feeds deserve special attention. The first class includes the antibiotics monensin, maduramicin and avoprocin as coccidostats in poultry and growth promoters in cattle. Monensin and maduramicin are toxic compounds and both compounds can be present in poultry litter in concentrations that are lethal to cattle fed the silaged poultry litter (*52*). Avoprocin was banned as it was associated with the development of vancomycin resistant bacteria in humans (*53*). However, the whole question of vancomycin resistance is complicated and more research is needed. The second class consists of steroidal hormones (estradiol, zearalanol, nortestosterone, and trenbolone). These substances have been banned, on rather slight scientific evidence, in most European countries. In the U.S., they are used extensively but their environmental fate is not known.

Antibiotic compounds in surface water and groundwater near swine and poultry farms and sewage treatment plants were studied by Meyer et al. (*7*), who detected

"one or more antibiotic compounds in water samples collected in the vicinity of swine and poultry AFOs [animal feeding operations] and in three surface-water samples collected downstream from sewage treatment plants. Most of the compounds were detected at concentrations that were less than 1.0 µg/L. In the vicinity of the swine AFOs in Iowa, one or more antibiotic compounds were detected in four of five ground-water samples, one of two tile-drain inlets, and three of four tile-drain outlets. The compounds detected included chlortetracycline, oxytetracycline, lincomycin, sulfamethazine, trimethoprim, sulfacimethoxine, and the dehydrated metabolite of erythromycin. In the vicinity of the poultry AFOs in Ohio, one or more antibiotic compounds were detected in 11 of 15 surface-water samples. The compounds detected included chlortetracycline, oxytetracycline, lincomycin, sulfadimethoxine, and trimethoprim. Tetracycline also was detected in three of four and trimethoprim in one of four ground-water samples collected near the poultry AFOs. From the 11 stream sites, one or more of the antibiotic compounds – chlortetracycline, trimethoprim, and the dehydrated metabolite of erythromycin – were detected in three samples collected downstream from sewage treatment plants."

Discharge of antibiotic drugs into the environment may create drug resistant bacteria. While many studies have addressed biological impacts of short-term high-level doses on non-target organisms (*54*), the effects of long-term low-level exposures to PPCPs are largely unknown.

It is possible to reduce steroidal hormones (estradiol, estrone, testosterone) from animal manures using simple low-technology solutions. Although stacking or silaging of manure does not substantially reduce the hormone content (45), preliminary experiments indicate that composting is a highly efficient (85-97%) method for reducing hormonal levels (C. Oshins, L.S. Shore, and L. Drinkwater, unpublished observations). Alternatively, hormones in runoff from manured fertilized fields can be prevented by the use of alum or buffer strips (55). As for sewage treatment plant effluents, sand filters should be effective to reduce hormonal content as sand has a high capacity for adsorbing steroid hormones (4). Although utilization of these technologies just to remove hormones may not be economically justified, there is ample justification for their use to remove pathogens and other pollutants that may damage human health and the environment.

With all the hormones, antimicrobials, and other pharmaceuticals routinely given to farm animals to increase production and to keep them healthy, there should be more concern about pharmaceutical residues in the manure. Issues to be addressed include how these residues survive animal waste treatment, composting, and storage, and how they may get into groundwater through leaching from manured fields and into surface water through surface runoff. This is one area where more research is critically needed. Aquaculture with its use of antimicrobials and disposal of spent water into surface water also must be considered (2).

Sewage Sludge

Sewage sludge is another solid waste often applied to agricultural fields for fertilizer (nitrogen) and soil conditioning. Because some hormones and other PPCPs are adsorbed by the sludge particles, land application of sludge could be another source of PPCP contamination of surface water and groundwater. Hormonally active agents like dioxin and nonylphenol, and other industrial chemicals have also been identified in sludge (56). Sludge application to land often is controlled by its nitrogen content to minimize nitrate contamination of groundwater and nitrate edge-of-field losses. Heavy metals like cadmium also could limit amounts of sludge that can be applied. No field work has been done on the effect of the wide use of sludge on the estrogen and testosterone content in the environment or on the plants themselves. However, aqueous extracts of sludge contained from 18 to 70 ng/L estrogen and 10 to 173 ng/L testosterone in a study by Shore et al. (43). Wenzel et al. (40) found that the average concentration in sludge for three estrogens was 59 μg/kg total solids for estrone, 28.9 μg/kg for estradiol, and 2.3 μg/kg for mestranol. Thus, land application of sewage sludge could be a potential source of hormone contamination of surface water via surface runoff, and of groundwater via infiltration of rain or irrigation water and deep percolation. As with chicken manure, estrogen could move mainly via surface runoff to surface water, and testosterone to both surface water and groundwater (4).

Discussion and Conclusions

Residuals of pharmaceuticals and personal care products (PPCPs) in sewage effluent are entering the surface water environment through direct discharges of variously treated sewage effluents into streams and lakes. PPCPs can then enter groundwater where the groundwater level is lower than the water level in affected streams or lakes, causing them to lose water to the underground environment. Other pathways to groundwater are via irrigation or artificial groundwater recharge systems that use sewage effluent or sewage contaminated water, and via septic tank leach fields. Pharmaceuticals used in animal production and present in animal waste could enter surface water via surface runoff and groundwater via infiltration and deep percolation from farms and manured fields. Health effects so far have been detected mainly in aquatic life (fish, amphibians) and animals up the food chain but not positively in humans, although there are significant indications of potential adverse effects (16). Even if the effects were known, eliminating PPCPs may be difficult and some form of source control and treatment may be a first step to minimize their presence and concentrations in surface water and groundwater. On the one hand, it may be argued that since the amounts of PPCPs ingested with drinking water are so small compared with the medical doses at which they are prescribed that adverse human health effects may be of no concern. On the other hand, since there is so little information on long-term and synergistic effects, there also are statements that PPCPs in water "should be avoided in principle" (23). For aquatic organisms, the exposure is complete immersion for 24 hours a day. Also, concentrations of the chemicals in organisms and animals increase up the food chain. Thus, there are real concerns about wildlife (15). PPCPs typically occur at nanograms per liter levels which seem completely insignificant considering that 1 ng/L is equivalent to 1 second in 31,000 years. On the other hand, 1 ng/L of a compound with a molecular weight of 300 still contains 2×10^{12} molecules per liter, which for endocrine disruptors still could bind to a lot of hormone receptors. Public opinion, as fostered through the media, also must be considered. Catchy headlines with words like "ecological disaster," "drugs in drinking water," or "AIDS-like symptoms" can easily stir up serious public concern to which scientists may be unable to respond adequately in the absence of adequate and reliable data. Insufficient information breeds concern, and concern leads to fear.

Historically, the U.S. has pursued a "straightforward and simple policy that no risk can be tolerated in the nation's food supply" and "that all food additives be proved safe before marketing and explicitly prohibits any food additive found to induce cancer in test animals" (57, and references therein) and, by implication, also in drinking water. Despite public objections on the grounds that this inhibited freedom of choice, this policy was rigorously enforced until the late 1970s when saccharine was discovered to cause cancer in rats. However, plans for taking saccharine off the market caused a serious public outcry against the federal government dictating what people could and could not eat. In response, the policy shifted to one of informing the public about risks and letting the people decide for themselves what they wanted to eat and drink. Thus, an informed public making its own decisions was the new policy, realizing that there is no such thing as a risk-free society. This also led to more studies of carcinogens naturally occurring in food and of risks associated with recreational activities and sports, risks in common human activities and environmental effects, occupational risks, and various cancer risks to show that life as a whole is not risk free (57). The matter of choice also is very important. Often, people are willing to accept higher risks in eating habits, sports, recreational and other activities which they chose to do, than in the quality of the food they buy or of the water that comes out of the tap over which they have no control.

The fear about carcinogens naturally occurring in food and drink or added artificially with food processing and in polluted water was somewhat attenuated by studies such as by Ames and Swirsky Gold (58), who showed that chemicals caused cancer in rodent bioassays not because they were carcinogenic, but because they were administered in such high doses that they caused cell damage in the test animals. Subsequent cell division to heal the damaged tissue then could produce mutations that caused malignant tumors. Thus, the linear response theory to extrapolate positive responses from high doses in the bioassays to low doses that are more realistic of exposures that can occur in real life is flawed because it does not recognize threshold concentrations below which positive responses do not occur (59). About half the chemicals found carcinogenic in such tests in actuality may not have been carcinogenic but rather produced cancer because of the high doses administered to the test animals (58), confirming Paracelsus' conclusion that it is the dose that makes the poison. For hormonally active compounds, however, dose-response relations may not be linear and may even be odd-shaped like an inverted "U" (16 and references therein). This makes extrapolation to different doses and exposures very problematic. Rodent bioassays also do not recognize genetic effects, as some chemicals caused cancer in mice but not in rats (60). Thus, if rodent bioassay results cannot be transferred to different rodent species, how can they be transferred to humans? However, this may not be true for hormones, which may function basically the same in all mammals (16).

Large systems such as the ecosystems we live in are inherently chaotic even though they may appear to be in equilibrium. Therefore any technological fix of one variable will change all the other variables in a totally unpredictable manner. It is, therefore, inadvisable to make large-scale changes to correct pollution without numerous small-scale studies. This requires long-term data base collection. Solutions based on small databases and extrapolated models are, therefore, not recommended. What is needed is ecological "common sense", e.g., if the fish are dying in the rivers used for drinking water, action must be taken immediately. If minor or localized disturbances are seen in wildlife ecology, the best course of action would be long-term data collection.

The true significance of PPCPs in the aquatic environment and in our water supplies is still a big question. Adverse effects on aquatic life and micro-organisms observed so far are serious enough to warrant more research, including effects on humans. Only if these effects are better understood can the public be sufficiently informed to make its own decisions. Because PPCPs play such an important role in the well being of people and animals, some adverse effects on aquatic organisms living in affected water and on people and animals drinking that water may have to be accepted, just as side effects of medical drugs are accepted. The question then is, what is acceptable and what can be done about it? Ideally, dose-response relationships are developed on which regulators can base appropriate maximum contaminant levels. Hormone disrupting chemicals, however, may not follow classical dose-response relations (16). Developing more biodegradable PCPPs is another avenue toward reducing their harmful impact. Phasing out a compound is, of course, an action of last resort. However, if there are serious enough concerns about a certain product, action may have to be taken before there is absolute scientific proof of harm. This is where eco-toxicology and eco-epidemiology become important (16).

The role of scientists in all this was already defined almost 400 years ago by Francis Bacon (61) who wrote "And we do also declare natural divinations (forecasting by natural observation) of diseases, plagues, swarms of hurtful creatures, scarcity, tempests, earthquakes, great inundations, comets, temperature of the years, and diverse other things; and we give counsel thereupon, what the people shall do for the prevention and remedy of them." Of course the list of "diverse other things" has greatly expanded over the years, and now definitely includes PPCPs! However, information about PPCPs in general and endocrine disruptors in particular is "limited, with sparse data, few answers, great uncertainties and a definite need for further research," as stated by Maczka et al. (5). Indeed, a lot more research needs to be done before scientists can give, in Bacon's words, counsel thereupon and what the people shall do.

References

1. Bouwer, H. "Integrated water management: emerging issues and challenges," *J. Agric. Wat. Manage.* **2000,** *45,* 217-228.
2. Daughton, C.G.; T.A. Ternes. "Pharmaceuticals and personal care products in the environment: agents of subtle change?" *Environ. Health Perspect.* **1999,** *107*(suppl 6), 907-938.
3. Rodgers-Gray, T.P.; S. Jobling; S. Morris; C. Kelly; S. Kirby; A. Janbakhsh; J.E. Haries; M.J. Waldock; J.P. Sumpter; C.R. Tyler. "Long-term temporal changes in the estrogenic composition of treatment sewage effluent and its biological effects on fish," *Environ. Sci. Technol.* **2000,** *34* (8), 1521-1528.
4. Shore, L.S., D. Correll, P.K. Chakroborty. "Fertilization of fields with chicken manure is a source of estrogens in small streams." In: Animal Waste and the Land-Water Interface (K. Steele, ed.), pp. 49-56. Lewis Publishers, Boca Raton, Florida. 1995.
5. Maczka, C., S. Pong, D. Policansky, R. Wedge. "Evaluating impacts of hormonally active agents in the environment," *Env. Sci. Technol. News,* March **2000,** 136A-141A.
6. National Research Council. Hormonally Active Agents in the Environment. National Academy Press, Washington, D.C. 1999.
7. Meyer, M.T., D.W. Kolpin, J.E. Bumgarner, J.L. Varnes, J.V. Daughtridge. "Occurrence of antibiotics in surface and groundwater near confined animal-feeding operations and wastewater treatment plants using radioimmunoassay and liquid chromotography/electrospray mass spectometry," presented at the 219th

National Meeting of the American Chemical Society, San Francisco, CA, 27 March 2000 (published in "Issues in the Analysis of Environmental Endocrine Disruptors", preprints of Extended Abstracts, 401, pp.106-107, 2000).

8. California State Department of Health Services. Water recycling criteria, title 22. Sacramento, California, 2000.

9. Asano, T., A.D. Levine. "Wastewater reclamation, recycling, and reuse: an introduction." In Wastewater Reclamation and Reuse, T. Asano, ed., Technomic Publishing Company, Lancaster, Pennsylvania. p. 1-56, 1998.

10. California State Department of Health Services, "NDMA in drinking water", CSDHS, Sacramento, California. 1998.

11. World Health Organization. Health Guidelines for the Use of Wastewater in Agriculture and Aquaculture, Technical Bulletin Series 77, WHO, Geneva. 1989.

12. Bouwer, H. "Artificial recharge of groundwater: systems, design, and management", Chapter 24 in Hydraulic Design Handbook, L.W. Mays ed., McGraw-Hill, New York, N.Y. p. 24.1-24.44, 1999.

13. Ayers, R.S., D.W. Westcot. "Water quality for agriculture." FAO Irrigation and Drainage Paper No. 29, Food and Agriculture Organization of the United Nations, Rome, 1985.

14. Baier, D.C., W.B. Fryer. "Undesirable plant responses with sewage irrigation," J. Irrig. Drainage Div., Am. Soc. Civil Engrs. **1973,** 99 No. IR2, 133-141.

15. Bouwer, H., E. Idelovitch. "Quality requirements for irrigation with sewage effluent," J. Irrig. and Drainage Div., American Society of Civil Engineers. **1987,** *113*(4), 516-535.

16. Colborn, T., D. Dumanoski, J.P. Myers. Our Stolen Future, Penguin Books USA Inc., New York, NY, 1996

17. Shore, L.S., Y. Kapulnik, M. Gurevich, S. Weninger, H. Badamy, M. Shemesh. "Induction of phytoestrogens production in Medicago sativa leaves by irrigation with sewage water," *Environ. Exper. Bot.* **1995,** *35*, 363-369.

18. Shore, L.S., Y. Kapulnik, B. Ben-Dov, Y. Fridman, S. Weninger, M. Shemesh. "Effects of estrogen and 17 β-estradiol on vegetative growth of Medicago sativa," Physiologia Plantarum **1995,** *84*(2), 217-222.

19. Guan, M., and J.G. Roddick. "Comparison of the effects of epibrassinolide and steroidal estrogens on adventitious root growth and early shoot development in mung bean cuttings," *Physiol. Plant.* **1998,** *73*, 246-431.

20. Bouwer, H., P. Fox, P. Westerhoff, J.E. Drewes "Integrating water management and reuse: causes for concern?" *Wat. Qual. Internat.* Jan/Feb **1999**, 19-22.

21. Bouwer, H. "Agricultural chemicals and groundwater quality," *J. Soil Wat. Conserv.* **1990,** *45*(2) 184-189.

22. Bouwer, H., R.L. Chaney, "Land treatment of wastewater." In Advances in Agronomy 26:133-176, C.B. Brady, ed. Academic Press, New York, NY. 1974.

23. Zullei-Seibert, N. "Your daily "drugs" in drinking water? State of the art for artificial groundwater recharge", Proc. Third Internat. Symp. on Artificial Recharge of Groundwater, Amsterdam, Jos H. Peters et al., eds. A.A. Balkema, Rotterdam, The Netherlands p. 405-407, 1998

226

24. Drewes, J.E., P. Fox. "Fate of natural organic matter (NOM) during groundwater recharge using reclaimed water," *Wat. Sci. Technol.* **1999**, *40*(9), 241-248.
25. Drewes, J.E., P. Fox. "Effect of drinking water sources on reclaimed water quality in water reuse systems," *Wat. Environ. Res.* **2000**, *72*(3), 353-362.
26. Singer, P.C., R.D.G. Pyne, M. AVS, C.T. Miller, C. Mojoonnier. "Examining the impact of aquifer storage and recovery on DBPs," *J. Am. Wat. Works Assoc.* **1993**, Nov., 85-94.
27. Bouwer, H. "Simple derivation of the retardation equation and application to preferential flow and macrodispersion," *Ground Wat. J.* **1991**, *1*, 41-46.
28. McCarty, P.L., B.E. Rittman, E.J. Bouwer, "Microbiological Processes Affecting Chemical Transformations in Groundwater", in G. Bitton and C.P. Gerba, eds., Groundwater Pollution Microbiology, John Wiley and Sons, New York, pp. 89-116, 1984.
29. Crook, J., Hultquist, R., Sakaji, R. "New and Improved Draft Ground-water Recharge Criteria in California." Proceedings Annual Conference American Water Works Association (AWWA), Denver, Colorado.
30. Drewes, J.E., Rauch, T., Rincon, M., Nellor, M., P. Fox, "A watershed guided approach for water quality assurance in indirect potable reuse systems: Experiences from field studies in Arizona and California," Proceedings Annual Research Conference of the Water Reuse Association, Monterey, California, 1999.
31. Bouwer, E.J., P.L. McCarty, H. Bouwer, R.C. Rice, "Organic contaminant behavior during rapid infiltration of secondary wastewater at the Phoenix 23rd Avenue Project," *Wat. Res.* **1984**, *18*, 463-472.
32. Bouwer, E.J., J. Weiss, C. O'Melia. "Water quality improvements during riverbank filtration at three Midwest utilities," Proc. International Riverbank Filtration Conference, 4-6 Nov 1999, Louisville, Kentucky, National Water Research Inst., Fountain Valley, California. p. 49-52, 1999.
33. Bouwer, E.J., M. Reinhard, P.L. McCarty, "Organic contaminant behavior during rapid infiltration of secondary wastewater at the Phoenix 23rd Avenue Plant," Technol. Report No. 264, Department of Civil Engineering, Stanford University, California, 1982.
34. Nellor, M.H., R.B. Baird, J.R. Smith "Summary of Health Effects Study: Final Report," County Sanitation Districts of Los Angeles County, Whittier CA, 1984.
35. Sloss, E.M., S.A. Geschwind, D.F. McCaffrey, B.R. Ritz. "Ground-water Recharge with Reclaimed Water: An Epidemiologic Assessment in Los Angeles County," 1987-1991, RAND, Corp. Santa Monica CA, 1996.
36. Sharpe, R.M., N.E. Skakkebaek, "Are oestrogens involved in falling sperm count and disorders of the male reproductive tract?" *Lancet* **1993**, *341*, 1392.
37. Colborn, T., F.S. Vom Saal, A.M. Soto, "Developmental effects of endocrine-disrupting chemicals in wildlife and human," *Environ. Health Perspect.* **1993**, *101*, 378.
38. Ternes, T. "Occurrence of drugs in German sewage treatment plants and rivers," *Wat. Res.* **1998**, *32*(11), 3245-3260.

39. Stumpf, M., Th. Ternes, K. Haberer, P. Seel, W. Baumann, "Nachweis von Arzneimittelrueckstaenden in Klaeranlagen und Fliessgewaessern," *Vom Wasser* **1996**, *86*, 291-303 (in German).
40. Wenzel, A., Th. Kuechler, J. Mueller. "Konzentrationen oestrogen wirksamer Substanzen in Umweltmedien. Report," Project sponsored by the German Environmental Protection Agency, Project No 216 02 011/11, 1998.
41. Stuer-Lauridsen F, M. Birkved, L.P. Hansen, H.C. Holten Lutzhoft, B. Halling-Sorensen, "Environmental risk assessment of human pharmaceuticals in Denmark after normal therapeutic use," *Chemosphere* **2000**, *40*, 783-793.
42. Routledge, E.J., D. Sheahan, C. Desbrow, G.C. Brighty, M. Waldock, J.P. Sumpter, "Identification of estrogenic chemicals in STW effluent. 2. In vivo responses in trout and roach," *Environ. Sci. Technol.* **1998**, *32*(11), 1559-1565.
43. Shore, L., M. Gurevich, M. Shemesh. "Estrogen as an environmental pollutant". *Bull. Environ. Contam. Toxicol.* **1993**, *51*, 361-366.
44. Sontheimer, H. "Trinkwasser aus dem Rhein?" Academia Verlag, Sankt Augustin (in German), 1991.
45. Steele, K., ed. Animal Waste and the Land-Water Interface. CRC Press, Boca Raton, Florida, 1995.
46. Sims, J.T., "Characteristics of animal wastes and waste amended soils: An overview of the agricultural and environmental issues". In Animal Waste and the Land-Water Interface, K. Steele, ed. Lewis Publishers, Boca Raton, Florida, 1-14, 1995.
47. Shemesh, M., L.S. Shore. "Effects of hormones in the environment on reproduction in cattle". In: Factors Affecting Net Calf Crops (M.J. Fields and R.S. Sand, eds.), pp. 287-298. CRC Press, Boca Raton, Florida, 1994.
48. Nichols, D.J., Daniel T.C., Moore A., Edwards D.R, D.H. Pote, "Runoff of estrogenic hormone 17 beta-estradiol from poultry litter applied to pasture," *J. Environ. Qual.* **1997**, *26*, 1002-1006.
49. Miles-Richardson, S.R., V.J. Kramer, S.D. Fitzgerald, J.A. Render, B. Yamini, S.J. Barbee, J.P. Giesy, "Effects of waterborne exposure of 17 beta-estradiol on secondary sex characteristics and gonads of fathead minnows (*Pimephales promelas*)," *Aquat. Toxicol.* **1999**, 47,129-145.
50. Shore, L.S., D.W. Hall, M. Shemesh, "Estrogen and testosterone in ground water in the Chesapeake Bay watershed." Dahlia Greidinger Inter. Symp. on Fertilization and the Environment", Technion, Haifa, Israel, 24-27 March, 1997. Pp. 250-255, 1997.
51. Shore, L.S., M. Shemesh. "Analysis of testosterone levels in chicken manure", *Israeli J. Vet. Med.* **1993**, *48*, 35-37.
52. Schlosberg, A, S. Perl, A. Harmelin, V. Hanji, M. Bellaiche, E. Bogin, R. Cohen, O. Markusfield-Nir, N. Shpigel, Z. Eisenberg, M. Furman, A. Brosh, A. Holzer, Y. Aharoni, "Acute maduramicin toxicity in calves," *Vet Rec.* **1997**, *140*, 643-646.
53. Wegener, H. C., F. M. Aarestrup, L. B. Jensen, A. M. Hammerum, F. Bager, "Use of Antimicrobial Growth Promoters in Food Animals and *Enterococcus*

228

faecium Resistance to Therapeutic Antimicrobial Drugs in Europe, Danish Veterinary Laboratory, Copenhagen, Denmark," *Emerg. Infect. Dis.* **1999**, *5*, 329-35.

54. Halling-Sorensen, B., S.N. Nielsen, P.F. Lansky, F. Ingerslev, H.C.H. Lutzhoft, S.E. Jorgenson, "Occurrence, fate and effect of pharmaceutical substances in the environment - A review," *Chemosphere* **1998**, *36*, 357-394.

55. Nichols, D.J., T.C. Daniel, D.R. Edwards, P.A. Moore, D.H. Pote. "Use of grass filter strips to reduce 17 beta-estradiol in runoff from fescue-applied poultry litter," *J. Soil Wat. Conserv.* **1998**, *53*, 74-77.

56. Benne, D.T. "Review of the environmental occurrence of alkylphenols and alkylphenol ethoxilates," *Wat. Qual. Res. J. Canada* **1999**, *34*(1), 79-122.

57. Hutt, P.B. "Unresolved issues in the conflict between inividual freedom and government control of food safety," *Ecotoxic. Environ. Safety* **1978**, *2*, 447-469.

58. Ames, B.N., L. Swirsky Gold. "Too many rodent carcinogens: mitogenesis increases mutagenesis," *Science* **1990**, *249*, 940-971.

59. Calabrese, E.J. "Animal extrapolation–a look inside the toxicologists' black box," *J. Environ. Sci. Technol.* **1987**, *21*, 612-623.

60. Lave, L.B., F. K. Ennever, H.S. Rosenkranz, and G.S. Omenn, "Information value of the rodent bioassay," *Nature* **1988**, *336*, 631-633.

61. Bacon, F. The New Atlantis (also "Solomon's House") 302 pp., **1626** (available: http://www.sirbacon.org/links/newatlantis.htm).

Acknowledgment
The authors gratefully acknowledge Dr. Herman Bouwer, Phoenix (Arizona), for his encouragement to write this paper. His advice, in-depth discussions, and his review have significantly enriched the paper. This paper is dedicated to him.

Ecotoxicological Issues

Chapter 13

Iodinated X-ray Contrast Media in the Aquatic Environment: Fate and Effects

T. Steger-Hartmann, R. Länge, and H. Schweinfurth

Schering AG, Research Laboratories, D-13342 Berlin, Germany

Iodinated X-ray contrast media have gained a prominent role in the present discussion of pharmaceuticals in the environment. Rather than for toxicological or ecological reasons this is mainly due to the fact that these substances are applied at high dosages and possess a high meta-bolic stability, which renders them relatively resistant to degradation. Therefore, in contrast to many other pharmaceuticals, comparatively large amounts are released into the environment. In this contribution, the existing data on degradation, ecotoxicology, and mammalian toxicology is summarized. A risk assessment based on the presented data leads to the conclusion that even though traces of iodinated X-ray contrast media or their degradation products may show up in surface or drinking water, a risk for aquatic life or human consumers' health is not discernible.

Introduction

Iodinated X-ray contrast media are pharmaceuticals that are mainly applied intravenously for the display of soft-tissue or vessels, which otherwise would give insufficient contrast in X-ray imaging, e.g., blood vessels (angiography), urinary tract (urography), gallbladder (choleography).

The substances consist of derivatives of the triiodobenzoic acid and can be divided into two groups: ionic X-ray contrast media with a free acid group and nonionic X-ray contrast media, where all the substituents bound to the benzene ring are nonionizable hydrophilic chains. The first iodinated X-ray contrast media marketed were ionic ones, but since 1984, when nonionic contrast media were first introduced to the United States, their market share has been continuously increasing due to their higher comfort and safety. Meanwhile more than 80% of all contrast studies are performed with nonionic contrast media in the U.S. Up to 200 g of X-ray contrast medium may be applied intravenously prior to a single radiologic session, a fact already illustrating, that extremely good tolerance is a prerequisite for these substances. The total amount of iodinated X-ray contrast media consumed in the U.S. was estimated by our sales department to be 1330 t for 1999. The substances are optimized for their biological inertness, i.e., no metabolism occurs during the passage through the body. Since they are highly hydrophilic, they are excreted unchanged via urine into wastewater. More than 80% of the applied dose is renally excreted within 24 h. Environmental fate and potential effects of these substances are presented in this paper.

Exposure

Predicted Environmental Concentration and Monitoring Data

According to the Center for Drug Evaluation and Research (U.S. FDA), the expected introduction concentration (EIC) of a pharmaceutical into the aquatic environment can be calculated on the basis of the following formula (1):

EIC-aquatic (ppb) = A x B x C x D where
A = kg/year used product (1339×10^3)
B = 1/liters per day entering municipal sewage treatment plants*
C = year/365 days
D = 10^9 µg/kg (conversion factor)
* 1.214×10^{11} liters per day entering municipal sewage treatment plants in the U.S.

Thus using the above-mentioned figure for yearly consumption, the EIC for X-ray contrast media, i.e., the wastewater concentration, is estimated to be 30 μg/L. If an average dilution factor for wastewater of 10 is taken into account (2), the estimated average U.S. surface water concentration can therefore be given as 3 μg/L. Analytical monitoring data for iodinated X-ray contrast media are not available for the U.S. For Germany, the predicted environmental concentration (PEC) in surface water was calculated as 2 μg/L for iopromide, a non-ionic X-ray contrast medium with the highest German market share. Considering the total contrast media consumption, a PEC of less than 8 μg/L is obtained (3), which is well in the range of the calculated value for the U.S. In regions of Germany where surface water is used for the production of drinking water, trace amounts of contrast media or their metabolites were detected in the latter, at first determined with sum-parameter analysis of adsorbable organic halogens (4). We were able to determine iopromide in the effluent of a Berlin sewage treatment plant with a median of 5 μg/L (3,5). Hirsch et al. found median values in effluents for various X-ray contrast media below 1 μg/L (maximum value: 15 μg/L iopamidol) (6). In surface waters, Ternes and Hirsch detected median concentrations of 0.23 μg/L for the ionic contrast medium diatrizoate and 0.49 μg/L for the non-ionic iopamidol (see 7). Summarizing the presently existing monitoring data, it can be concluded that the calculated PEC generally represents a worst case scenario, which only occurs in regions with locally high contrast media consumption, especially when the sewage treatment plant discharges its effluents into small rivers or creeks.

Degradation of Iodinated X-ray Contrast Media

Standard tests, such as described under OECD guideline 301, on the biodegradability of various representatives of contrast media were conducted in our laboratories. They showed that these substances are generally not readily biodegradable. Less than 10% of the substances were mineralized in the course of the 28-day incubation period. Subsequently we chose iopromide (N,N'-bis (2,3-dihydroxypropyl)-2,4,6-triiodo-5-methoxyacetylamino-N-methylisophthal-amide, see Figure 1, left) as a model compound for further investigations.

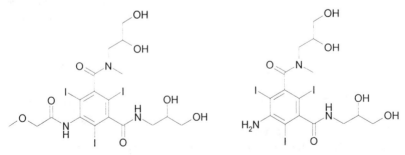

Figure 1: The structure of the nonionic X-ray contrast medium iopromide (left) and its main metabolite 5-amino-N,N'-bis(2,3-dihydroxypropyl)-2,4,6-triiodo-N-methylisophthalamide (right), which was identified in a biodegradation test

In a test for inherent biodegradability (Zahn-Wellens test, OECD guideline 302B), iopromide showed a 35% degradation on the basis of DOC (dissolved organic carbon) reduction within the study period of 28 days (Figure 2). A prolongation did not increase degradation. Less than 15% of iodine was released during the degradation process as confirmed by concomitant AOX (adsorbable organically bound halogens) analysis, thus indicating the appearance of stable metabolites. Iopromide had no toxic or inhibitory effects on the wastewater bacteria, which was shown by concomitant incubation of iopromide and the reference substance diethylene glycol (1:1, toxicity control).

In a test simulating the sewage treatment process on laboratory scale (modified Coupled Units test, OECD 303A) we identified one metabolite of iopromide degradation, the free aromatic amine of iopromide (IUPAC: 5-amino-N,N'-bis(2,3-dihydroxypropyl)-2,4,6-triiodo-N-methylisophthalamide,   see Fig. 1) which appeared after approximately three weeks of sludge adaptation and was not further degraded.

However, in a long-term degradation experiment over 23 weeks, simulating the situation in nutrient-poor aquatic environments with and without illumination, the initial metabolite disappeared in the dark cultures, and smaller highly hydrophilic intermediates turned up, which were no longer amenable to HPLC analysis with the established system (Figure 3). The degradation under light was low, probably because lower wavelengths responsible for photolysis (see below) were excluded by the glass vessels. Nevertheless, intensive growth of green algae was observed, thus indicating that the metabolite had no inhibitory effect on algal growth. In the dark, a different biocoenosis developed, including bacteria able to degrade the iopromide metabolite further. Degradation was significantly retarded in dark cultures, which were initially sterilized (data not shown).

Parallel experiments on photodegradation of the sewage metabolite with a standardized setup and a light-source simulating sun-light spectrum (8) showed that it decayed with a considerably shorter half-life than the parent compound iopromide (Table I), which we already reported previously (5). Photodegradation was accompanied by deiodination. Sensitivity of X-ray contrast media to photolysis was already shown for a number of other compounds (9). However, half-lives were not calculated in these experiments.

Kalsch (10) was able to show that iopromide was degraded in a river water/sediment system. In the environmentally relevant concentration range, the determined half-lives were inversely correlated with the initial iopromide concentration, i.e., the degradation occurred faster at higher initial iopromide concentrations. It can therefore be concluded that several mechanisms (e.g., photolysis, step-wise biodegradation in different compartments) may account for the disappearance of iodinated X-ray contrast media in the aquatic environment

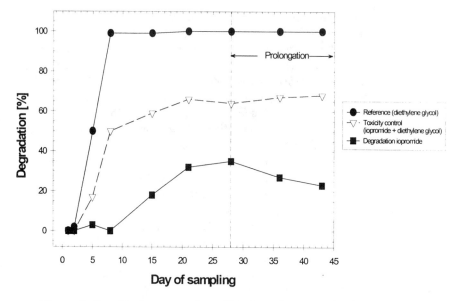

Figure 2. Graphical presentation of biodegradation on the basis of DOC reduction [%] in the test for inherent biodegradability of iopromide (Zahn-Wellens test); initial concentrations: 100 mg DOC/L for iopromide and diethylene glycol, 200 mg/L for toxicity control

Table I. Environmental half-lives of iopromide and its main metabolite at different latitudes and seasons, calculated using data from the photo-degradation experiment and the GCSOLAR software (EPA)

Substance	$t_{1/2}$ at low irradiation (fall, latitude 60° North)	$t_{1/2}$ at high irradiation (summer, latitude 20° North)
Iopromide	607.2 h (=25.3 days)	42.5 h
Free aromatic amine of iopromide, ZK 39216 [a]	25.9 h	4.9 h

[a] ZK is the abbreviation of the central compound file of Schering AG ("Zentralkartei")

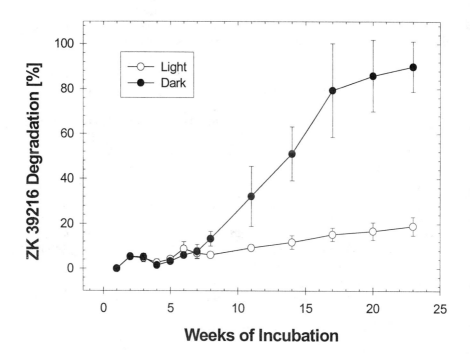

Figure 3. Percentage degradation of the iopromide metabolite (free aromatic amine of iopromide, ZK 39216) in a prolonged degradation assay with and without illumination; The values represent the mean ± standard deviation (n=4); initial concentration of ZK 39216: 92 mg/L

Effects

Ecotoxicity

To characterize the environmental hazard of different ionic and nonionic contrast media (amidotrizoic acid, iohexol, iotrolan, iopromide), we conducted a series of short-term (acute) and long-term (chronic) ecotoxicological tests with fish (*Danio rerio*, *Leuciscus idus*), water flea (*Daphnia magna*) and green algae (*Scenedesmus subspicatus*). More detailed results of the ecotoxicological testing were already published recently (*3, 5*). Summarizing these data, it can be stated that no effects were observed in the short-term tests even at the highest tested concentrations of 10 g/L. In the long-term toxicity tests with fish (*Danio rerio* Early-Life-Stage test) and water fleas (*Daphnia magna*, reproduction test) at concentrations up to 100 mg/L and 1 g/L, respectively, also no adverse effects were detected.

In the discussion of pharmaceuticals residues in the environment, it has often been argued that ecotoxicological data are, if at all, only available for the parent substances but not for metabolites. As described above, we were able to identify a metabolite in a laboratory test system, which we subsequently characterized for its ecotoxicity. Furthermore, we investigated theoretical metabolites that might occur by further side-chain cleavage of the parent molecule, but which could not be confirmed in model systems for analytical reasons. These substances were available to us, because they also occur as impurities during the production process of X-ray contrast media.

The no-observed-effect-concentrations (NOEC) determined in these investigations are given in Table II. The toxicity data for the metabolites is generally in the range of the parent compounds, i.e., the cleavage of side-chains during biodegradation does not appear to result in intermediates that are more toxic to aquatic life than the parent compounds.

Mammalian Toxicity and Genotoxicity

Due to the high dosage of contrast media needed for achieving good contrast of soft tissue, the media are optimized for extremely good tolerability in the mammalian organism. These animal toxicity data originally obtained during the drug development process may also be used to assess the risk that might result for humans by the uptake of drinking water containing traces of X-ray contrast media.

No acute toxic effects were observed after single-dose applications of non-ionic contrast media (iotrolan, iopromide) given either orally or intravenously in doses of ≥ 26 g/kg (rats) or ≥ 32 g/kg (mice). Repeated-dose toxicity studies with rats and dogs using doses of up to 5.3 g/kg of ionic contrast media (amidotrizoate, ioglicinate, iotrixinate) applied intravenously five times a week over the course of five weeks showed no morphological or functional changes of toxicological relevance. Nonionic contrast media (iopromide, iotrolan) were tested with even higher doses of 7.7 g/kg without such effects (Table III). In studies on reproduction toxicity with rats and rabbits, ionic or nonionic contrast media injected intravenously showed no embryolethal or teratogenic effects up to 5.3 and 7.7 g/kg, respectively (Table IV).

Extrapolation of these data to humans leads to the conclusion also supported by clinical experience that the intravenous administration of 310 g of ionic contrast medium or 460 g nonionic contrast medium may still be regarded as safe. This is far more than what is needed for reasonable contrast imaging. The rare adverse effects reported after contrast medium application were either due to misapplication (e.g., paravasal injection) or to pseudoallergic reactions, which can be mainly attributed to the high osmotic pressure (*11, 12*). Both effects are without any relevance for the concentration range expected in the environment.

Mutagenicity is also routinely tested during the drug development process. Besides the parent compound, the mutagenicity of the above-mentioned metabolites was also investigated because these substances may appear as impurities in the final drug. From a structure-activity view point, the metabolites were of particular interest because all of them contain free aromatic amino groups, thus raising a mutagenicity alert. However, in all investigated test systems for mutagenicity (Ames test, HPRT test, chromosome aberration test, micronucleus test in vitro and in vivo) both the parent compounds as well as the metabolites/intermediates proved to be negative, i.e. non-mutagenic (Table V).

For human risk assessment, it must be borne in mind that the bioavailability of iodinated X-ray contrast media after oral uptake is below 2% (Figure 4) (*13*). The low bioavailability is a consequence of the high hydrophilicity of the compounds. From these data as well as from the low $logP_{ow}$ it can be concluded that bioaccumulation does not occur either in aquatic species or in humans.

The above-mentioned animal data were all obtained after intravenous application, ensuring quantitative bioavailability. If humans experience exposure to environmental residues of X-ray contrast media, it would be dermally (washing, swimming) or orally (via surface or drinking water). Therefore, the extrapolation from the animal data to human safety includes an additional safety margin. Hence, it can be concluded that incorporation of trace amounts of iodinated X-ray contrast media via surface or drinking water is of no risk for the consumer.

Table II. NOEC-results (mg/L) of ecotoxicological testing of potential or proven X-ray contrast media metabolites

Metabolites (IUPAC name)	Algae S. subspicatus (72 h)	Crustacean D. magna (acute; 48 h)	Fish D. rerio (acute; 48 h)	Crustacean D. magna (chronic; 21 d)
ZK 39216 (5-Amino-N,N'-bis(2,3-dihydroxypropyl)-2,4,6-triiodo-N-methylisophthalamide)	250	1000	92[a]	n.d. (fish: 100)
ZK 39211 (5-Amino-N-(2,3-dihydroxypropyl)-2,4,6-triiodo isophthalamic acid)	10000	10000	10000	300
ZK 3513 (5-Amino-2,4,6-triiodoisophthalic acid)	10000	10000	10000	300

[a] for this test, effluent from the laboratory sewage treatment plant with a ZK 39216 concentration of 92 mg/L was directly used for the fish test

Table III. Toxicity after repeated intravenous application (3-5 weeks, 5-7 times per week)

Substance	Results
Ionic X-ray contrast media	
Amidotrizoate	no serious health effects at doses of up to
Ioglicinate	2.8 g iodine/kg (= 4.5 g contrast media/kg)
Iotrixinate	in rats or dogs
Nonionic X-ray contrast media	
Iopromide	no serious health effects at doses of up to
Iotrolan	3.7 g iodine/kg (= 7.7 g contrast media/kg)
	in rats or dogs
Metabolites, Intermediates	
ZK 39216	no signs of toxic effects in dogs at a dose of 30 mg/kg given over 2 weeks
ZK 39211	LD50 after single intragastric or dermal application is > 2 g/kg)

Table IV. Reproductive toxicity after repeated intravenous application through days 6-15 (Rats) or 6-18 (Rabbits) postcoital

Substance	Results (Embryolethality/Teratogenicity)
Ionic X-ray contrast media	not observed at doses of up to
Amidotrizoate	3.0 g iodine/kg (= 4.8 g contrast media/kg)
Ioglicinate	3.0 g iodine/kg (= 5.3 g contrast media/kg)
Iotrixinate	2.8 g iodine/kg (= 4.5 g contrast media/kg)
Nonionic X-ray contrast media	not observed at doses of up to
Iopromide	3.7 g iodine/kg (= 7.7 g contrast media/kg)
Iotrolan	3.0 g iodine/kg (= 6.9 g contrast media/kg)[a]

[a] in rabbits, a dose-dependent increase in embryo-lethality was observed starting at 1.9 g iotrolan/kg

Table V. Results of genotoxicity testing in various test systems

Substance	*Results*
Ionic X-ray contrast media Amidotrizoate Ioglicinate Iotrixinate	**Negative** in Ames test (-/+S9), micronucleus assay (mice, human lymphocytes), and in tests for chromosomal aberrations
Nonionic X-ray contrast media Iopromide Iotrolan	**Negative** in Ames test (-/+S9), UDS test, micronucleus assay (mice), dominant lethal test (mice)
Metabolites, Intermediates or Impurities ZK 39216	**Negative** in Ames test (-/+S9), HGPRT test (V79), micronucleus assay (mice), and in tests for chromosomal aberrations
ZK 39211 ZK 3513	**Negative** in Ames test (-/+S9)

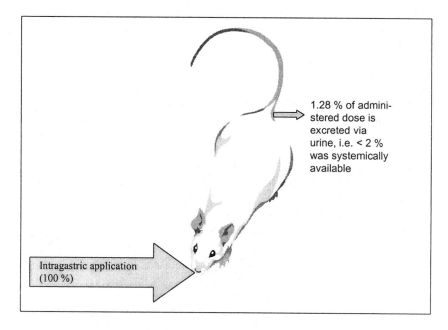

Figure 4. After intragastric application less than 2 % of the applied X-ray contrast medium iopromide is systemically available

Conclusion

An environmental exposure with iodinated X-ray contrast media may be expected in locales with intensive use and short water cycles in the lower microgram per liter range. The NOECs for aquatic organisms were determined to be in the upper milligram per liter range. For mammalians, the no-observed-effect levels were even higher, in the gram range after intravenous administration. Consequently, if the expected exposure concentrations are compared with the toxicity data, neither a risk for the aquatic environment nor for human consumers' health due to the release of iodinated X-ray media is predicted.

References

1. FDA/CDER. Guidance for Industry – Environmental Assessment of Human Drug and Biologics Applications. CMC 6 (Revision I). July 1998, p. 4.
2. European Commission. Assessment of Potential Risks to the Environment Posed by Medicinal Products for Human Use (Excluding Products Containing Live Genetically Modified Organisms): Phase I Environmental Risk Assessment, III/5504/94-Draft. 1995, p. 8.
3. Steger-Hartmann, T.; Länge, R.; Schweinfurth, H. Environmental Risk Ecotox. Environ. Saf. 1999, 42, 274-281.
4. Schulze-Rettmer, R. Halogenorganische Verbindungen in gewerblichen und industriellen Abwässern. Korrespondenz Abwasser 1998, 8, 1423-1433.
5. Steger-Hartmann, T.; Länge, R.; Schweinfurth, H. Vom Wasser 1998, 91, 185-194.
6. Hirsch, R.; Ternes, T.; Kratz, K.-L.; Wilken, R.-D. Occurence of X-ray Diagnostics in German Rivers and Ground Waters. 9th Annual Meeting of SETAC-Europe, 25-29 May 1999.
7. Daughton, C.G.; Ternes, T Environ. Health Perspect. 1999, 107 (suppl. 6), 907-938.
8. An Assessment of Test Methods for Photodegradation of Chemicals in the Environment (Technical report No. 3). Lemaire, J.; Campbell, I.; Hulpke, H.; Guth, J.A.; Merz, W.; Philp, J.; von Waldow, C.; Eds. ECETOC. Brussels, 1981.
9. Hundesrügge, T. Pharmaceutical Products in the Environment. Arzneimittel in der Umwelt - Weg des Röntgenkontrastmittels Iopentol. Krankenhauspharmazie 1998, 5, 245-248.
10. Kalsch, W. Sci. Total Environ. 1999, 225, 143-153.

11. Krause, W.; Press, W.R.; Frenzel, T. Preclinical Testing of Iopromide. 1st Communication: Pharmacological Evaluation. *Arzneim.-Forsch./Drug Res.* **1994,** 44 (III), Nr. 10, 1162-1166.

12. Krause, W.; Schöbel, C.; Press, W.R. Preclinical Testing of Iopromide. 2nd Communication: Toxicological Evaluation. *Arzneim.-Forsch./Drug Res.* **1994,** 44 (III), Nr. 11, 1275-1279.

13. Mützel, W.; Speck, U.; Weinmann, H.-J. In *Contrast Media in Urography, Angiography and Computerized Tomography*; Taenzer,V.; Zeitler, E; Eds.; Thieme-Verlag: Stuttgart, New York, 1981, pp 85-90.

Chapter 14

Multidrug–Multixenobiotic Transporters and Their Significance with Respect to Environmental Levels of Pharmaceuticals and Personal Care Products

D. Epel[1] and T. Smital[2]

[1]Hopkins Marine Station of Stanford University, Ocean View Boulevard,
Pacific Grove, CA 93950
[2]Department for Marine and Environmental Research, Ruder Boskovic Institute,
Bijenck AJ 54, 1000 Zagreb, Croatia

What might be the consequences to aquatic organisms of low levels of pharmaceuticals and personal care products (PPCPs) in the aquatic environment? One possibility is interference with efflux transporters, which protect aquatic organisms from a wide variety of potentially toxic substances. These transporters act to efflux putative toxicants out of the cell and organism. In this article we review the properties of these transporters as regards substrate specificity and substrate affinity, protective role, and the modulation of activity by pharmaceutical products. We conclude that although PPCPs could potentially modulate these transporters, it is unclear whether this modulation can occur at the low PPCP levels present in the environment. Work on this question is urgently needed, and we suggest directions for this research.

Organisms use a variety of cellular mechanisms to avoid or repair the effects of toxicants. Historically, most research has focused on detoxification or conversion of foreign compounds once they enter the cell. These detoxification systems usually involve an initial alteration (phase I, typically an oxidation) of the xenobiotic molecule (*1*). The molecule can then be conjugated to create a more soluble form (phase II, ref *2*), and the altered molecule is then typically secreted by the cell to eventually leave the organism (*2,3*). These detoxification systems are poised to act on xenobiotics, as indicated by their high inducibility in response to chemical stress (*4*). However, these systems can create problems for the cell since some of the detoxification/degradation products can be even more toxic than the parent compound (*5*).

An alternative mechanism, which avoids the problems of creating new toxicants in the cell, is utilization of a newly discovered protective activity provided by multidrug/multixenobiotic transport proteins. These transporters, exemplified by the P-glycoprotein (Pgp), provide a first line of defense against toxicants by acting as efflux transporters which prevent the toxicant from even entering the cell in the first place (*6-8*). The P-glycoprotein-like transport proteins are present in many aquatic organisms and minimize the exposure of these organisms to toxicants. The protection offered by these transporters is referred to as multixenobiotic resistance (MXR, ref *6*) since, as will be described below, the transporters work on a wide variety of chemically unrelated xenobiotics.

However, the main disadvantage of the toxicant avoidance system provided by the Pgp which is also the reason this chapter forms part of this book is its ability to be compromised by the presence of alternate substrates such as specific PPCPs. This weakening of the defense arises because the transporters act on a wide variety of chemical compounds (as will be described later) and a consequence of this non-specificity is that the system can become saturated/overwhelmed by the presence of these potential substrates. The result of such inhibition can be enhanced intracellular accumulation of other more toxic xenobiotics frequently present in the aquatic environment.

The major question that we will therefore address in this article is whether the release of pharmaceuticals into the environment can act as Pgp-competitive substrates and thus compromise this protective system. To address this question, we will first present current data on the properties of this transporter system and then consider the question of whether the levels of PPCPs in the environment could affect the protection provided by the Pgp.

The Efflux Transporters

Members of the Pgp family have been identified in all eukaryotic cells from protozoa through plants, fungi, and animals (*9*). The protein is a member of the ABC (ATP-binding cassette) family of transport proteins and most of its functions appear to be related to a role as an efflux transporter. This transport mode can be used for numerous cell functions ranging from transport of physiological substances such as cholesterol and peptides to transport of toxic substances out of the cells (*8, 10-12*).

The role of the Pgp in transporting xenobiotics out of cells is well studied and well documented. The interest and reason for this large amount of work on the Pgp arose from a correlation between the expression of the Pgp in tumor cells and the acquisition of the tumor cell resistance to chemotherapeutic drugs. Numerous studies showed that this acquired resistance to chemotherapeutic drugs coincided with the expression of a glycosylated protein (the Pgp) in the membrane of these resistant tumor cells. This resistance was also associated with reduced accumulation of the chemotherapeutic drugs within the cells. Evidence that the Pgp is involved in this reduced drug accumulation came from transfection experiments in which cells not expressing the Pgp were transfected with the gene. Successful transfection resulted in acquisition of drug resistance by these cells and also the phenotype of reduced accumulation of the xenobiotics within the cytoplasm. This efflux transporter mechanism was further confirmed by studies in which the transporter was inserted into vesicle preparations. The transporter in these membrane vesicles utilized ATP to pump substrates from the medium into the intravesicular space (above reviewed in ref *8-10*).

Proteins functionally related to the multidrug transporters are also present in bacteria. A fascinating finding is that expression of one of these bacterial effluxers (LmRA) in human cells resulted in a phenotype virtually identical with that of Pgp-mediated drug resistance (*13*). These results then indicate that this type of efflux transporter is an evolutionarily ancient system and that one of its roles is to subvert the effects of toxic compounds in the organism's environment.

Roles of the Efflux Transporters

Various genetic manipulation studies similarly support a role for the transporter in xenobiotic protection. First are studies on the soil nematode *Caenorhabditis elegans* (*14*), in which mutants of the Pgp family showed increased sensitivity to potential toxic substances such as colchicine. Conversely, over-expression of members of this family provided increased protection against xenobiotics.

A second critical study comes from studies on knockout mice for the *MDR*1 gene (coding for the Pgp responsible for drug-resistance in mice). The phenotype of these mice was apparently normal. However, exposure to xenobiotics (such as pesticides) that had no affect on wild-type animals resulted in death of the knockout mice (*15*).

The apparently normal phenotype of the knockout mice suggests that the only role of the Pgp is in toxicant evasion. However, there is similarly good evidence for roles of the Pgp in normal cell function. For example, Pgp is involved in immune responses (*16*) and also in transfer of lipids across the membrane bilayer (*17*) suggesting that the efflux transporter may have additional roles in transferring substances across the plasma membrane that are not related to toxicant evasion. Why these roles are not apparent in the mouse knockouts is unknown but could relate to redundancy of pathways or to the possibility that the knockout mice were not stressed sufficiently to reveal these alternative functions.

Modulation of Activity of the P-Glycoprotein Transporters

One of the most interesting properties of the Pgp is its ability to efflux a wide variety of substrates. Unlike the immune system in which each antibody is exquisitely specific for each antigenic compound, the Pgp effects the efflux of a wide variety of compounds whose major common denominator appears to be the shared property of moderate hydrophobicity (*18*). This is a highly "economical" approach since most organisms possess genes coding for only a few of these Pgp-like transport proteins and this small inventory apparently suffices to provide a first line of defense against most xenobiotics (*10, 13*).

A list of potential substrates is shown in Table I. The list is biased towards pharmacological agents since most work has been done with these compounds. There is some information on environmental substrates such as pesticides, and these are also indicated in the table.

Table I. Some substrates and modulators of the Pgp.

Calcium channel blockers	verapamil, diltiazem, nifedipine
Calmodulin antagonists	trifluoperazine, chlorpromazine, thioridazine
Anti-hypertensive agents	reserpine
Vinca alkaloids	vincristine, vinblastine
Steroids	progesterone
Anti-arrhythmics	amiodarone, quinidine
Anti-parasitics	quinacrine, quinine
PKC-inhibitors	staurosporine
Immuno-suppressants	cyclosporin A, PSC-833
Anti-estrogens	tamoxifen, toremifene
Pesticides	pentachlorophenol, dachtal

The lack of substrate specificity and resultant wide range of substrates that can be handled by the transporter is indeed remarkable and particularly relevant to its involvement with the PPCPs. Research on mechanisms of transport action that can account for this wide range of substrates indicates that there are two and possibly three distinct substrate binding sites on the transport protein (*19, 20*). These are referred to as the H and R sites named in relation to specific binding of rhodamine (R site) and Hoechst 33342 (H site). There is also evidence of a separate binding site for some steroids (*20*).

Another aspect of transport activity relative to substrates is that the affinity of the transporter varies for different substrates; some substrates bind to the transporter at very low concentrations whereas other substrates require much higher levels. The implication vis-a-vis environmental protection is that the transporter will protect the cell efficiently against high-affinity substrates but poorly against low-affinity substrates.

A caveat with this simple idea, however, is that some of these high-affinity substrates can also inhibit the transporter activity such that they abrogate the protective function so that other xenobiotics that normally would be excluded from the cell can now enter; an example is verapamil, which is a potent inhibitor of Pgp activity (*21*). Other substances can inhibit Pgp by affecting the P-glycoprotein ATPase; cyclosporin might work in this fashion (*21*). Finally, some compounds act indirectly by affecting enzymes that regulate the phosphorylation state of the transporter; examples are protein kinase C (PKC) inhibitors such as staurosporine (*23*). If staurosporine action is indeed through inhibition of PKC, then P-glycoprotein might require rapidly turning over phosphorylation mediated by protein kinase C.

Finally, activity is affected in mixtures of different substrates. Assuming none of these inhibits the activity, this modulation arises from competitive interaction between these various substrates such that the efflux/exclusion of one substrate is enhanced over another, effectively inhibiting or reducing the effectiveness of the Pgp. This regulation of activity by substrates is referred to as **modulation** and the regulating substrates are referred to as chemical **modulators**. If the PPCPs are modulators or inhibitors, their presence in the environment could affect/inhibit the transport activity and thus allow chemicals to enter the cell that would normally have been kept at bay by the transport functions of the P-glycoprotein.

Efflux Transporters, PPCPs and Aquatic Organisms

Predictions on any environmental consequences of the PPCPs on the Pgps requires information on whether 1) these transporters are present and important for xenobiotic protection in aquatic organisms; 2) whether the PPCPs are present at high enough levels to be modulators or inhibitors, and 3) whether the affinity characteristics of these substrates can compromise the first line of defense provided by the efflux transporters. These questions will be addressed in the next sections.

Presence of Pgp-Like Proteins in Aquatic Organisms

Work pioneered by Kurelec and his colleagues in the early 1990s first demonstrated the presence of Pgp-like molecules in aquatic organisms (reviewed in ref *6*). This list has now been extended to a wide variety of aquatic animals, including sponges, mussels, clams, fish, and echiuroid worms (*6-8, 24*).

Evidence for activity encompasses a wide variety of assays and techniques. One of the initial demonstrations was an ATP-dependent binding of substrates to membrane vesicles (*25*) that was inhibited by verapamil (a high-affinity substrate for the Pgp). Subsequent work showed the presence of the protein on the basis of immunological similarity to the mammalian Pgp (*6, 27-33*) using monoclonal antibody C219, which is generated against a highly conserved epitope on the Pgp (similar epitope present in plants, fungi, and animal cells).

Further evidence comes from activity studies showing that many moderately hydrophobic substrates are excluded from cells but that these compounds can accumulate in the cytoplasm at much higher rates in the presence of Pgp antagonists such as verapamil. This has been shown in a variety of ways such as directly measuring uptake of vinblastine and other substrates (*6, 27-33*). Another assay — which can be used with single cells such as oocytes or thin cell sheets such as in mussel gills — is to measure the uptake of a fluorescent substrate, such as rhodamine (*26,31*). Under normal conditions, rhodamine is excluded from the cell but if inhibitors or modulators of the transporter are added, rhodamine now enters the cell. The kinetics of this uptake can be easily followed by fluorescence microscopy and image processing equipment. An important modification of the dye approach is to measure dye efflux in a non-invasive manner as developed by Smital and Kurelec (*34*). Activity studies have also been done with pollutants. Cornwall et al., for example, measured the transport of the herbicide DCPA (dimethyl 2,3,5,6-tetrachloro-1,4-benzenedicarboxylate) in the absence and presence of verapamil. The presence of verapamil resulted in a more than two-fold increase in herbicide uptake by mussel gills (*31*).

Another proof of function/presence of the Pgp has been provided by photo-affinity labeling studies. The embryos of the mud-dwelling worm *Urechis caupo* showed Pgp activity based on transport studies with rhodamine B and the cross-reactivity of a membrane protein with the C219 antibody (*26*). More definitive proof was obtained by photo-affinity labeling of the membrane proteins with a substrate of the transporter and immunoprecipitation with an antibody directed against the mammalian Pgp (*26*). The experiment involved exposure of membrane fractions to an I^{125} labeled and photo-activated, cross-linkable derivative of forskolin (a Pgp substrate). Several protein bands were labeled in the plasma membrane by this compound, and one of these could be immunoprecipitated with a Pgp-precipitating antibody. The molecular weight of the I^{125} labeled immunoprecipitated protein corresponded to the Pgp and to the C219-reactive protein (*26*). Thus a substrate of the transporter — forskolin — binds to a Pgp-like molecule.

Definitive evidence for the presence of the Pgp-like protein in these aquatic organisms has come from isolation of genes from three aquatic organisms, which are closely related to the mammalian Pgp. Partial sequences have now been obtained from the mussel *Mytilus galloprovincialis*, the aforementioned echiuroid worm *Urechis caupo*, and the winter flounder *Pleuronectes americanus*. The partial sequences for these three glycoproteins and comparison with the human Pgp are shown in Figure 1 (refs *35-37*).

```
                     10         20         30         40         50
human p-gp       1 ypsrkevkil kglnlkvqsg qtvalvgnsg cgksttvqlm qrlydptegm
flounder p-gp    1 ---------- ---------- ---------- ---------- ----------
Urechis p-gp     1 typsrpdnqv lkgisfkvnh gemvavvgps gggkstmvnl ierfydpdsg
Mytilus p-gp     1 ---------- ---------- ---------- ---------- qrfydpdagq
                                           +      +  +                +
                     60         70         80         90        100
human p-gp      51 vsvdgqdirt invrflreii gvvsqepvlf attiaeniry grenvtmdei
flounder p-gp   51 ---------- ---------- ---------- ---------- ----------
Urechis p-gp    51 fiylgntdvr llnanwfrrq lamvgqepil fatsirdnia ygkdatieqi
Mytilus p-gp    51 vlldgnnikd lnlnwlrqni gvvsqepvlf gctiaenirl gnpnatitei
                   +                      +          +           +    +
                    110        120        130        140        150
human p-gp     101 ekavkeanay dfimklphkf dtlvgergaq lsgqgkqria iaralvrnpk
flounder p-gp  101 ---------- -------kf dtlvgdrgtq msggqkqrva iaralvrnpk
Urechis p-gp   101 eevarsanah dfiadfaegy dtmvgergvr lsggqkqrva iarallmdpa
Mytilus p-gp   101 eqaakqanah dfikslpqsy ntlvgergaq lsggqkqrva iaralirdpr
                   +++        ++           + ++  ++    ++++++++ ++++++  +
                    160        170        180        190        200
human p-gp     151 illldeatsa ldteseavv- qvaldkarkg rttiviahrl ..........
flounder p-gp  151 illldeatsa ldaesetivv qealdqaskg rtciivahrl ..........
Urechis p-gp   151 vllldeatsa ldaesehlv- qeaidramnn rtvlviahrl ..........
Mytilus p-gp   151 illldeatsa ldseseniv- qealekarqg rttlviahrl
                   +++++++++ ++ +++   +   +++++ +  + ++      ++++
```

Figure 1. *Partial amino acid sequences from **Mytilus galloprovincialis** (mussel, ref 35), **Urechis caupo** (echiuroid worm, ref 36), **Psuedopleuronectes** (fish, winter flounder, ref 37) and **Homo sapiens** (human, accession # PO8183). Conserved amino acids shown in the last row as a +. Where no sequence data available, indicated as a minus (-) sign.*

The above evidence then indicates that Pgp-like transport molecules are present in aquatic organisms. The evidence includes transport activity similarities, presence of related immunoreactive molecules similar to the mammalian Pgp, cross-linking studies showing that substrates of this transporter can be cross-linked to a protein that precipitates with the Pgp antibody and finally related gene sequences. This presence is not surprising; given the occurrence of the gene from bacteria through plants, fungi, and animals, the transporter gene and protein are probably ubiquitous.

Our preliminary data, however, indicates that the expression of this gene varies dramatically in different animals. In *Urechis* embryos and mussel gills, for example, the protein is present in very high titer whereas it is in low titer in sea urchin embryos (see later). Figure 2 depicts a western blot for C219-reactive material in the eggs of *Urechis caupo* and compares this with the amount of Pgp in a drug-resistant human tumor cell line (ref *26*). The same amount of protein was applied to each lane of the gel and as can be seen, the amounts of immunoreactive material are remarkably similar. Thus, the amount of protein present in the *Urechis* embryo plasma membrane is equivalent to that of a highly drug-resistant tumor cell. This is a large amount of expression and as shown in the next section is more than adequate to keep out toxicants.

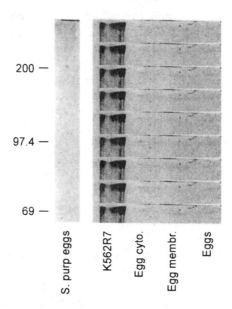

Figure 2. *Western blot of Pgp in an echiuroid worm (**Urechis**) and comparison with the Pgp in a Pgp-expressing drug resistant human tumor cell line K562R7 (from ref 26). Note that the intensity of the Pgp band in **Urechis** is comparable to the drug-resistant tumor cell. There is no detectable Pgp in the eggs of the sea urchin **S. purpuratus** (lane 1).*

Sea urchin embryos do not express large amounts of C219-reactive protein, but do evidence protection against some xenobiotics that is abrogated by Pgp inhibitors (see below). This could mean that these low Pgp levels are adequate for protection, or that there might be other types of efflux transporters present in these embryos (and in other aquatic organisms also).

Other Types of Efflux Transporters

Additional efflux transporter candidates are the multidrug resistant protein referred to as MRP (multidrug resistance protein), also a member of the ABC transport family and similar to the Pgp except in having an additional transmembrane component (38). Some of the modulators that affect this transporter can inhibit activity of the Pgp, so differential diagnosis on the basis of effects of these shared substrates is problematic. The major MRP role appears to be the efflux of organic anions, weak acids, and conjugates of xenobiotics such as glutathione adducts (39). Most of the research has been on mammalian systems, and there are few probes useful for identifying this protein in aquatic animals, although an antibody directed against the rabbit MRP does cross-react with a fish MRP-like protein (40). Availability of such probes is essential to determine the role of the MRP in xenobiotic protection, if any, in non-mammalian organisms.

Another efflux transporter that might be relevant to aquatic organisms is LRP (lung resistance-related protein), whose action is implicated in some cases of drug resistance in cancer chemotherapy (41). Again, probes for non-mammalian organisms are not yet available. Finally, there is a widely distributed family of organic anion transporters, and these could similarly be involved in xenobiotic resistance (40). Their relevance to the transport and evasion of xenobiotics has been studied in fish (40), but not extensively in other aquatic organisms.

Protective Role of Pgp in Aquatic Organisms

There is extensive evidence that the Pgp provides a protective function for aquatic organisms similar to that described above for mammals and terrestrial nematodes. One line of evidence comes from studies describing protection from the microtubule depolymerizing drug vinblastine in embryos of *Urechis caupo* (26). As shown in Figure 3, cell division in these embryos is insensitive to vinblastine at the indicated concentrations (up to 0.5 µM).

However, in the presence of vinblastine plus verapamil (a Pgp-inhibiting drug), the embryos no longer can divide, indicating that vinblastine can now accumulate in the cell and prevent cell division (verapamil alone has no affect). A second line of evidence has been carried out by Kurelec and his colleagues looking at protection from DNA damage by the Pgp in mussel gills. To demonstrate how xenobiotics that are good Pgp substrates may competitively inhibit the pumping out of other xenobiotics, mussels (*M. galloprovincialis*)

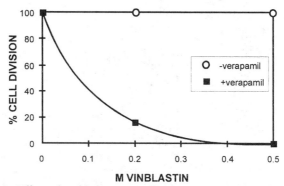

M VINBLASTIN

Figure 3. *Effect of vinblastine on cell division in Urechis (from ref 26). As seen, there is no effect of vinblastine on cell division at the indicated concentrations (in the absence of verapamil), but there is a very strong effect when the Pgp inhibitor verapamil is present. Verapamil alone has no effect on cell division.*

were exposed to radiolabelled vincristine (^{3}H-VCR) in the presence of Diesel-2 oil. The presence of this pollutant enhanced the accumulation of the radioactivity by 3-fold, or to the level equivalent to enhanced accumulation caused by 8.5 µM verapamil (Figure 4A). Similarly, an indirect chemosensitizer of MXR, the PKC inhibitor staurosporine (1 µM), enhanced by 7-fold the accumulation of vincristine (Figure 4A).

In the case of exposure to carcinogenic xenobiotics, such enhanced accumulation could correspondingly enhance mutagenic and carcinogenic effects. This speculation was confirmed by two experiments. First, *Dreissenia polymorpha* mussels exposed to water spiked with 2-aminofluorene (AF) in the presence of 10 µM cyclosporin A (a direct, Pgp-ATPase-inhibitor) enhanced the production of mutagens in exposure medium 460% compared with mussels exposed to AF without cyclosporin A (Fig. 4B). Second, staurosporine (0.5 µM), which can inhibit the Pgp in a fresh water clam, *Corbicula fluminea,* made the clams much more susceptible to acetylaminofluorene (AAF) toxicity. As shown in Figure 4C, single strand breaks in the DNA were now seen at a 10-fold lower concentration of the AAF (0.01 µM) in the presence of the Pgp inhibitor. In summary, the above results indicate that a verapamil-sensitive process, most likely the Pgp transporter, protects these organisms from the detrimental affects of added xenobiotics.

254

There do exist, however, verapamil-sensitive mechanisms to exclude toxicants from cells that might not be related to the Pgp. Mentioned above was the low titer of C219-reactive protein in sea urchin embryos. Yet, these eggs are

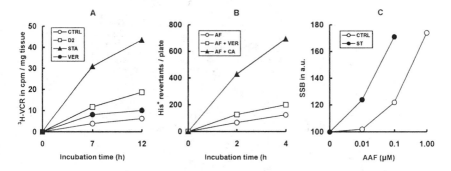

Figure 4. *Protection by the Pgp in various molluscs. Figure 4A depicts the effects of Diesel-2-oil (D2), 1 μM staurosporine (STA), or verapamil (VER) on H³ vincristine accumulation in mussel gill tissue (Mytilus galloprovincialis). Diesel oil is presumably acting as an inhibitor or competitive substrate, and staurosporine is acting as an indirect inhibitor. Figure 4B depicts the effects of the Pgp inhibitor cyclosporine (CA, 10 μM) or verapamil (VER) on 2-aminofluroene induction of mutagens in the mussel Dreissenia polymorpha. Figure 4C compares the effects of 0.5 μM staurosporine (STA) on DNA-strand breaks [SSB, expressed as arbitrary units (a.u.)] induced by 2-acetylaminofluorene (AAF); as seen, the effect is enhanced 10-fold by the presence of this Pgp inhibitor (from ref 6).*

protected from the effects of vinblastine by a verapamil-sensitive mechanism similar to that depicted in Figure 3 for *Urechis* embryos (*42*). As noted above, this could result from the activity of a non-Pgp efflux transporter. Irrespective of the mechanism, these types of experiments indicate that drug exclusion mechanisms can be affected by the presence of other types of drugs, such as verapamil.

Another possibility is that low levels of Pgp seen in the sea urchin embryo might be adequate for protection and that the higher levels seen in mussel gills or *Urechis* embryos (and drug-resistant tumor cells) are an aberration. Thus, we see exclusion of rhodamine dye from *Urechis* or mussel cells at rhodamine levels up to 2.5-5.0 μM. This contrasts with the sea urchin, which excludes dye only up to 0.25 μM, a 10-fold lower level (*42*). Perhaps the small amount of Pgp present is only effective at excluding low levels of xenobiotics. If so, the opposite case is seen in mussels or the mud-dwelling worm embryos, which have high Pgp titer and exclude higher concentrations of drugs. This over-expression may be necessary for more complete protection from xenobiotics for organisms that are potentially exposed to high concentrations of toxicants as with filter-feeders or mud-dwellers

Affinity of the Transporter for Various Substrates

The critical question relevant to this chapter is whether the Pgp can be affected by environmental levels of PPCPs (e.g., see first chapter of this book, by Daughton). Although there is yet no specific information on this question, there are good data on the affinity of the Pgp in aquatic organisms towards various chemicals. This data, from studies on mussel gills and *Urechis* embryos, has been obtained by exposing these cells or tissues to varying concentrations of rhodamine B and looking at accumulation of the fluorescent dye in the cells in the presence and absence of Pgp inhibitors or competitive Pgp substrates (*26, 33*). Figure 5 shows this type of study done with verapamil. As seen, uptake of the fluorescent dye is enhanced by verapamil at rhodamine levels as low as 1 nanomolar.

The enhancement increases with rhodamine concentration, with a peak at 10^{-7} M but enhancement decreases with higher levels and is no longer evident above 2.5 micromolar. These results are also depicted in Figure 6 as ratios of uptake of rhodamine in the presence or absence of verapamil, which more graphically indicates the exclusion of rhodamine in the range between 10^{-9} and 10^{-6} M. Below this range there is no exclusion, and above the micromolar level rhodamine entry is seemingly independent of the efflux transporter. These results then indicate that the transporter helps to exclude this compound at the lower levels (less than micromolar). The apparent ineffectiveness of the transporter at the higher dye levels suggests that the transporter is saturated by rhodamine and that at saturation the transport activity makes no difference — the transporter is working at maximum capacity and cannot effect a reduction in concentration.

Similar types of experiments have been done with other substrates to estimate the affinity of the transporter to various compounds. The uptake of 1 μM rhodamine was measured in the presence of different levels of putative substrates, assuming that these substrates are competing with rhodamine. Toomey and Cornwall and colleagues (*26, 31*) looked for the concentration of competing substrate where the rhodamine accumulation was increased to 50% of the maximum value (maximum value being the accumulation of rhodamine in the presence of verapamil). This data, shown in Table II, indicates a wide range of concentrations over which the transporter can exclude rhodamine in the

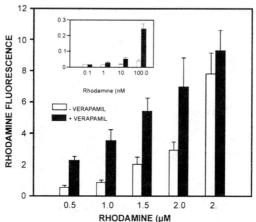

Figure 5. *Concentration dependence of rhodamine accumulation/exclusion from mussel gill tissue in the absence or presence of verapamil (from ref 31). As seen, rhodamine accumulation is enhanced in the presence of verapamil between 1 x 10^{-9} and 2.5 x 10^{-6} M. Below or above this level there is no effect of verapamil. We interpret this finding as indicating that the Pgp is excluding rhodamine accumulation between these concentrations (1 x 10^{-9} and 2.5 x 10^{-6} M). Below 10^{-9} M, the pump apparently is not operative, probably because of affinity for substrate. Above 2.5 x 10^{-6} M, the pump is also ineffective, most likely because the pump is saturated, either because of affinity considerations or because the rate of diffusion into the cell cannot be balanced by the activity of the Pgp pumping out of the cell.*

presence of these various compounds. Some compounds will not affect the exclusion of rhodamine until the concentration is very high (60 μM for acetylaminofluorine) whereas in other cases very low levels of compound sufficed (<1 μM verapamil).

The above affinity measurements, however, are rife with assumptions. For example, this work was carried out before it was realized that there are at least two binding sites for substrates. These measurements, made with rhodamine B, would therefore only assess competition for the R site and not the H site. Some of the higher affinity constants observed (such as for acetylaminofluorene) could therefore relate to these compounds being substrates/modulators of the H site as opposed to the R site. Another concern is that the classification of substances as substrates or inhibitors of the Pgp activity could result from effects on cellular ATP levels or from increases in membrane permeability, which could be consequences of exposure to toxic chemicals.

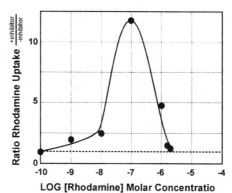

Figure 6. *Ratio of uptake of rhodamine vs rhodamine + verapamil. This recalculation of the data of Figure 4 indicates the limited range over which the Pgp can act. See legend to Figure 5 for possible explanations.*

More extensive studies are needed to determine which compounds might modulate the Pgp of aquatic organisms. Studies on the various PPCPs at their environmental levels are particularly relevant to the questions addressed in this monograph. One question is whether these compounds can affect the transporter at nanomolar levels; another is whether the PPCPs can affect the accumulation of other substrates, especially ones that might be toxic.

A very interesting environmental study carried out by Kurelec and his colleagues looked at the affects of hospital wastewaters on the accumulation of rhodamine in mussel gills (Kurelec and Smital, unpublished data). These studies, shown in Figure 7, indicate that wastewaters near hospitals contain substances that increase rhodamine accumulation, indicating that modulators from this hospital — presumably medically related products such as pharmaceuticals — could augment the uptake of rhodamine into gill cells. These results then clearly indicate the presence of Pgp modulators in these hospital waters. These results also fit with measurements of Pgp modulating substances from landfills, which similarly contain PPCPs. Dichloromethane-methanol extracts of solid waste taken at the municipal landfill near Zagreb (Croatia) revealed significant potential to inhibit the Pgp in NIH 3T3 mouse fibroblasts transfected with the human *MDR*1 gene (*43*).

Table II. Substrates of the drug transporter of
Mytilius and *Urechis*

Pharmacologicals	$(EC_{50}, \mu M)$
Verapamil	0.25
Cyclosporin	0.4
Dideoxyforskolin	0.6
Forskolin	2.2
Quinidine	5.0
Emetine	11
Promethazine	18
Vinblastine	40
2-Acetylaminofluorene	80
Pesticides	
Chlorbenside	~5
Dacthal	~5
Pentachlorophenol	~3
Sulfallate	~3

Source: Unpublished observations of R. Cornwall and B. Toomey

Conclusions and Overview

As we hope this review has shown, the Pgp transporter provides a potent, protective mechanism against xenobiotics in aquatic organisms. This mechanism may be particularly important in the aqueous environment since the cells of aquatic organisms are constantly exposed to whatever xenobiotics are present around them. Their exposure then is not episodic, as would be the case in terrestrial organisms, which are primarily exposed during feeding. Rather, aquatic organisms are literally immersed in a media containing potentially detrimental substances. This would be a particular problem for filter feeders or fish, which are constantly passing large amounts of water through their feeding apparatus or gills. The presence of efflux transporters in such organisms then makes sense, and these transporters may be particularly important to enable such lifestyles.

The critical question is whether the pharmaceuticals and personal care products, typically present at 10^{-9} M levels in fresh water situations, can affect these transport systems. One question relates to the ability of this diverse group of chemicals to act as chemical modulators. If they modulate Pgp at these low

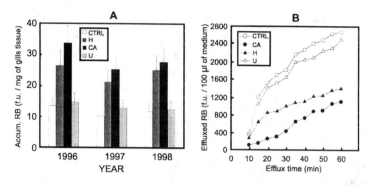

Figure 7. *The effect of water near a hospital sewage site on rhodamine (RB) uptake in gill tissue of M. galloprovincialis (Unpublished data of Kurelec and Smital). **A**. Rhodamine accumulation in gills tested in water samples from the hospital site (H) and compared with water from an unpolluted site (U) and with cyclosporin A (CA, an inhibitor of the Pgp). Samples taken during one of the indicated years. **B**. Efflux of rhodamine from gills in the presence of waters from a hospital site (H), an unpolluted site (U), or in the presence of cyclosporin A (CA). The efflux of the dye is a more direct measure of the transport activity. In control and unpolluted site, the efflux of dye is very rapid (upper two curves). In the presence of a Pgp inhibitor (CA) or water from the hospital site (H), the efflux is considerably reduced; f.u. refers to fluorescence units.*

levels, then they certainly could adversely affect the protective role of the transporters. For example, if these substances bind to the transporters with high affinity, they could allow other xenobiotics which would normally be excluded by the cells to now accumulate in the cytoplasm. If these other xenobiotics are harmful, they could then detrimentally affect the organism since the organism's first line of defense has been compromised. Another possibility is of course that particular PPCPs might themselves be harmful, and their entry into cells of aquatic organisms could then be directly harmful.

Obviously more research is needed to answer these questions. Most important will be a cataloging of the affinities of PPCPs for the multixenobiotic transporters of representative aquatic organisms. Better methods are needed for this. One approach as noted above is measuring the effects of candidate substrates on the uptake of model substrates, such as rhodamine B or Hoechst 33342. This can be easily done with single-cell assays or with more rapid throughput assays *(34)* with whole mussels. Another approach could be the sensitive and rapid assay described by Sarkadi et al. *(44)*, which can be used to characterize the MXR inhibiting potential of environmental samples. This assay measures the property of environmental samples to stimulate (verapamil-like), or to inhibit (cyclosporin A-like) the vanadate sensitive *MDR*1-ATPase activity in membrane vesicles isolated from Sf9 insect cells infected with a recombinant baculovirus carrying the human *MDR*1 gene. Another approach is to look at the effect of environmental samples on the transport of easily assayed substrates in Pgp-expressing mammalian cells in culture.

A problem with the latter two approaches, which use mammalian Pgp, is that these comparisons may not be relevant for aquatic organisms since Pgp primary structure and hence transport properties are not completely conserved. The observed 50-60% homology at the amino acid level in fish, worms, and mussels indicates divergence in structure and presumably function, and this could be reflected in differing substrate preferences. Also, one has to look at both fresh water versus marine environments as the differing properties of these media could also affect the transport functions with different substrates.

Another question is how these transporters are affected by the presence of multiple competing substrates in the environment. The PPCPs could comprise dozens of putative compounds at nanomolar levels in the environment. What is the nature of interaction between all these different chemicals? Will this plethora of substrates pose unique problems for aquatic organisms? The problems of complex mixtures is just beginning to be appreciated in environmental toxicology and recent analyses of this problem using "isobole" approaches might allow dissection of this potentially complex area (45). Alternatively, complex mixtures may pose less significant problems to aquatic organisms that live in organic or chemically enriched environments that have naturally elevated levels of Pgps. The numerous plants, animals, and bacteria that co-exist in this environment might all be producing xenobiotics, and the efflux transporters might therefore be pre-adapted to handle the resultant multiple chemicals simultaneously. The PPCPs might just be an added but unimportant burden to these pre-adapted organisms.

Additionally and importantly, there are certainly alternate mechanisms than the Pgp transporters for excluding compounds from cells. We referred above to the MRP, LRP, and organic anion transporters. These might also be present in aquatic organisms and need to be similarly studied in the future so as to ascertain the consequences of multiple chemicals in the environment.

A final note of skepticism but also concern needs to be made about the consequences of the nanomolar levels of these PPCPs. The skepticism is whether such low levels, even if they could affect the Pgp transporters, would affect the target organisms. The concern is that there indeed could be effects, but not the sort that stand out in short-term toxicity studies. Rather, there could be subtle effects of these compounds, effects that can only be detected in long-term studies on individual species or even on communities. This type of determination will be a true challenge for the future. The design of reporter systems in test organisms to assess these low level effectors is urgently needed.

Dedication and Acknowledgments: This article is dedicated to the memory of Dr. Branko Kurelec, a great scientist and pioneer in ecotoxicology who spent his last years building a scientific basis for the role of the efflux transporters as a first line of defense against xenobiotics in the aquatic environment.

The work by DE was funded in part by the National Sea Grant Program, National Oceanic and Atmospheric Administration, US Department of Commerce under project number R/CZ-143 through the California Sea Grant College. The views expressed herein are those of the authors and do not necessarily reflect the views of NOAA or any of its subagencies. The US government is authorized to reproduce and distribute for governmental purposes. Work by TS was supported by the Ministry of Science and Technology of the Republic of Croatia, Project No 00981510.

References

1. Livingstone, D.R. *Biochem. Soc. Trans.* **1990**, *18*, 15-19.
2. Commandeur, J.N.M.; Stijntjes, G.N.; Vermeulen, N.P.E. *Pharmacol. Rev.* **1995**, *47*, 271-330.
3. Sijm, D.T.H.M.; Opperhuizen, A. In *Handbook of Environmental Chemistry;* Hutzinger, O; Springer: Berlin, Germany, 1989, 2E, pp 163-235.
4. Timbrell, J.A. *Principles of Biochemical Toxicology;* 2nd ed; Taylor and Francis: London, 1991.
5. Goksøyr, A.; Förlin, L. *Aquat. Toxicol.* **1992**, *22*, 287-312.
6. Kurelec, B. *Crit. Rev. Toxicol.* **1992**, *22*, 23-43.
7. Epel, D. *Comp. Biochem. Physiol. A* **1998**, *120*, 23-28.
8. Bard, S.M., *Aquat. Toxicol.* **2000**, *48*, 357-389.
9. Bosch, I; Croop, J. M. *Cytotechnology*, **1998**, *27*, 1-30.
10. Gottesman, M.M.; Pastan, I. *Annu. Rev. Biochem.* **1993**, *62*, 385-427.
11. Endicott, J.A.; Ling, V. *Annu. Rev. Biochem.* **1989**, *58*, 137-171.
12. Higgins, C.F. *Cell* **1994**, *79*, 393-395.
13. van Ween, H.W.; Callaghan, R.; Soceneantu, L.; Sardini, A.; Konings, W.N.; Higgins, C.F. *Nature* **1998**, *391*, 291-295.
14. Broeks, A.; Gerrard, B.; Allikmets, R.; Dean, M.; Plasterk, R.H.A. *EMBO J.* **1996**, *15*, 6132-6143.
15. Borst, P.; Schinkel, A.H. *Eur. J. Cancer* **1996**, *32A*, 985-990.

16. Randolph, G.J., Beaulieu, S., Pope, M. Sugawara, I., Hoffman, L., Steinman, R.M., Muller, W.A. *Proc. Natl. Acad. Sci., USA* **1998**, *95*, 6924-6929.
17. Luker, G.D., Nilsson, K.R., Covey, D.F., Piwnica-Worms, D. *J. Biol. Chem.,* **1999**, *274*, 6979-6991.
18. Pearce, H.L., Safa, A.R., Bach, N.J., Winter, M. A., Cirtain, M. C., Beck, W. T. *Proc. Nat. Acad. Sci. USA* **1989**, *86*, 5128-5132.
19. Shapiro, A.; Ling, V. *Eur. J. Biochem.* **1998**, *254*, 181-188.
20. Shapiro, A.B., Fox, K; Ping, L.; Ling, V. *Eur. J. Biochem.* **1999**, *259*, 841-850.
21. Tsuruo, T.; Iida, H.; Sukagoshi, S.; Sakurai, Y. *Cancer Res.* **1981**, *41*, 1967-1972.
22. Rao, U.S.; Scarborough, G.A. *Mol. Pharmacol.* **1994**, *45*, 773-776.
23. Yu, G.; Ahmad, D.; Aquino, A.; Fairchild, C.R.; Trepel, J.B.; Ohno, S.; Suzuki, K.; Tsuruo, T.; Cowan, K.H.; Glaser, R.I. *Cancer Commun.* **1991**, *3*, 181-189.
24. Minier, C.; Eufemia, N.; Epel, D. *Biomarkers* **1999**, *4*, 442-454.
25. Kurelec, B.; Pivcevic , B. *Biochem. Biophys. Res. Comm.* **1989**, *164*, 934-960.
26. Toomey, B.H.; Epel, D. *Biol. Bull.* **1993**, *98*, 197-203.
27. Eufemia, N., Epel, D. *Aquatic Toxicol.* **2000**, *49*, 89-100.
28. Minier, C.; Akcha, F.; Galgani, F. *Comp. Biochem. Physiol. C* **1993**, *106*, 1029-1036.
29. Minier, C.; Moore, M.N. *Mar. Ecol. Prog. Ser.* **1996**, *142*, 165-173.
30. Galgani, F.; Cornwall, R.; Toomey, B.H.; Epel, D. *Environ. Toxicol. Chem.* **1995**, *15*, 325-331.
31. Cornwall, R.; Toomey, B.H.; Bard, S.; Bacon, C.; Jarman, W.M.; Epel, D. *Aquatic. Toxicol.* **1995**, *31*, 277-296.
32. Bonfanti, P.; Colombo, A.; Camatini, M. *Chemosphere* **1998**, *37*, 2751-2760.
33. Köhler, A.; Lauritzen, B.; Jansen, D.; Böttcher, P.; Teguliwa, L.; Krüner, G.; Broeg, K. *Mar. Environ. Res.* **1998**, *46*, 411-414.
34. Smital, T. , Kurelec, B. *Environ. Toxicol. Chem.* **1997**, *16*, 2164-2170.
35. Amino Acid Sequence Accession # AAD42898
36. Amino Acid Sequence Accession # AAA79094
37. Chan, K.M.; Davies, P.L.; Childs, S.; Veinot, L.; Ling, V. *Biochim. Biophys. Acta.* **1992**, *1171*, 65-72
38. Cole, S.P.C.; Bhardway, G.; Gerlach, J.H.; Mackie, J.E.; Grant, C.E.; Almquist, K.C.; Stewart, A.J.; Kurz, E.U.; Duncan, A.M.V.; Deely, R.G. *Science* **1992**, *258*, 1650-1654.
39. Jedlitschky, G.; Leier, I.; Buchholz, U.; Center, M.; Kepler, D. *Cancer Res.* **1994**, *54*, 4833-4836.

40. Miller, D.S.; Masereeuw, R.; Henson, J.; Karnaky, K.J. *Am. J. Physiol.* **1998**, *275,* R697-R705.
41. Shepper, R.J.; Broxterman, G.L.; Scheffer, G.L.; Kaijk, P.; Dalton, W.S.; van Heijnengen, H.M.; van Kalken, C.K.; Slovak, M.V.; de Vries, E.G.E.; van der Kalk, P.; Meier, C.J.C.M.; Pinedo, H.M. *Cancer Res.* **1993**, *53,* 1475-147
42. Eufemia, N.; Narayanan, D; Xing, J; Epel, D. **2000**, Stanford University, unpublished results
43. Ahel, M.; Mikac, N.; Æosoviæ, B.; Prohiæ, E.; Soukup, V, *Wat. Sci. Technol.* **1998**, *37,* 203-210.
44. Sarkadi, B.; Price, E.M.; Bouchers, R.C.; Germann, U.A.; Scarborough, G.A. *J. Biol. Chem.* **1992**, *267,* 4854-4858.
45. Broderius, S.J.; Kahl, M.D.; Hoglund, M.D. *Environ. Toxicol. Chem.* **1995**, *14,* 1591-1605.
46. Waldmann, P.; Piovcevic, B.; *Biochem. Biophys. Res. Comm.* **1989**, *164,* 934-960.

Chapter 15

Antidepressants in Aquatic Organisms: A Wide Range of Effects

Peter P. Fong

Department of Biology, Gettysburg College, 300 North Washington Street, Gettysburg, PA 17325

Selective serotonin reuptake inhibitors (SSRIs) and tricyclics are commonly prescribed antidepressants in the developed world. The increasing use of such antidepressants coupled with their possible release in municipal sewage effluent creates the potential for physiological and behavioral modification of aquatic organisms. Antidepressants often act by modulating and mimicking the effect of the neurotransmitter serotonin. Since serotonin regulates a wide range of physiological systems, including vasoconstriction in trout, retinomotor activity in sunfishes, larval metamorphosis in snails, reproduction in a variety of molluscs, and ciliary beating in protozoans, drugs that mimic its action could have salient effects on these and other organisms. Laboratory data have shown that several groups of aquatic organisms respond physiologically to applied antidepressants. This review summarizes our current knowledge of the effects of SSRIs, tricyclic antidepressants, and serotonin on the physiology and behavior of aquatic organisms.

Introduction

Clinical depression affects more than 18 million Americans and one in eight needs treatment for depression during his or her lifetime. Many of these individuals take prescription antidepressants to treat their condition. Commonly prescribed antidepressants are of three classes: Monoamine oxidase inhibitors (MAOIs), tricyclic antidepressants, and selective serotonin reuptake inhibitors (SSRIs). These three classes include several popular drugs indicated for a number of conditions besides depression, including obsessive-compulsive disorder, panic disorder, social phobia, and attention deficit disorder. Of these three classes, tricyclics and SSRIs work by blocking reuptake of neurotransmitters, especially norepinephrine, dopamine, and serotonin. Tricyclics were introduced in the 1950s, and SSRIs were developed and first marketed in the 1980's. Since then, the latter have become extremely popular; for example, since its introduction in 1986, fluoxetine (Prozac) has been prescribed for over 34 million people in more than 100 countries, including more than 17 million people in the United States alone. A newer SSRI, fluvoxamine (Luvox) is currently prescribed for over 12 million people worldwide.

Clinical depression can be caused by low levels of the neurotransmitter serotonin (5-hydroxytryptamine; 5-HT) in the brain (*1*). Women, who may experience higher incidence of depression than men, have a lower rate of 5-HT synthesis than men (*1*). 5-HT reuptake transport proteins return released 5-HT back to the presynaptic terminal where the transmitter is repackaged into synaptic vesicles. The SSRIs increase 5-HT neurotransmission by blocking these reuptake transporters (*2*). Since 5-HT itself does not cross the blood-brain barrier, oral administration of 5-HT is not a practical treatment for clinical depression. However, SSRIs do cross the blood-brain barrier. Moreover, since their mechanism of action increases 5-HT neurotransmission, application of SSRIs can also mimic the action of 5-HT.

Tricyclics have a longer history than SSRIs, and even though not as popular today as SSRIs, some, such as imipramine (Tofranil), amitriptyline (Elavil), desipramine (Norpramine), and clomipramine (Anafranil), still enjoy wide use for a variety of disorders including panic disorder, social phobia, and attention deficit disorder. The current human medical interests in SSRIs are in the treatment of not only depression, but also for obsessive compulsive behaviors associated with Tourette's Syndrome (*3*), obesity in non-insulin dependent diabetics (*4*), and convulsive seizures (*5*).

Since 5-HT is an ancient and ubiquitous biochemical found in all phyla of animals thus far examined (*6*) as well as plants such as bananas (*7*), our knowledge of the physiological systems serotonin acts upon or modulates is quite extensive. Moreover, since both tricyclics and SSRIs can act to mimic the action of 5-HT by blocking its reuptake, a number of recent studies have utilized these drugs to clarify the nuances of 5-HT neurotransmission in the nervous systems of several animal taxa (see *8* for a review of serotonin modulation of behavior in a number of taxonomic groups).

We have no data on the quantities of antidepressants released into the environment or their resulting concentrations (*9*). However, there has been a recent dramatic surge in the number of office visits for treatment of depression and a corresponding increase in the number of prescriptions for antidepressants in the U.S. (*10*). Moreover, recent studies (*9*) have shown that even though concentrations of some drugs detected in the effluent of municipal water treatment plants are probably too low to affect humans, aquatic organisms are likely to be impacted. Furthermore, laboratory experiments applying antidepressant drugs to aquatic organisms have shown that they can have neurohormone-like effects (e.g., stimulating reproductive processes) akin to those elicited by environmental estrogen mimics. Since antidepressants mimic the action of the biogenic monoamine serotonin in a number of species, this review will attempt to summarize our current knowledge of effects of serotonin, and both tricyclic antidepressants and SSRIs, on the physiology and behavior of aquatic animals.

Effects of 5-HT and SSRIs on fishes, amphibians, and aquatic mammals

In humans, 5-HT mediates or modulates a number of activities in addition to depression including sleep, appetite, arousal, and anxiety (*11*). It is a classic vasoconstrictor in mammals (*12*). Although there are a fair number of aquatic mammals and a large number of amphibians, fishes are the most speciose and numerous aquatic vertebrates, and thus it is no surprise that 5-HT controls several piscine physiological systems.

In terms of reproduction, serotonin inhibits steroid-induced meiotic maturation of ovarian follicles in the common estuarine mummichog, *Fundulus heteroclitus* (*13, 14*). Specific blockers of serotonin receptors induced meiotic maturation indicating that the response is mediated by a serotonin receptor (*15*). During the spawning season, the mummichog is known to spawn every two weeks during the new and full moons. During ovulation and spawning, levels of pituitary serotonin decrease and levels of hypothalamic serotonin increase, and therefore levels of brain serotonin are critical for the fish to coordinate its biweekly spawning pattern.

By contrast, in goldfishes, serotonin injections produce an increase in serum gonadotropin in male and female pre-spawn fish, and fluoxetine potentiates the serotonin-induced increase (*16*). Furthermore, in the Atlantic croaker, concomitant administration of serotonin and luteinizing hormone-releasing hormone (LHRH) elevated the level of gonadotropin higher than that by LHRH alone. Fluoxetine alone slightly enhanced the effect of serotonin, but the elevation was completely blocked by the serotonin receptor antagonist ketanserin (*17*).

In salmonids, serotonin affects or modulates a number of physiological systems. For example, in trout, serotonin has vasoconstrictive effects on the vasculature of the gill filaments (*18*). This vasoconstriction can be blocked by the serotonin receptor antagonist methysergide. Furthermore, serotonin causes a decrease in blood pressure in trout dorsal aorta and a corresponding decrease in arterial oxygen tension (*19*). In salmon, diets deficient in tryptophan (the precursor amino acid to serotonin) cause a spine deformation known as scoliosis. Administration of 100 mg of 5-hydroxytryptophan (another serotonin precursor, not to be confused with 5-HT) per 100 g diet reduced the incidence of scoliosis from 65 to 1.7 % (*20*). Furthermore, oral administration of serotonin to young salmon reduced the incidence of scoliosis (*21*).

Serotonin is present in the retinas of both teleosts and cartilaginous fishes, and is apparently involved in retinomotor movements by modulating the release of another monoamine, dopamine. Dearry and Burnside (*22*) reported that serotonin and GABA affect cone retinomotor movements by affecting the release of dopamine, serotonin by stimulation, and GABA by inhibition of dopamine release in the green sunfish. In the goldfish retina, tritiated paroxetine binds to isolated retinal cells, and in skate's retinas, both amacrine and bipolar cells stain immunopositive for serotonin and compete for extracellular serotonin by using distinct uptake mechanisms (*23*). Zimelidine, one type of SSRI, blocked serotonin uptake by bipolar cells but had no effect on amacrine cells, whereas fluoxetine failed to block serotonin uptake into bipolar cells and in some cases increased uptake (*24*).

In amphibians, injections of 5-hydroxytytryptophan (a serotonin precursor) stimulates the dispersion of melanin in dermal melanophores of red-spotted newts (*Notophtalmus viridescens*). Inhibitors of serotonin synthesis blocked melanin dispersal (*25*). Serotonin blocks progesterone-induced oocyte maturation in *Bufo viridis* and *Xenopus laevis*, and serotonin antagonists trigger maturation (*26*). In *Xenopus* oocytes, serotonin induces an inward ion current through channels, and this current can be blocked by the tricyclic antidepressant imipramine, which the authors suggest block 5-HT_{1C} receptors.

Finally, there is very little information of effects of 5-HT or antidepressants in aquatic mammals. However, some studies have shown serotonin to be important in inhibiting predatory aggression in captive and wild mammals, including a semi-aquatic mammal, the mink. Studies (*27, 28*) showed that serotonin precursors increased levels of 5-HT in the mink nervous system with a concomitant inhibition of predatory attacks on rats. Furthermore, a similar predatory reduction response has been recorded in rats, mice, minks, and silver foxes when brain serotonin levels are increased, and it is suggested that serotonin represents a dietary response factor that regulates predatory behavior in carnivores.

Effects of 5-HT and antidepressants on aquatic invertebrates

Among aquatic invertebrates, 5-HT has wide ranging effects from induction of spawning and oocyte maturation in clams and mussels (*29, 30, 31*), mediation of aggression in crayfish (*32*), regeneration of cilia in protozoans (*33*), induction of rhythmic contractions in coelenterates (*34*), induction of negative phototaxis in bryozoan larvae (*35*), and induction of metamorphosis in hydrozoan larvae (*36*).

SSRIs mimic the action of 5-HT in a number of aquatic species (molluscs, crustaceans). In others (some crustaceans), they antagonize the 5-HT-induced action. The mode of action of SSRIs such as fluoxetine in aquatic invertebrates may be via inhibition of serotonin reuptake transporters (for example in lobsters, *37*), or they can bind directly to 5-HT (and other) receptors (for example in frog embryos; *38, 39, 40*). Our present knowledge of the effects of antidepressants on physiology and behavior of aquatic invertebrates come from work on mainly molluscs and crustaceans since they are economically and ecologically important, and because they are model organisms for the study of physiological processes in invertebrates.

Molluscs: Bivalves

Because it is a serious biofouling pest in fresh water, the zebra mussel, *Dreissena polymorpha* has been the focus of a number of studies on the regulatory mechanisms of oocyte maturation and spawning. Serotonin is an internal regulator of oocyte maturation and germinal vesicle breakdown in *D. polymorpha* and a number of other bivalve species (*41, 42, 43*). Spawning as well is under serotonergic control, and application of exogenous serotonin and compounds that bind to serotonin receptors results in spawning of both males and female zebra mussels and other bivalve species (*29, 43, 30*). In zebra mussels, application of the SSRIs Prozac (fluoxetine) and Luvox (fluvoxamine) results in spawning of both sexes (*44*). Fluvoxamine is particularly potent, inducing spawning in 100% of both sexes at 10^{-5} and 10^{-6} M. Spawning was significantly induced in males at concentrations as low as 10^{-9} M and in females at 10^{-7} M. The lowest concentration of fluvoxamine to induce spawning was 10^{-8} M for females (40%) and 10^{-10} M for males (20%). Gametes spawned in fluvoxamine (10^{-5} M and lower) were viable and swimming trochophore larvae were formed within 20 hours. Fluoxetine was also an effective spawning inducer, causing 100% of males to spawn at 5×10^{-6} M. Fluoxetine significantly induced spawning in males at concentrations as low as 5×10^{-8} M and in females at 5×10^{-6} M. Interestingly, both of these SSRIs induce spawning at concentrations that are orders of magnitude lower than serotonin itself, and at high concentrations of fluoxetine (10^{-4}-10^{-3} M), there is a sharp reduction in spawning percentage indicating a potential toxic effect. Paroxetine (Paxil) was less potent at inducing spawning and only did so in males, but its effectiveness was reduced at high concentration (10^{-4} M).

As SSRIs induce spawning in zebra mussels, tricyclic antidepressants block it. Hardege et al. (*45*) found that the tricyclics imipramine and desipramine significantly blocked 5-HT-induced spawning in males, but not in females. However, clomipramine blocked spawning in both sexes.

The observation of SSRI induction of reproductive processes was also shown by Fong et al. (*46*) who studied the release of juveniles by fresh water fingernail clams. Fluoxetine and paroxetine mimicked the action of 5-HT by inducing release of juvenile *Sphaerium striatiunum*. Fluoxetine potentiated 5-HT-induced juvenile release. However, similar to the observations of Hardege et al. (*45*), 5-HT-induced release of juveniles was inhibited by simultaneous application of desipramine, clomipramine, or imipramine. In further support of the inhibitory actions of tricyclics on bivalve reproductive processes, Juneja et al. (*47, 48*) found that imipramine, clomipramine, and desipramine all inhibited 5-HT-induced germinal vesicle breakdown and calcium uptake by oocytes of the marine clam *Spisula solidissima*. It should be noted that in the case of both serotonin and SSRI-induced parturition in *Sphaerium* spp., it is very common for immature juveniles to be released during parturition as well as larger, mature juveniles.

In the common sea mussel *Mytilus galloprovincialis*, 5-HT (as well as other neuroactive compounds) acts as an inducer of larval metamorphosis from the pediveliger (infant mussel, fraction of a millimeter long with gills and foot) to the juvenile stage (*49*). Interestingly, these authors also found that yohimbine, a compound that binds to both adrenergic and serotonergic receptors also induces larval metamorphosis in *Mytilus*. Yohimbine is a bioactive ingredient in over-the-counter herbal preparations, some of which have been used for hundreds of years as a purported treatment for impotence.

In the freshwater unionidae, a group of bivalves that is extremely diverse in eastern North America and to which a number of endangered species belong, serotonin has salient effects on at least two physiological systems. Serotonin induces release of glochidia larvae in *Utterbackia imbecillis* and *Pygandon cataracta* (*50*). In *Ligumia subrostrata*, serotonin regulates sodium influx across gills, and the response is mediated by cyclic AMP (*51*).

Finally, in the marine bivalve *Macoma balthica*, fluoxetine was used to stimulate spawning of both male and female clams (*52*). These clams were stimulated to spawn even though the population itself was well past the normal window of reproductive time. This indicates that fluoxetine is particularly potent, yet not toxic since viable gametes and larvae were produced from these spawnings.

Molluscs: gastropods

Serotonin regulates a wide range of physiological responses in gastropods. There is a large and varied literature on detection and localization of serotonin, as well as effects, especially in opisthobranchs, for example *Aplysia* (*53, 54*).

Serotonin stimulates ciliary beating in cells of several animal taxa from frogs (*55*) and echinoderms (*56*) to molluscs, including gastropods (*57, 58*). In embryos of the freshwater gastropod *Helisoma trivolis*, serotonin stimulates ciliary activity and thus also the rotation of embryos within egg masses. Endogenous serotonin is released from serotonergic neurons that directly activate ciliated cells, stimulating their cilia to beat and causing embryo rotation (58, 59). I have observed greatly accelerated rotation of encapsulated veliger larvae of the marine opisthobranch (*Phidiana crassicornis*) after addition of serotonin (personal observation). This serotonin-induced cilia-driven rotational behavior in gastropod embryos can be mimicked by SSRIs. Uhler et al. (60) showed that the SSRI fluvoxamine elicited a significant dose-dependent (10^{-6} M to 10^{-4} M) increase in embryo rotation as that seen with 5-HT. Furthermore, paroxetine and fluoxetine both caused an increase in rotation at 10^{-6} M and 10^{-5} M, but reduced rotation rate below that of baseline at 10^{-4} M, indicating a possible toxic effect at higher concentrations.

Serotonin also appears to play a role as an inducer of larval metamorphosis in marine gastropods. In the nudibranch *Hermissenda crassicornis*, a variety of neurotransmitter substances can induce metamorphosis, including serotonin (61). Larvae of the mud snail *Ilyanassa obsoleta* metamorphose with either bath-applied or injected serotonin (62). Metamorphosis can also be induced by injected fluoxetine (10^{-6} M), and the percentage of metamorphosed larvae is not significantly different from those induced by bath-applied serotonin. Furthermore, fluoxetine potentiates low concentrations of serotonin to produce an additive effect (62).

Arthropods: crustaceans

The effects of serotonin and other neurotransmitters on behaviors in crustaceans are well known. Crayfishes and lobsters are model organisms for the study of crustacean neuroendocrinology, and a fairly large literature exists for these groups (63). Moreover, the effects of SSRIs on crayfish reproductive processes have been investigated. Serotonin was detected in crustaceans over 40 years ago (64). Several behaviors in crustaceans, especially crayfishes and lobsters, have been shown to be regulated by serotonin. Aggressive postures characterized by rigid flexion of the legs and abdomen with chelipeds open and facing forward can be induced by hemocoelomic injection of serotonin in both crayfishes and lobsters. Injection of octopamine causes the opposite responses, consisting of extension of both the chelipeds and tail fan in a subordinate fashion (32). Subsequently, fluoxetine was tested in order to tease apart the mechanisms by which serotonin exerts its action. Interestingly, acute administration of fluoxetine does not mimic the action of serotonin. This result mirrors that which has been found in humans in that the antidepressant action of fluoxetine requires prolonged administration. Fluoxetine, in combination with serotonin reduces the aggressiveness achieved by application of serotonin alone. Thus, serotonin reuptake is important in modulating the effects of serotonin (65, 66).

Serotonin also has salient effects on reproductive processes in crayfishes by stimulating the release of ovary-stimulating hormone (and see 67 for a review of biogenic amine detection and functions in crustaceans). Kulkarni et al. (*68*) found that injected serotonin increased the size of ovaries and oocytes in the crayfish *Procambarus clarkii* compared with controls. Moreover, the compound fenfluramine, which has some SSRI-like characteristics, and fluoxetine both mimicked and potentiated the action of serotonin, in all cases increasing the size of ovaries and oocytes compared with controls. Similarly, in the fiddler crab, *Uca pugilator*, ovarian development can be stimulated by serotonin (*69*). As in crayfishes, the SSRIs fenfluramine and fluoxetine both mimicked and potentiated the action of serotonin by significantly increasing the size of ovaries and oocytes compared with controls (*70*). Furthermore, in the case of fiddler crabs, testicular development can be stimulated by serotonin, fenfluramine, and fluoxetine (*71*).

Serotonin and SSRIs also modulate the dispersion of red pigment in shrimp and fiddler crabs. In the both of these crustaceans, pigmentation is directly under control of two neurohormones — red pigment concentrating hormone and red pigment dispersing hormone (RPDH). In the shrimp, *Macrobrachium ohione*, serotonin elicits pigment dispersion by triggering the release of RPDH. Fenfluramine and fluoxetine mimic the serotonin-induced effect (*72*). Serotonin acts in a similar fashion in fiddler crabs, and the SSRIs fluvoxamine and fluoxetine not only enhance pigment dispersion, but they slow pigment aggregation induced by dopamine (*73*).

Since serotonin mediates a variety of behaviors in crustaceans, it is no surprise that parasites of some crustaceans can alter the distribution of serotonin in the nervous system and therefore alter the behavior of the crustacean host, enhancing transmission of the parasite. Such is the case for the acanthocephalan worm *Polymorphus paradoxus* that parasitizes the common freshwater amphipod *Gammarus lacustris*. Amphipods infected by the larval stages of *P. paradoxus* show abnormal behavior of swimming to the water surface and clinging to an object. Amphipods injected with serotonin show a similar abnormal behavior. This behavior makes the amphipods more likely to be consumed by ducks, the definitive host of the parasitic worm. This serotonin-regulated behavior is probably related to the normal clinging behavior of males during precopulatory grasping of females (*74*).

Other aquatic invertebrates

Although most of the work on serotonin and SSRIs in aquatic invertebrates has focused on molluscs and arthropods, other important groups of invertebrates have also been studied. 5-HT has been detected in the larvae and embryos of starfishes (*75, 76*). In starfish oocytes, serotonin potentiates the maturation-inducing effect of the hormone 1-methyladenine, and serotonin antagonists block maturation of oocytes (*26, 77*). Thus the role of serotonin in starfish oocytes appears to be opposite to that in amphibian oocytes. In sea urchin embryos, 5-HT increases the speed of forward swimming possibly by increasing cililary activity (56). On the other hand, a long-lasting increase in muscular activity of sea urchin larvae is also induced by low (10^{-7} M) concentrations of serotonin (78). The action of serotonin on muscular activity in sea urchin larvae may be regulated by second messenger systems involving cyclic AMP and calcium. Serotonin antagonists increase intracellular free calcium in growing embryos of sea urchin larvae (79). Of critical importance could be the role of serotonin in the regulation of cell divisions in early embryonic stages of sea urchins. Serotonin is present in sea urchin oocytes and blockers of serotonin receptors (gramine and metergoline) cause a significant delay in cell divisions (80). The observed delay can be prevented by application of serotonin or serotonin precursors. As in serotonin-induced muscular activity, calcium and cyclic AMP appear to be involved in the signal transduction pathway.

A variety of common aquatic worms are strongly affected by serotonin. In the nematode *Caenorhabditis elegans*, serotonin mediates egg laying behavior. Egg laying is a random process; worms fluctuate between active egg laying periods and an inactive non-laying period. The transitional switch is regulated by serotonin released from two pairs of motor neurons, and individual egg laying events are triggered by acetylcholine, also released by these neurons (81).

In leeches, feeding and swimming behaviors are regulated by serotonin. In the medicinal leech *Hirudo medicinalis*, biting, salivation, and contraction of pharyngeal muscles are all controlled by serotonin (82, 83). Bath-applied serotonin causes leeches to bite more frequently and ingest more blood than control leeches (84). Application of reserpine (a serotonin depleter) inhibits *Hirudo* from biting for 4 months (85) and decreases food intake in a carnivorous leech *Haemopis marmorata* (86). Swimming in leeches is also regulated by serotonin in the ventral nerve cord. Medicinal leeches swim spontaneously upon exposure to serotonin (87, 88).

The oligochaete *Lumbricus terrestris* has a circadian periodicity of locomotion. The serotonin precursor 5-hydroxytryptophan decreases locomotor activity and does so with circadian rhythm. Fluoxetine mimics the action of 5-hydroxytryptophan (89). Earthworms are not aquatic, but there are a number of aquatic oligochaete species that could be similarly affected.

In protozoans, serotonin stimulates phagocytosis (90), and it regenerates cilia in *Tetrahymena thermophila* (33, 91).

Discussion

Both serotonin and SSRIs, which can mimic the action of serotonin, have wide-ranging effects on the physiology and behavior of aquatic organisms (Table I). It is difficult to determine if any of the physiological systems mentioned above would be negatively impacted by concentrations of antidepressants released into the environment. However, based on laboratory experiments, some harmful effects could be expected. These include increased vasoconstriction of gill blood vessels and subsequent reduction in blood pressure in rainbow trout, important sport fish known to require clean, highly oxygenated streams. Stimulation of reproductive processes in fishes, bivalves, and crayfishes could have a negative impact at the population level if environmental conditions (temperature, food supply, abundance of predators) for gametes, larvae, and juvenile stages are not optimal, or if only one gender of a species is affected when the antidepressant-induced stimulation occurs. There is laboratory evidence for induction of premature births of fingernail clams exposed to serotonin and SSRIs (personal observation). Furthermore, in the bivalve *Macoma balthica*, since fluoxetine can induce out-of-season release of gametes that would normally be resorbed, the adults would lose energy, and the larvae from such fertilizations may have insufficient planktonic food for proper development. The induction of release of glochidia larvae in the freshwater unionids by SSRIs could have significant ecological effects. Normally, unionids release their glochidia when fishes (e.g., sunfishes) or other aquatic vertebrates pass over them. The glochidia then attach to their vertebrate hosts, becoming parasitic, and also hitching a ride as a means of dispersal before eventually dropping off.

On the other hand, tricyclic antidepressants exert an inhibitory effect on spawning in freshwater bivalve molluscs and this too could have negative effects on maintenance of their populations. The unionids are a group of bivalves which have several endangered species. Glochidial release induced by pharmaceuticals when appropriate hosts are not present could reduce their numbers and add to the already serious decline of many populations of native freshwater bivalves in North America (*92, 93*).

Thus, it is possible that water-borne antidepressants could be taken up by aquatic organisms through gills, tentacles, or the general body surface, enter the blood stream, and then have neurohormone-like effects similar to those caused by environmental estrogen mimics (*94, 95*).

Another critical and yet potentially unseen effect of antidepressants in the aquatic environment could be their modulation of ciliary activity at all levels of the food chain. 5-HT stimulates ciliary activity and cilia regeneration in a number of organisms (*60, 91, 96*). Most aquatic animals have organs or tissues containing ciliated cells. With the exception of crustaceans, many aquatic invertebrate larval stages are heavily ciliated. Thus, exposure to serotonin-mimicking drugs could seriously affect locomotion of larvae (*56*). There are also countless numbers of ciliated protozoans in the benthos and plankton of aquatic systems. These protozoans play an important role as producers, consumers, and decomposers in the food chain. Given that serotonin and SSRIs increase activity of ciliated cells and the large number of ciliated protozoans in aquatic systems, release of SSRIs could have marked effects on these populations and on the food chain. Furthermore, since ciliated surfaces of adult

Table I. Effects of serotonin (5-HT), tricyclic antidepressants, and SSRIs on selected aquatic organisms.

Animal	Drug	Selected Effects	References
Fishes			
Mummachog	5-HT	Inhibits steroid-induced maturation of ovarian follicles	13, 14
Goldfish	5-HT	Stimulates serum gonadotropin in both sexes	16
	fluoxetine	Potentiates the 5-HT-induced increase in serum gonadotropin	
Atlantic Croaker	5-HT	Elevates level of gonadotropin when given together with luteinizing hormone	17
	fluoxetine	Enhances 5-HT-induced gonadotropin increase	
Rainbow Trout	5-HT	Constricts blood vessels of gill filaments; decreases pressure in dorsal aorta	18, 19
Salmon	5-HT	Reduces incidence of scoliosis	
Amphibians			
Frogs, toads	5-HT	Blocks progesterone-induced oocyte maturation	26
Mammals			
carnivores	5-HT	Reduces predatory activity in minks	27, 28
Molluscs			
Bivalves			
Clams, mussels	5-HT	Induces oocyte maturation and spawning (many species); release of juveniles	30, 41, 42, 43
Mussels (*Mytilus* sp.)	5-HT	Induces metamorphosis of larvae	49
Zebra mussels (*Dreissena*)	fluoxetine, fluvoxamine, paroxetine,	Induces spawning	44
	tricyclics (imipramine, desipramine, clomipramine)	Blocks 5-HT-induced spawning	45

Table I. *Continued*

Fingernail clams (*Sphaerium*)	fluvoxamine, paroxetine, fluoxetine	Induces release of juveniles	*46*
		Potentiates subthreshold concentrations of 5-HT	
	tricyclics	Blocks 5-HT-induced juvenile release	personal observation
Unionids	5-HT	Regulates sodium influx; induces release of glochidia	*50*
Surf Clams (*Spisula*)	tricyclics	Blocks 5-HT-induced oocyte maturation and uptake of calcium into oocytes	*47, 48*
Baltic Clam (*Macoma*)	fluoxetine	Induces spawning	*52*
Gastropods			
Helisoma	5-HT	Stimulates ciliary activity in embryos	*58, 59*
Biomphalaria	5-HT, fluoxetine, fluvoxamine, paroxetine,	Stimulates ciliary activity in embryos	*60*
Mud snail (*Ilyanassa*)	5-HT, fluoxetine	Induces metamorphosis of larvae	62
Crustaceans			
Crayfish	5-HT	Induces dominant postures; stimulates release of ovary-stimulating hormone	*32, 65* *66, 68*
	fluoxetine	Reduces 5-HT-induced dominant postures; stimulates release of ovary-stimulating hormone	
Fiddler crab	5-HT	Stimulates ovarian and testicular development	*69, 70,* *71, 73*
	fluoxetine	Stimulates dispersion of red pigment and slows dopamine-induced aggregation of red pigment in chromatophores	
	fenfluramine		
Shrimp	fluoxetine, fenfluramine	Stimulates dispersion of red pigment in chromatophores	*72*
Echinoderms			
Sea urchins	5-HT	5-HT antagonists increase intracellular free calcium in developing embryos; induces	*56, 78, 79*

Continued on next page

Table I. *Continued*

Animal	Drug	Selected Effects	References
		muscular activity in larvae; induces swimming in embryos	
Starfish	5-HT	Potentiates maturation-inducing effect of 1-methyladenine in oocytes	*26, 77*
Annelids			
Leeches	5-HT	Induces feeding and swimming	*82, 83* *87, 88*
Ciliated Protozoans	5-HT	Regenerates cilia; stimulates phagocytosis	*33, 90* *91*

tissues (e.g., gills and feet of bivalves and snails) require energy for ciliary movement, the energetic cost to increased activity could compromise animals during periods of environmental stress. Clemmesen and Joergensen (97) showed that serotonin increased both ciliary beat frequency and energy expense in gills of the mussel *Mytilus edulis*.

The wide ranging effects of antidepressants on aquatic organisms makes predictions of future impacts difficult. Still, release of such antidepressants, and indeed, of pharmaceuticals and personal care products is both an unknown and until now, an unsuspected source of chemical pollutants in the environment. One of the goals of the United States Environmental Protection Agency's Strategic Plan 2000 is to identify such unsuspected risks (9). Thus, further testing for target pharmaceuticals needs to be seriously undertaken at the federal, state, and local levels in order to elucidate how widespread discharge of such drugs and personal care products is. Effects of such target chemicals need to be ascertained by biochemical (e.g., analysis of expression of stress protein genes), cell physiological, and whole animal assays.

It should be noted that even though most of the impacts of their discharge will probably be felt greatest by aquatic organisms, humans living downstream from sewage treatment plants could be affected. Wells communicate with ground water, and ground water can be contaminated by sewage discharge. Therefore, future studies should also be concerned with levels of pharmaceuticals in not only municipal water systems, but in private water systems.

Acknowledgments

I thank Kimberly Lellis and Caroline Philbert for help finding pertinent references.

References

1. Nishizawa, S.; Benkelfat, C; Young, S.N.; Leyton, M.; Mzengeza, S.; De Montigny, C. Blier, P.; Diksic, M. *Proc. Natl. Acad. Sci. USA* **1997**, *94*, 5308-5313.
2. Fuller, R.W. *Life Sci.* **1994**, *55*, 163-167.
3. Eapen, V.; Trimble, M.R.; Robertson, M.M. *Biol. Psychiatry* **1996**, *20*, 737-743.
4. Daubresse, J.C.; Kolanowski, J.; Krzentowski,G.; Kutnowski, M.; Scheen, A.; Vangaal, L. *Obesity Res.* **1996**, *4*, 391-396.
5. Pasini, A.; Tortorella, A; Gale, K. *Brain Res.* **1996**, *724*, 84-88.
6. Fujii, K.; Takeda, N. *Comp. Biochem. Physiol.* **1988**, *89 C*(2), 233-239.

278

7. Udenfriend, S.; Lovenberg, W.; Sjoerdsma, A. *Arch. Biochem. Biophys.* **1959**, *85*, 487-490.
8. Weiger, W. Biol. Rev. **1997**, *72*, 61-95.
9. Daughton, C.G.; Ternes, T.A. *Environ. Health Perspect.* **1999**, *107*(suppl. 6), 907-938.
10. Pincus, H.A.; Tanielian, T.L.; Marcus, S.C.; Olfson, M.; Zarin, D.A.; Thompson, J.; Magno Zito, J. *J. Am. Med. Assoc.* **1998**, *279*(7), 526-31.
11. Heninger, G.R. *Proc. Natl. Acad. Sci. USA* **1997**, *94*, 4823-4824.
12. Rapport, M.M.; Green, A.A.; Page, I.H. *J. Biol. Chem.* **1948**, *176*, 1243-1251.
13. Cerda, J.; Reich, G.; Wallace, R.A.; Selman, K. *Mol. Reprod. Dev.* **1998**, *49*(3), 333-341.
14. Cerda, J.; Subhedar, N.; Reich, G.; Wallace, R.A.; Selman, K. *Biol. Reprod.* **1998**, *59*(1), 53-61.
15. Cerda, J.; Petrino, T.R.; Greenberg, M.J.; Wallace, R. *Molecul. Reprod. Develop.* **1997**, *48*(2), 282-291.
16. Somoza, G.M.; Yu, K.L.; Peter, R.E. *Gen. Comp. Endocrinol.* **1988**, *72*(3), 374-382.
17. Khan, I. A.; Thomas, PS. *Gen. Comp. Endocrinol.* **1992**, *88*, 388-96.
18. Sundin, L.; Nilsson, G.E.; Block, M.; Lofman, C.O. *Am. J. Physiol.* **1995**, *268*, 1224-1229.
19. Fritsche, R.; Thomas, S.; Perry, S.F. *J. Exp. Biol.* **1992**, *173*, 59-73.
20. Akiyama, T.; Murai, T.; Nose, T. *Bull. Jap. Soc. Sci. Fish/Nissuishi.* **1989**, *52*, 1249-1254.
21. Akiyama, T; Murai,T.; Mori, K. *Bull. Jap. Soc. Sci. Fish/Nissuishi.* **1986**, *52*, 1255-1259.
22. Dearry, A.; Burnside, B. *J. Neurochem.* **1986,** *46*(4), 1022-1031.
23. Tohda, M.; Takasu, T.; Nomura, Y. *Eur. J. Pharmacol.* **1989**, *166*, 57-63.
24. Schuette, E.; Chappell, R.L. *Int. J. Neurosci.* **1998**, *95*(1-2), 115-132.
25. Miller, L.J. *Life Sci.* **1989**, *44*, 355-359.
26. Buznikov, G.A.; Nikitina, L.A.; Galanov, A.; Malchenko, L.; Trubnikova, O. *Int. J. Dev. Biol.* **1993**, *37*, 363-364.
27. Nikulina, E.M.; Popova, N.K. *Aggressive Behavior* **1988**, *14*, 77-84.
28. Nikulina, E.M. *Neurosci. Biobehav. Rev.* **1991**, *15*, 545-547.
29. Hirai, S.; Kishimoto, T.; Kadam, A.L.; Kanatani, H.; Koide, S.S. *J. Exp. Zool.* **1988**, *245*, 318-321.
30. Ram, J.L.; Crawford, G.W.; Walker, J.U.; Mojares, J.J.; Patel, N.; Fong, P.P.; Kyozuka, K. *J. Exp. Zool.* **1993**, *265*, 587-598.
31. Fong, P.P.; Wall, D.M.; Ram, J.L. *J. Exp. Zool.* **1993**, *267*, 475-482.
32. Livingstone, M.S.; Harris-Warrick, R.M.; Kravitz, E.A. *Science.* **1980**, *208*, 76-79.
33. Rodriguez, N.; Renaud, F.L. *J. Cell Biology.* **1980**, *85*, 242-247.
34. Carlberg, M.; Anctil, M. *Comp. Biochem. Physiol.* **1993**, *106* C(1), 1-9.

35. Pires, A.; Woollacott, R.M. *Biol. Bull.* **1997**, *192*, 399-409.
36. McCauley, D.W. *Dev. Biol.* **1997**, *190*, 229-240.
37. Huber, R.; Orzeszyna, M.; Pokorny, N.; Kravitz, E.A. *Brain Behav. Evol.* **1997**, *50* (suppl. 1), 60-68.
38. Ni, Y.G.; Miledi, R. *Proc. Natl. Acad. Sci. USA.* **1997**, *94*, 2036-2040.
39. Garcia-Colunga, J.; Awad, J.N.; Miledi, R. *Proc. Natl. Acad. Sci. USA* **1997**, *94*, 2041-2044.
40. Maggi, L.; Palma, E.; Miledi, R.; Eusebi, F. *Molec. Psychiatry.* **1998**, *3*(4), 350-355.
41. Fong, P.P.; Kyozuka, K.; Abdelghani, H.; Hardege, J.D.; Ram, J.L. *J. Exp. Zool.* **1994**, *269*, 467-474.
42. Guerrier, P.; LeClerc-David, C.; Moreau, M. *Dev. Biol.* **1993**, *159*, 474-484.
43. Gibbons, M.C.; Castagna, M. *Aquaculture* **1984**, *40*, 189-191.
44. Fong, P.P. *Biol. Bull. Mar. Biol. Lab. Woods Hole* **1998**, *194*, 143-149.
45. Hardege, J.D.; Duncan, J.; Ram, J.L. *Comp. Biochem. Physiol.* **1997**, *118* C, 59-64.
46. Fong, P.P.; Huminski, P.T.; D'Urso, L.M. *J. Exp. Zool.* **1998**, *280*, 260-264.
47. Juneja, R.; Ueno, H.; Segal, S.J.; Koide, S.S. *Neurosci. Lett.* **1993**, *151*, 101-103.
48. Juneja, R.; Ito, E.; Koide, S.S. *Cell Calcium* **1994**, *15*, 1-6.
49. Satuito, C.G.; Notoyama, K.; Yamazaki, M.; Shimizu, K.; Fusetani, N. *Fish. Sci.* **1999**, *65*, 384-389.
50. Dimock, R.V.; Strube, R.W. *Am. Zoologist* **1994**, *34*, 96.
51. Dietz, T.H.; Steffens, W.L.; Kays, W.T.; Silverman, H. *Can. J. Zool.* **1985**, *63*(6), 1237-1243.
52. Honkoop, P.; Luttikhuizen, P.C.; Piersma, T. *Mar. Ecol. Prog. Ser.* **1999**, *180*, 297-300.
53. Gosselin, R.E. *J. Cell. Comp. Physiol.* **1961**, *58*, 17-25.
54. Kandel, E.F.; Swartz, J.H. *Science* **1982**, *218*, 433-433.
55. Maruyama, I.; Inagaki, M.; Momose, K. *Eur. J. Pharmacol.* **1984**, *106*, 499-506.
56. Mogami, Y.; Watanabe, K.; Ooshima, C. *Comp. Biochem. Physiol.* **1992**, *101C*, 251-254.
57. Audesirk, G.; McCaman, R.E.; Dennis Willows, A.O. *Comp. Biochem. Physiol.* **1979**, *62C*, 87-91.
58. Diefenbach, T.J. ; Kowhncke, N.K.; Goldberg, J.I. *J. Neurobiol.* **1991**, 22, 922-934.
59. Goldberg, J.L.; Koehncke, N.K.; Christopher, K.J.; Neumann, C.; Diefenbach, T.J. *J. Neurobiol.* **1994**, *25*, 1545-1557.
60. Uhler, G.C.; Huminski, P.T.; Les, F.T., Fong, P.P. *J. Exp. Zool.* **2000**, *286*, 414-421.

61. Avila, C.; Tamse, C.T.; Kuzirian, A.M. *Invert. Rep. Dev.* **1996**, *29*, 127-141.
62. Couper, J.M.; Leise, E.M. *Biol. Bull. Mar. Biol. Lab. Woods Hole* **1996**, *191*, 178-186.
63. Sandeman, D.C.; Sandeman, R.E.; Aitkin, A.R. *J. Comp. Neurol.* **1988**, *269*, 465-478.
64. Welsh, J.H.; Moorehead, M. *J. Neurochem.* **1960**, *6*, 146-169.
65. Huber, R.; Delago, A. *J. Comp. Physiol. A.* **1998**, *182*, 573-583.
66. Huber, R.; Smith, K.; Delago, A.; Isaksson, K.; Kravitz, E.A. *Proc. Natl. Acad. Sci. USA* **1997**, *94*, 5939-5942.
67. Fingerman, M.; Nagabhushanam, R.; Sarojini, R.; Reddy, P.S. *J. Crust. Biol.* **1994**, *14*(3), 413-437.
68. Kulkarni, G.K.; Nagabhushanam, R.; Amaldoss, G.; Jaiswal, R.G.; Fingerman, M. *Invert. Reprod. Devel.* **1992**, *21*(3), 231-240.
69. Richardson, H.G.; Deecaraman, M.; Fingerman, M. *Comp. Biochem. Physiol.* **1991**, *99C*, 53-56.
70. Kulkarni, G.K.; Fingerman, M. *Comp. Biochem. Physiol.* **1992**, *101 C*(2), 419-423.
71. Sarojini, R.; Nagabhushanam, R.; Fingerman, M. *Comp. Biochem. Physiol.* **1993**, *106C*(2), 321-325.
72. Hanumante, M.M.; Fingerman, M. *Comp. Biochem. Physiol.*, **1983**, *74* C(2), 303-309.
73. Fingerman, M.; Hanumante, M.M.; Fingerman, S.W. *Comp. Biochem. Physiol.* **1981**, *68*(2), 205-211.
74. Helluy, S.; Holmes, J.C. *Can. J. Zool.* **1990**, *68*(6), 1214-1220.
75. Moss, C.; Burke, R.D.; Thorndyke, M.C.; Brownlee, C. *J. Mar. Biol. Assoc. U.K.* **1993**, *74*, 61-71.
76. Chee, F.; Byrne, M. *Biol. Bull. Mar. Biol. Lab. Woods Hole* **1999**, *197*, 123-131.
77. Buznikov, G.A.; Nikitina, L.A.; Galanov, A.; Malchenko, L. *Biol. Membr.* **1993**, *6*, 1403-1404.
78. Gustafson, T. *Comp. Biochem. Physiol.* **1991**, *98C*, 307-315.
79. Shmukler, Y.B.; Buznikov, G.A.; Whitaker, M.J. *Int. J. Dev. Biol.* **1999**, *43*, 179-182.
80. Renaud, F.; Parisi, E.; Capasso, A.; De Prisco, P. *Dev. Biol.* **1983**, *98*, 37-46.
81. Waggoner, L.E.; Zhou, G.T.; Schafer, R.W.; Schafer, W.R. *Neuron* **1998**, *21*, 203-214.
82. Lent, C.M.; Dickinson, M.H.; Marshall, C.G. *American Zoologist* **1989**, *29*, 1241-1254.
83. Lent, C.M. *Brain Res. Bull.* **1989**, *14*, 643-655.
84. Lent, C.M.; Dickinson, M.H. *J. Comp. Physiol.* **1984**, *154A*, 457-471.

85. O'Gara, B.A.; Chae, J.; Latham, L.B.; Friesen, W.O. *J. Neurosci.* **1991**, *11*, 96-110.
86. Goldburt, V.; Sabban, B.A.; Kleinhaus, A.L. *Behav. Neural. Biol.* **1994**, *61*, 47-53.
87. Mangan, P.S.; Curran, G.A.; Hurney, C.A.; Friesen, W.O. *J. Comp. Physiol.* **1994**, *175A*, 709-722.
88. Brodfuehrer, P.D.; Debski, E.A.; O'gara, B.A.; Friesen, W.O. *J. Neurobiol.* **1995**, *27*, 403-418.
89. Burns, J.T.; Gryskevich, C.L.; Artman, S.A.; Hon, D.M.; Huffner, J.A.; Jones, B.R. *J. Interdiscip. Cycle Res.* **1992**, *23*, 218-219.
90. Quinones-Maldonado, V.; Renaud, F.L. *J. Protozool.* **1987**, *34*, 435-438.
91. Darvas, Z.; Arva, G.; Csaba, G.; Vargha, P. *Acta Microbiol.* (Hung.) **1988**, *35*, 45-48.
92. Bogan, A.E. *J. Shellfish Res.* **1996**, *15*, 484.
93. Neves, J. *J. Shellfish Res.* **1997**, *16*, 345.
94. Sonnenschein, C.; Soto, A.M. *Steroid Biochem. Molec. Biol.* **1998**, 65, 143-150.
95. Sumpter, J.P. *Toxicol. Lett.*, **1998**, *102-103*, 337-342.
96. Castrodad, F.A.; Renaud, F.L.; Ortiz, J.; Phillips, D.M. *J. Protozool.* **1988**, *35*, 260-264.
97. Clemmesen, B.; Joergensen, C.B. *Marine Biology* **1987**, *94*, 445-449.

Chapter 16

Veterinary Medicines and Soil Quality: The Danish Situation as an Example

John Jensen

Department of Terrestrial Ecology, National Environmental Research Institute, P.O. Box 314, Vejlsoevej 25, DK-8600, Silkeborg, Denmark

Veterinary medicines — biologically active substances designed to harm organisms such as bacteria, viruses, and parasites — are also potentially hazardous for non-target species, such as those in soil. Some, medicines, for example antibiotics and anthelmintics, are used in large amounts and may enter the soil environment through animal excrement. Consumption in 1997 of antibiotics in Denmark exceeded 150,000 kg, whereas the quantity of antiparasitics used was unknown. Harmful effects on indigenous soil organisms have been observed in controlled experiments. In-situ effects may therefore also occur, and more information on fate and effects of veterinary medicines is still needed to assess long-term ecological consequences. On the basis of current knowledge, it is, however, concluded that the current use of veterinary medicine in Denmark is not likely to pose a significant risk to the overall quality of agricultural soils.

Introduction

Technological and scientific development in agriculture has made it possible to significantly increase production of food in post-war time. However, the requirement for improvements in agricultural technology to feed an increasing population is accompanied by increasing concern for the health of consumers and the environment. The increased crop production as a result of pesticide use and the subsequent discovery of pesticides or metabolites in grounwater is a good example of this. Also animal production has undertaken large changes during the last decades. Larger and more modern husbandries have optimized the spatial use. However, the close contact between animals increases the risk of diseases spreading. High hygienic safety in farms cannot prevent the need for medical intervention. Therefore, a significant consumption of veterinary medicines and growth promoters is found wherever modern husbandry practices are found.

Veterinary medicinal products are authorized for use by regulatory authorities if they comply with scientific criteria on quality, efficacy, and safety. The primary historical concerns have been occupational health during production and handling, and the ultimate safety to the treated animal and the consumer. Only more recently has the environmental risk of veterinary medicinal products become a matter of increasing public concern and legal requirements. This paper is an attempt, on the basis of the current knowledge, to outline the overall problems associated with the release of veterinary drugs to the terrestrial environment. The use of veterinary medicines in Denmark, a small European country of 5 million inhabitants and an intensive agricultural sector, will serve as an example. This paper is not intended as an all-inclusive review of the topic. For more detailed compilations on some of the aspects of this chapter, please refer to other reviews (e.g., 1,2). The bulk of information currently published on veterinary drugs in the environment concerns antibiotics and antiparasitics. This review is therefore primarily devoted to these two groups of drugs.

Use and consumption of veterinary medicine in Denmark

Large amounts of veterinary medicines are used worldwide to either prevent or to cure diseases within livestock and pets. Although a small country, Denmark has a relatively intense livestock production with approximately 20, 0.7, and 115 million slaughtered pigs, cattle, and poultry, respectively, in 1997. The Danish Plant Directorate annually publishes data on the consumption of growth promoters and feed-administrated drugs (Fig.1), whereas detailed information on the consumption of veterinary medicines for therapeutic

284

purposes is less accessible. The 1996 and 1997, consumption figures for a number of high-volume veterinary medicinal products were assembled and analyzed by the Royal Danish School of Pharmacy and the Danish Environmental Research Institute. The use of selected groups of substances is presented in Table I. The consumption is based on the number of recorded prescriptions. A more detailed description of data can be found in a draft report from the Danish EPA and Danish Medicines Agency (3).

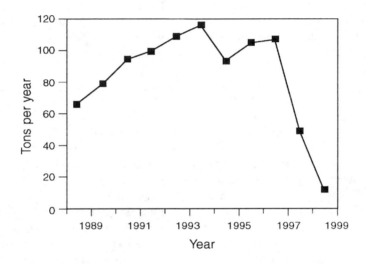

Figure 1. The annual use of antimicrobial growth promoters in Denmark. Data from the Danish Plant Directorate (3)

The use of antibiotics in Denmark has changed significantly during the last 5 years (Fig. 1). Avoparcin, previously a very popular antibiotic in Denmark and the rest of EU, was banned in 1995 due the risk of inducing cross-resistance against other antibiotics used for humans. Due to the risk of cross-resistance, virginamycin was banned by the EU in January 1998. In 1999, EU banned zinc bacitracin, spiramycine, tylosine, carbadox and olaquindox as growth promoters in animal feed. These bans, and a voluntary agreement for a gradual reduction in the use of antibiotics as growth promoters between the farmers, the feed producers and the authorities in Denmark, have caused a significant reduction in the use of antibiotics (Table II, Fig. 1). For example, all use of antimicrobial growth promoters for pigs was to stop by January 2000. Unfortunately no data are available to judge whether the use of antibiotics for therapeutic treatments

Table 1. Selected groups of veterinary medicines used for therapeutic purposes in Denmark 1996 and 1997 (kg active substance prescribed per year). The most common substances in each group and their relative fraction of the total use are shown in parentheses. Data from (3).

Substances/Therapeutic use [CAS RN]	1996	1997
Antibiotics (therapeutic)	**42,968**	**48,530**
tetracycline/chlortetracycline [60-54-8/57-62-5]	(23%)	(23%)
benzylpenicillin (penicillin G) [61-33-6]	(15%)	(15%)
Alimentary tract/metabolism	**14,212**	**14,743**
calcium gluconate [299-28-5]	(54%)	(52%)
(neomycin B sulfate [1405-10-3]	(24%)	(27%)
Antiparasitics (therapeutic)	**242**	**236**
sulfaclozine sodium [102-65-8]	(100%)	(100%)
CNS active compounds	**237**	**230**
metamizol sodium (dipyrone) [5907-38-0]	(83%)	(77%)
butanilicaine phosphate [3785-21-5]	(11%)	(13%)
Hormones	**27.3**	**28.7**
medroxyprogesterone acetate [71-58-9]	(55%)	(49%)
prednisolonacetate [52-21-1]	(30%)	(37%)

Table II. The use of antibitic growth promoters and coccidiostatics in Denmark from 1996 to 1999 (kg active substance sold per year). The most common substances (in 1996) and their relative fraction of the total sale are shown in parentheses. Data from (3) and the Danish Plant Directorate.

Substances [CAS RN]	1996	1997	1998	1999
Antibiotic growth promoters	**105,548**	**107,179**	**49,294**	**12,283**
tylosin [1401-69-0]	(65%)	(58%)	(27%)	(15%)
olaquindox [23696-28-8]	(13%)	(16%)	(58%)	(76%)
virginiamycin [11006-76-1]	(5%)	(10%)	(2%)	(0%)
Coccidiostatics	**19,400**	**17,037**	**18,292**	**25,493**
salinomycin sodium [55721-31-8]	(23%)	(25%)	(43%)	(35%)
metichlorpindol/methylbenzoquate (chlopidol/nequinate) [2971-90-6 /13997-19-8]	(25%)	(23%)	(5%)	(<1%)
monensin sodium [17090-79-8]	(18%)	(22%)	(20%)	(34%)

increased in the same period. However, from 1996-1999 the consumption of coccidiostatics increased by 31% to now approximately 25,500 kg per year (Table II).

The use of anthelmintic substances like avermectines is presumably relatively large. However, due to the regulation of these compounds in Denmark, the sale of antiparasitic compounds was only registered and compiled if specifically used for therapeutic treatment. In 1997, prescriptions of 236 kg of sulfaclozine was the only registered use of antiparasitic drugs. In the future, a far better estimate of the Danish use of antiparasitic substances can be made because all use of antiparasitic substances in livestock production has needed a prescription as of August 1999.

Release and effects of veterinary medicines in the terrestrial environment

Compared with industrial chemicals, the exposure routes of veterinary medicinal products are relatively easy to identify and relate to specific field scenarios. Veterinary medicines may be spread to the terrestrial environment, either directly when using the drugs (minor), by disposal (minor), or by subsequent excretion from the animals (major). The dominating pathway of environmental release in the terrestrial compartment is by amendment of arable soil with manure or slurry. Likewise, drugs used in aquaculture may reach soil in cases where sediment/sludge from fish farms are spread on arable land. If the drug is hydrophobic, it will mainly adsorb to particulate matter. In contrast, if the substance is hydrophilic, it will be dissolved in the aqueous fraction of the manure/slurry, and the risk of run off during application is thereby higher. When used therapeutically for animals outside confined areas, these substances (e.g., hormones, antibiotics, antiparasitic drugs) will be directly deposited via urine or feces on the field; resultant terrestrial exposures could therefore result from high, localized concentrations. Information on several of the pathways to soil for an administered drug are lacking or inadequate.

Most medicines are at least partly metabolized before being eliminated in the urine or feces. Metabolism changes the physical and chemical behavior of the substance, as it typically makes the metabolites more water soluble than the parent compounds. If more than 20% of the drug is excreted as a single metabolite, the environmental risk also needs to be assessed for the metabolites.

There are examples of a natural conversion of metabolites back to parent compounds. Chloramphenicol glucoronide and N-4-acetylated sulfadimidine were converted back to the parent chloramphenicol (56-75-7) and sulfadimidine (sulfamethazine; 57-68-1), respectively, in samples of liquid manure (4).

Fate of veterinary drugs in manure and soil

In Denmark, regulations prescribe a minimum storage capacity of nine months for manure production.Very little is known about the occurrence and fate of veterinary drugs in manure when stored prior to soil amendment. The antibiotic tylosin rapidly degrades or at least disappears from the aqueous phase under conditions partly resembling the situation in manure tanks (5). Its half-life in the aqueous phase was less than three days. Future studies will investigate whether the disappearance was due to degradation or sorption to solids and colloids. Shore et al. measured testosterone and estrogen in manure from American chickens indicating an incomplete degradation of hormones in manure (6).

Limited information is also available to judge the degradation rates of antibiotics when mixed into the soil after tilling. Ceftiofur sodium (104010-37-9; a wide-spectrum cephalosporin antibiotic) is quickly degraded to non-degradable metabolites in different soils, with half-lives between 22 and 49 days (7). Soil sterilization inhibits the degradation of the substance, suggesting that microorganisms may be responsible for the degradation. Donoho found that monensin (17090-79-8) applied as growth promoter for chickens and cattle, was degradable in manure and soil (8). Warman and Thomas found chlortetracyclines (CTC) in soil amended with chicken manure (9). However, the recovery rate was generally low, and they could not detect any CTC in manure-to-soil ratios greater than 1:10. Tetracyclines are generally difficult to extract from soil material (10), which could account for the low recovery rates.

Sorption and mobility studies may give an indication of the potential for biodegradation. Substances with high sorption to minerals or organic material in soils or manure are likely to have slow degradation rates as they are unavailable for degradation by microorganisms. Raboelle and Spliid (10) observed significant difference in the mobility of four antibiotics. Tylosin and especially oxytetracycline (OTC), with Kd values of more than 100 and 1000, respectively, were highly immobile in soil, whereas metronidazole and olanquindox was fully recovered in the leachate. They did not find any correlation between sorption of the four test antibiotics, and their respective physico-chemical characteristics. Efromycin (56592-32-6), an antibiotic used as a growth promoter for pigs, was immobile in five soils of various properties (11). It may be concluded that veterinary drugs with high absorption cabability are likely to be relatively persistent, at least under the anaerobic and methanogenic conditions prevalent in manure tanks, but probably also in soils with high clay and organic matter content.

Anthelmintics are used against parasites in ruminants. Most information is available for ivermectin, an antiparasitic drug developed in the early 1980s for treatment of cattle, pigs, horses, and sheep. It is excreted almost entirely in feces. Chiu et al. showed that 60-80% of radiolabelled ivermectin was excreted as ivermectin or metabolites in the feces over the first week after an injection dose of 0.3 mg kg^{-1}, and that more than 60% was excreted during the first three days (12). Less than 1% was excreted through the urine. Halley et al. found by using radiolabeled ivermectin that parent compounds comprised approximately 40-45% of the total radioactivity in dung from steers, and 60-70% and 40% in dung from sheep and pigs, respectively (13); the remaining components were primarily polar metabolites. Sommer and Nielsen found, two days after injection with 0.2 mg kg^{-1} body weight, a maximum concentration in dung from cattle of 3.8 mg kg^{-1} d.w (14). After 7 and 17 days, the concentrations in fresh dung dropped to 1.6 and 0.3 mg kg^{-1}, respectively. However, ivermectin excretion may continue for five weeks after a single injection (15).

Ivermectin and other antiparasitic drugs may be given to cattle and sheep by sustained release boli, which release the drug over an extended period of time, often encompassing the entire grazing season. This may be of concern, as it will prolong the time of exposure for dung-living organisms. Ivermectin has been shown to be persistent in dung voided from treated cattle. It is not easily degraded or washed out by rainfall. It was not possible to detect any decrease in the ivermectin concentration in dung pats during an entire experiment (45 days) in the field under temperate conditions (16). Ivermectin is also persistent under tropical conditions (14, 16). These findings are in agreement with the insecticidal properties observed for aged dung (17) and the fact that avermectins were determined to be immobile in three different soils (18).

Effects of veterinary drugs in the terrestrial environment

Antibiotics

When evaluating effects on the microbial community, it is meaningful to keep in mind that target organisms vary between antibiotics. Antibiotics may have a broad spectrum of activity or designed to be either specifically active against gram-negative or gram-positive bacteria. An indigenous community of bacterial and fungal populations is very complex. Together with the micro-fauna, they carry out and safeguard important processes in the cycling of nutrients. Processes affected may include those driven solely by a few bacterial species, or more general processes such as decomposition of organic matter, which is a cooperation between a conglomerate of different types of microorganisms. The overall effect of antibiotics on total numbers of micro-organisms or soil respiration may tell very little about the potential risk to specific soil functions as these measures generally have low sensitivity. Not only bacterial, but also initially fungal numbers may be reduced by streptomycin at concentrations of 3000 mg kg^{-1}, whereas 1000 mg kg^{-1} caused no changes in the total numbers (*19*). Streptomycin applications of 10-100 mg kg^{-1} only result in temporary reductions of bacterial and fungal numbers (*20*). Soil concentrations of 1000-3000 mg kg^{-1} are not environmetal relevant as the level of antibiotic in soil after normal application of manure will be far below these concentrations.

More sensitive or vulnerable measures are the so-called "bottleneck" processes that are maintained by a few, metabolically restricted groups of microorganisms. A proper cycling of nutrients is crucial for a good soil quality and essential for maintaining a sustainable use of agricultural soils. Nitrogen is one of the most important nutrients in agricultural systems. The two-step oxidation of ammonium to nitrate (nitrification) is driven by only two groups of bacteria, which consist of a few gram-negative genera, e.g., *Nitrosomonas* and *Nitrobacter*. Effects on the nitrification processes by gram-negative antibiotics like aminoglycosides and quinolones or broad spectrum antibiotics like sulfonamides and tetracyclines could therefore be critical in the nitrogen cycling. Few studies, however, have recorded the effects of antibiotics on the potential ammonium oxidation in natural soils. Warman studied the effects of amprolium (coccidiostat, 121-25-5) and aureomycin (chlorotetracycline, 57-62-5, antibiotic) on the nitrification of poultry manure-amended soil (*21*). No nitrification effect was found at concentrations of 22.5 mg kg^{-1}. At high concentrations, streptomycin did not affect the entire bacterial population, but appeared to reduce only a small (but important) portion, that is those bacteria that nitrify ammonia (*20, 22*). Patten et al. did not find any effect on nitrogen mineralization in field soil or the growth of corn seedlings in greenhouse experiments (*23*). The experiments of Patten et al. were conducted by mixing soil with manure containing 5.3 mg chlortetracycline and 11.3 mg oxytetracycline kg^{-1}. The application rates were 4.2-8.4 g manure per kg soil. The maximum soil concentrations of 0.045 mg CTC and 0.095 mg OTC kg^{-1} were hence significantly lower than the concentrations used by Ingham et al. (*20, 22*). Beare et al. used biocides to experimentally control the decomposer organisms in field experiments on conventional and no-tillage agro ecosystems (*24*). For a bactericide, they used oxytetracycline. Nitrogen fluxes in the litter layer were determined in buried litter bags and surface soil. The exclusion studies showed that fungi had somewhat greater influence on the decomposition of surface litter, while bacteria were more important in the decomposition of buried litter. Oxytetracycline at a dose of 15 g m^{-2} depressed the total bacterial biomass by approximately 50-60%. The decomposition of surface litter declined 25% in the no-tillage systems, whereas the decomposition rates were reduced 35% by oxytetracycline in the litter bags buried in the conventional fields.

Resistance

In combination with the direct toxic effects on micro-flora, attention has been drawn to the possibility that antibiotics and antibiotic-resistant bacteria may cause undesirable changes in natural populations of microbiota. Selection of resistant bacteria in the presence of antibiotics is an already recognized problem, which may threaten future effectiveness of antimicrobial treatment for livestock and humans. Development of resistant bacteria in nature may be caused by antibiotic residues forming a selective pressure on soil bacteria or by possible donation of resistance genes by transient bacteria from manure or sludge to the natural bacteria flora. Transient antibiotic-resistant bacteria may exchange DNA in natural environments with even distantly related bacteria via the processes of transformation (uptake of soluble DNA), conjugation (transfer by cell-to-cell contact), and transduction (transfer by bacteriophages) (3). Gene transfer has mainly been investigated in laboratory experiments, and less is known concerning transfer in natural habitats (25). It has been demonstrated in habitats such as water, wastewater, sewage, sediment, and soil (3, 26, 27, 28, 29). It is not known, however, whether the pool of resistance genes offered by transient bacteria (e.g., pathogenic bacteria in slurry and sewage-sludge) is important compared with the background level of resistance genes originating in the soil bacteria. Some antibiotics, such as tetracyclines, are known to enhance gene transfer (30), but the relevance of this observation in relation to transfer in natural environments is unknown. The observation may cause concern, however, as it may be problematic that antibiotics provide a selective environment and at the same time promote the spread of the gene responsible for the selective advantage. There remain a number of open questions — What are the environmental and public health consequences of the presence of resistant bacteria in sewage treatment plants or manure tanks? How and at what rate can these bacteria transfer their resistance genes to the naturally occurring microbiota after discharge onto arable land?

Non-target organisms

The effect of antibiotics on soil organisms other than bacteria, such as plants and soil fauna, has also been studied. Batchelder has tested the effects of the antibiotics chlortetracycline and oxytetracycline on plants. In greenhouse studies, pinto beans (*Phaseolus vulgaris*) were grown in aerated nutrient media (*31*) and in soil watered with oxytetracycline solutions (*32*). Growth and development of roots was markedly decreased as concentrations were increased from 0 to 160 mg L^{-1} in solution. At the lowest test concentration of 10 mg L^{-1}, dry weights and root dry weights were significantly lower than the control. Plant mortality also increased with increasing concentrations of OTC and all plants died at 160 mg L^{-1}. Pinto beans grown in soil were also severely affected by antibiotics when watered with solutions containing 160 mg OTC L^{-1}. The mechanism behind the observed phytotoxicity is likely to be due to formation of complexes between OTC and essential ions, such as calcium. Calcium deficiency symptoms including chlorosis, necrosis, and curling of leaves were observed. Ingham et al. found that streptomycin applications of 10-100 mg kg^{-1} soil resulted in the death of all crops within a two-to-four week period (*20*). One year later, the streptomycin-treated areas still elicited poor plant growth. Ingham and Coleman found that streptomycin significantly affected plant health at high concentrations (1000-3000 mg kg^{-1}) (*19*). Whether this was a direct phytotoxic effect or an indirect effect on plants mediated by effects on the soil bacteria was not fully elucidated. However, in both cases only minor initial effects on the overall number of bacteria and fungi were observed. The soil concentrations of antibiotics were in both studies far above the concentrations likely to be found after soil application of manure from medicated animals.

Baguer et al. exposed three soil-dwelling invertebrate species to tylosin and oxytetracycline in the laboratory (*33*). The two antibiotics were not fatal for earthworms (*Aporrectodea caliginosa*) and springtails (*Folsomia fimetaria*) at concentrations below 5000 mg kg^{-1}, whereas an LC50 of 3,381 mg kg^{-1} was estimated for the enchytraeids (*Enchytaeus crupticus*). Reproduction was more sensitive, with EC10 and EC50 starting at 134 and 2,520 mg kg^{-1}, respectively. The low toxicity observed may partly be due to a strong binding of tylosin and especially OTC in soil (*10*). The study by Baguer et al. suggests that direct effect of antibiotics on soil fauna is not likely at environmentally relevant concentrations. However, soil micro and macro fauna may be impacted indirectly by changes in their food web. Soil ecosystems contain many interactions both in spatial and temporal scales within food webs. Due to its high complexity, all interactions are far from well described or understood at present, and links between the community structure and essential soil functioning are not always straightforward.

Simple functional food webs for soil ecosystems may be described by a bacterial or a fungal-driven pathway. The microbiota may hence consist of bacterial-feeding nematodes and protozoa, fungi-feeding nematodes, springtails, and mites, and predatory/omnivorous species such as nematodes, mites, spiders, and ants. The food web responses to perturbations are not fully known or understood (34). The high degree of omnivory typical for below-ground food webs (35) may blur the use of theories based on direct interactions between prey and predator. The literature contains a number of examples of community structure and food web interactions resulting from changes in environmental conditions (36). The perturbations include the use of herbicides (37, 38), insecticides (39), fungicides (19), bactericides (19, 39), and agricultural intensification (37, 40). If the bacterial or fungal populations are strongly altered as a result of antibiotics, the feeding of microbivore species like springtails, mites, enchytraeids, nematodes, and even plants can be impacted. Beare et al., for example, found that the population dynamics of bacterivorous nematodes were strongly linked to their bacterial food source (24), whereas protozoa were unaffected, a tendency also observed by Ingham et al. (22).

Despite many difficulties, it must therefore be recommended to include a food-web approach in ecotoxicological risk assessment of antibiotics (as well as for other toxic substances). Changes in one trophic level may eventually lead to alteration at another level. To support classical ecotoxicological testing, experiments including different trophic levels are, therefore, recommended to provide a hierarchical risk assessment in soil (41-43).

Anthelmintics

In the 1980s, a new class of compounds known as macrocyclic lactones were introduced to the market. This class includes avermectins (doramectin [117704-25-3], abamectin [71751-41-2], and ivermectin [70288-86-7]), all broad-spectered antiparasitics) and milbemycins (moxidectin [113507-06-5]). Most ecotoxicological studies available today concern this group of anthelmintics. The development of anthelmintics with a wide range of target pathogens has resulted in a more efficient and economically feasible solution to parasite control. However, the risk of affecting non-target organisms has increased as the spectrum of activity of the drugs has increased. In 1987, Wall and Strong reported that residues of ivermectin in cattle dung had lethal effects on beneficial dung-living insects and hence delayed degradation of dung pats (44). The pats were more or less intact 100 days after deposition. Large numbers of papers later confirmed the insecticidal properties of avermectin (17, 45, 46). Research has shown that the duration of effects depends on the non-target species (beetles, flies, earthworms), form of drug application, and livestock species. By studying the number and development of immature dung beetles (*Onthophagus gazella*), Sommer and Nielsen showed a nearly 100% larvicidal effect of ivermectin (1.6 mg kg^{-1}, d.w.) in dung collected one week after treatment (14). In dung voided 17 days after treatment, only half of the larvae were able to survive, despite the ivermectin concentration dropping to 0.3 mg kg^{-1} (d.w.). Madsen et al. observed the effects of ivermectin on two species of flies in dung from treated heifers (*17*). No chemical analysis of the dung was performed, but results from the laboratory studies showed that the face fly (*Musca autumnalis*) was more sensitive to ivermectin than the house fly (*Musca domestica*), and that dung excreted 40 days after treatment still killed half of the fly larvae.

Ivermectin and abmectin are also toxic to earthworms. Gunn and Sadd showed that ivermectin caused mortality and reduced growth and cocoon production in the compost worm (*Eisenia fetida*) at concentrations similar or lower to the levels found in dung pats of treated animals (*47*). All worms died at soil concentrations above 20 mg kg^{-1} (LC50=15.7 mg kg^{-1}), whereas growth and reproduction was reduced at 8 and 4.7 mg kg^{-1}, respectively. The relatively high toxicity of ivermectin to earthworms documented by Gunn and Sadd diverge somehow to the result from a few other studies concerning effects on earthworms. Halley et al. also exposed compost worms to ivermectin in the lab and found an LC50 of 314 mg kg^{-1} (*13*). The large difference between the results of these two tests may be explained by differences in pH and the formulation of ivermectin (*47*). A 25%, but not statistically significant reduction, in the total mass of earthworms was observed when exposed 98 days to dung from ivermectin-treated animals (*48*). Wislocki et al. exposed earthworms to abamectin and found at 28 days an LC50 of 28 mg kg^{-1} (*49*, c.f. 50) From the studies above, it is concluded that ivermectin and other avermectins are highly toxic to dung-living insects and toxic to earthworms.

Information on other anthelmintics is scarce. Moxidectin (113507-06-5), a member of the milbemycins, was harmful to face fly larvae, effecting a reduction of 30-90% during the 10-week experiment (*51*). Other papers have not found any significant toxicity of moxidectin to non-target insects. The toxicity of moxidectin was 1/64th that of abamectin for larvae of the dung beetle (*Onthophagus gazella*) and the buffalo fly (*Haematobia irritans*) (*52*). No significant difference between numbers of larval Coleoptera and Diptera in control samples versus moxidectin-treated dung was observed (*53*). Wardhaugh et al., did not find any reduction in survival of larvae of the bush fly and the housefly in dung from moxidectin-treated cattle (*54*). The mean number of brood balls and emergence of adult dung beetles (*Euoniticellus intermedius* and *Onthophagus gazella*) was unaffected by moxidectin (*55*). Madsen et al. studied the effect of a number of anthelmintics on the degradation of dung pats and the presence of earthworms. Ivermectin caused mortality among dung fly larvae and prolonged the degradation of dung pats, whereas the remaining anthelmintics metrifonatum (trichlorofon [52-68-6]), levamisoli chloridum, pyranteli citras, and fenbendazolum did not have any effect on the degradation of pats and survival of dung flies (*48*). None of the substances had any significant effect on the total biomass of earthworms.

Other drugs

Antibiotics and anthelmintics are the most widely used groups of veterinary medicines, but by no means the only ones. Even less is known about the other groups of medicines. In the EU, as opposed to the U.S., hormones can not be used as growth promoters. Hormones are a group of substances with a very high potential for environmental side effects. In light of the strong ongoing debate about endocrine disrupters (*56*), environmental risk of veterinary use of hormones should be evaluated whenever more information becomes available. Shore et al. found that growth of alfalfa was affected by steroidal estrogen in the concentration range found in sewage water used for irrigation of agricultural fields in especially arid and semi-arid regions (*57, 58*).

Veterinary medicines and the impact on soil quality

Medicines used for livestock, unless mineralized within the animal, will end up in the soil environment. Residues of veterinary medicinal products may be deposited on arable land or pastures as an unintended constituent of manure and slurry. Veterinary medicine products are in general highly biologically active substances, i.e., they are designed to pass through membranes, and in the case of antibiotics and antiparasitics, for example, to have a toxic action towards target organisms. Manure/slurry as fertilizers on arable land may thus pose a risk for the environment. This risk must be balanced against the beneficial use/recycle of manure/slurry nutrients, an important aspect of the concept of a self-sustainable agriculture. It is therefore important to assess whether the long-term use of veterinary medicines in the agricultural sector leads to deterioration of soil quality. To assess changes in soil quality within the context of sustainable land use, however, is by no means straightforward. Even to define soil quality may cause large divergence among experts. Doran and Parkin (59) suggest the following definition of good soil quality "The capacity of a soil to function within ecosystem boundaries to sustain biological productivity, maintain environmental quality, and promote plant and animal health".

Essential parts of the global carbon, nitrogen, phosphorus, and water cycles are carried out in the soil compartment largely through microbial and faunal interactions with organic and inorganic material. If use of veterinary medicines in the agricultural sector poses a risk to this soil functioning, it may become an environmental problem in the long run. The use of manure is closely associated with arable land, of which large areas are intensively managed today. Many of the available management tools have a large influence on the soil as habitat for animals and plants. Pesticide spraying, use of organic and inorganic fertilizers, tilling, compaction, and sewage sludge application are all practices affecting the soil ecosystem. Very often these operations lead to reduced biodiversity. Although not fully validated, it is generally anticipated that soil biodiversity is positively associated with stability and good soil quality (*40*). The potential effects of veterinary drugs in manure should, therefore, not only be balanced with the large benefits of increased animal welfare and recycling of nutrients, but also compared with the general effects of agricultural practices on soil ecosystems.

Many obstacles must be overcome when assessing the ecological consequences of veterinary drug use. First of all, it is necessary to get a good accounting of the actual consumption of different groups of drugs and to identify all relevant exposure routes. When sufficient information is present, methods for predicting the environmental concentrations are available (*60, 61*). Secondly, information on toxic effects of drugs on single species, food webs, and soil functioning should be generated. The fact that the consumption in Denmark of antimicrobial growth promoters declined by 80% in two years has without doubt reduced the release of antibiotics to the soil environment significantly. Information about fate during manure storage and after soil application is still needed to improve the assessment of environmental exposure. Whereas the consumption of antibiotics in Denmark has been well documented, the actual use of anthelmintics is not yet known in detail. As of August 1999, all use of antiparasistic substances in livestock production requires a prescription in Denmark.

For assessing the effects of antibiotics on soil quality, much more information is still needed. Nevertheless, the last couple of years have provided us with a better understanding of where potential problems may arise. In 1997, several national projects concerning fate and effects of antibiotics in the environment were initiated within the *Center of Sustainable Land Use & Management of Contaminants: Carbon and Nitrogen* (http://www.landuse.dk/uk/index.htm) and funded by the Danish Environmental Research Program. Several of these studies are published in a special issue of the journal "Chemosphere" (*62*).

The effects of anthelmintics, such as the avermectins, to non-target organisms are well documented in a number of studies and may have implications for soil quality. First of all, the effects on dung insects can prolong the degradation of dung pats. The disintegration of pats is typically initiated by dung beetles, which aerate the dung making it more attractive to earthworms, nematodes, and other invertebrates that subsequently disperse the pat into the soil (*63*). The ecological consequences of avermectins may, however, not be limited to the direct effects on dung insects. Although not fully verified, secondary effects could occur with wildlife such as birds that routinely disrupt cowpats while searching for insects (*64*). This may not only include dung-feeding birds but also soil-probing birds during the summer where soil invertebrates are limited in the dry soils. By occasionally shifting foraging methods, these birds may stay in these areas under otherwise unfavorable conditions.

The ecological consequences of using anthelmintics may be viewed on both local and global scales. Globally less than 15% of all cattle are treated with avermectins. The majority of manure from most husbandry practices will hence not contain avermectins (65). This makes it unlikely that anthelmintics will cause significant global problems at present (2). On a regional scale, however, overall consequences may differ. Drug administration and health care may differ to some extent from region to region. Furthermore, the primary colonist of dung pats changes according to climate and regions. In Europe and North America, flies and scarabs are important for degradation, whereas in more arid areas like Spain, Australia, and South America, scarabaine beetles have a more important role (66). Locally, the use of special sustained-release boli may represent a higher risk to populations of dung insects. Anthelmintic therapy, however, is normally targeted at young cattle with undeveloped immune systems. It has been estimated that pats excreted by this group of animals normally will comprise up to 20% of all dung produced by a beef herd and less than 5% by a dairy herd (63). Although some studies may indicate that dung from avermectin-treated animals is especially attractive for dung beetles (67), it is likely that the remaining pats will provide refuge for the beetles and in this sense prevent extinction. Ecological models have predicted that in typical cattle farming systems the cumulative insect mortality would rarely exceed 25% (67). Whether this is enough to cause significant long term ecosystem disruptions remains unclear.

Conclusions

Many veterinary medicines are potentially hazardous for soil organisms. Some, such as antibiotics and anthelmintics, are used in large amounts and may enter the soil environment via the use of manure as organic fertilizers or the dropping of cow dung in the field. The observed effects of antibiotics in the field or laboratory are generally found at least one order of magnitude below environmentally realistic concentrations, whereas effects of avermectins have been identified in controlled small-scale studies at environmentally realistic dung concentrations. Data on long-term regional effects in the field are, however, scarce if not totally lacking for both groups of substances. If these compounds change the overall rates of dung degradation or affect the nitrogen mineralization, this will affect nutrient cycling and in the long run soil fertility.

Relatively long-term effects on plant health have been observed in the field at high concentrations of antibiotics. These and other findings may indicate that more information about fate and effects in the environment is necessary before we can judge whether veterinary drugs reduce the quality of our soils. Furthermore, although the probability of antibiotics in manure posing a risk for development of resistance in human pathogens is low, very little is known about the consequences if populations of resistant bacteria develop in agricultural soils.

As a precautionary principle, all use of veterinary drugs should be based on a management strategy balancing cost and benefits. The fact that all new veterinary medicinal products launched in EU after 1 January 1998 shall undergo an environmental risk assessment before marketing will help to include ecological considerations in such cost-benefit analyses. The benefits from veterinary medicines are generally so many that their prohibition or restriction in use would have to be given an extraordinarily careful examination. In cases where potential ecological risk has been identified, one step forward would be to consider whether less hazardous substances in a therapeutic group could be used to accomplish the same veterinary goal. Alternatively, minor restrictions or changes in medication or handling may overcome the problems.

Denmark is a good example for how a mixture of central registration of sales, voluntary agreements, and bans without large economic costs has led to a significant reduction in the general use of antibiotics and a full halt for the most problematic ones. Although these actions primarily were driven by risk to human health, they may serve as models if significant ecological problems are identified for groups of veterinary medicinal products.

References

1. Halling-Sørensen, B.; Nielsen, S. N.; Lanzky, P. F.; Ingerslev, F.; Lützhøft, H. C. H.; Jørgensen, S. E. "Occurrence, fate and effects of pharmaceutical substances in the environment - A review." *Chemosphere* **1998,** *36*(2), 357-393.

2. McKellar, Q. A. "Ecotoxicology and residues of anthelmintic compounds." *Veterin. Parasitol.* **1997,** *72*(3-4), 413-435.

3. Halling-Søresen, B., Jensen, J., Nielsen, S.N. Environmental assessment of veterinary medicinal products in Denmark. *Draft Report.* Danish Environmental Protection Agency, Copenhagen, Denmark.

4. Berger, V. K.; Petersen, B.; Büning-Pfaue, H. Persistenz von Gülle-arzneistofffen in der nahrungskette. *Archiv für Lebenmittelhygiene* **1986,** *37,* 85-108.

298

5. Loke, M. L.; Ingerslev, F.; Halling-Sorensen, B.; Tjornelund, J. "Stability of Tylosin A in manure containing test systems determined by high performance liquid chromatography." *Chemosphere* **2000**, *40* (7), 759-765.

6. Shore, L.; Shemesh, M.; Cohen, R. "The role of estradiol and estrone in chicken manure silage in hyperestrogenism in cattle." *Austral. Veterin. J.* **1988**, *65* (2), 68.

7. Gilbertson, T.; Hornish, R. E.; Jaglan, P. S.; Koshy, K. T.; Nappier, J. L.; Stahl, G. L.; Cazers, A. R.; Nappier, J. M.; Kubicek, M. F.; Hoffman, G. A.; Hamlow, P. J. "Environmental fate of ceftiofur sodium, a cephalosporin antibiotic - role of animal excreta in its decomposition." *J. Agric. Food Chem.* **1990**, *38*(3), 890-894.

8. Donoho, A. L. "Biochemical studies on the fate of monensin in animals and the environment." *J. Anim. Sci.* **1984**, *58* (6), 1528-1539.

9. Warman, R. P.; Thomas, R. L. "Chlortetracycline in soil amended with poultry manure." *Can. J. Soil Sci.* **1981**, *61,* 161-163.

10. Raboelle, M.; Spliid, N. H. "Sorption and mobility of metronidazole, olaquindox, oxytetracycline and tylosin in soil." *Chemosphere* **2000**, *40* (7), 715-722.

11. Yeager, R. L.; Halley, B. A. "Sorption/desorption of 14C [efromycin] with soils." *J. Agric. Food Chem.* **1990, *38*,** 883-886.

12. Chiu, S.-H. L.; Green, M. L.; Baylis, F. P.; Eline, D.; Rosegay, A.; Meriwater, H.; Jacob, T. A. "Absorption, tissue distribution, and excretion of tritium-labeled Ivermectin in cattle, sheep and rat." *J. Agric. Food Chem.* **1990**, *38,* 2072-2078.

13. Halley, B. A.; Jacob, T. A.; Lu, A. Y. H. "The environmental impact of the use of ivermectin: Environmental effects and fate." *Chemosphere* **1989**, *18* (7-8), 1543-1564.

14. Sommer, C.; Nielsen, B. O. "Larvae of the dung beetle Onthophagus gazella exposed to lethal and sublethal Ivermectin concentrations." *J. Appl. Entomol.* **1992**, *114,* 502-509.

15. Strong, L. "Avermectins - a review of their impact on insects of cattle dung." *Bull. Entomol. Res.* **1992**, *82* (2), 265-274.

16. Sommer, C.; Steffansen, B. "Changes with time after treatment in the concentrations of ivermectin in fresh cow dung and in cow pats aged in the field." *Veterin. Parasitol.* **1993**, *48*, 67-73.

17. Madsen, M.; Nielsen, B. O.; Holter, P.; Pedersen, O. C.; Jespersen, J. B.; Jensen, K. M. V.; Nansen, P.; Gronvold, J. "Treating cattle with ivermectin: Effects on the fauna and decomposition of dung pats." *J. Appl. Ecol.* **1990**, *27*, 1-15.

18. Gruber, V.; Halley, B. A.; Hwang, S. C.; Ku, C. C. "Mobility of avermectin B-1a in soil." *J. Agricul. Food Chem.* **1990**, *38* (3), 886-890.

19. Ingham, E. R.; Coleman, D. C. "Effects of streptomycin, cycloheximide,

fungizone, captan, carbofuran, cygon, and PCNB on soil microorganisms." *Microb. Ecol.* **1984**, *10*, 345-358.

20. Ingham, E.R.; Parmelee, R.; Coleman, D. C.; Crossley, D. A. "Reduction of microbial and faunal groups following application of streptomycin and captan in Georgia no-tillage agroecosystems." *Pedobiologia* **1991**, *35(5)*, 297-304.

21. Warman, P. R. "The effect of amprolium and aureomycin on the nitrification of poultry manure-amended soil." *Soil Sci. Soc. Am. J.* **1980**, *44*, 1333-1334.

22. Ingham, E. R.; Trofymow, J. A.; Ames, R. N.; Hunt, H. W.; Morley, C. R.; Moore, J. C.; Coleman, D. C. "Trophic interactions and nitrogen cycling in a semi-arid grassland soil." *J. Appl. Ecol.* **1986**, *23*, 615-630.

23. Patten, D. K.; Wolf, D. C; Kunkle, W. E.; Douglas, L. W. "Effect of antibiotics in beef cattle feces on nitrogen and carbon mineralisation in soil and on plant growth and composition." *J. Environ. Qual.* **1980**, *9*(1), 167-172.

24. Beare, M. H.; Parmelee, R. W.; Hendrix, P. F.; Cheng, W. X.; Coleman, D. C.; Crossley, D. A. "Microbial and faunal interactions and effects on litter nitrogen and decomposition in agroecosystems." *Ecol. Monographs* **1992**, *62*(4), 569-591.

25. Davies, J. "Inactivation of antibiotics and the dissemination of resistance genes." *Science* **1994**, *264* (5157), 375-382.

26. van Elsas, J. D.; Trevors, J. T.; Starodub, M. E. "Bacterial conjugation between pseudomonads in rhizospehere of wheat." *FEMS Microbiol. Ecol.* **1988**, *53*, 299-306.

27. Top, E.; Desmet, I.; Verstraete, W.; Dijkmans, R.; Mergeay, M. "Exogenous isolation of mobilizing plasmids from polluted soils and sludges." *Appl. Environ. Microbiol.* **1994**, *60* (3), 831-839.

28. Mach, P. A.; Grimes, D. J. "R-plasmid transfer in a wastewater treatment plant." *Appl. Environ. Microbiol.* **1982**, *44*, 1395-1403.

29. Fernandez-Astorga, A.; Dearanguiz, A. F.; Pocino, M.; Umaran, A.; Cisterna, R. "Conjugal transfer of R-plasmids to and from enterobacteriaceae isolated from sewage." *J. Appl. Bacteriol.* **1992**, *72* (5), 381-385.

30. Salyers, A.; Shoemaker, N. B. "Broad-host-range gene-transfer - Plasmids and conjugative transposons." *FEMS Microbiol. Ecol.* **1994**, *15*, 15-22.

31. Batchelder, A. R. "Clortetracycline and oxytetracycline effects on plant growth and development in liquid cultures." *J. Environ. Qual.* **1981**, *10*, 515-518.

32. Batchelder, A. R. "Chlortetracycline and oxytetracycline effects on plant growth and development in soil systems." *J. Environ. Qual.* **1982**, *11*, 675-678.

300

33. Baguer, A. J.; Jensen, J.; Krogh, P. H. "Effects of the antibiotics oxytetracycline and tylosin on soil fauna." *Chemosphere* **2000**, *40* (7), 751-757.

34. Wardle, D. A. "Impacts of disturbance on detritus food webs in agro-ecosystems of contrasting tillage and weed management practices." *Adv. Ecol. Res.* **1995**, *26,* 105-184.

35. Gunn, A.; Cherrett, J. M. "The exploitation of food resources by soil meso-invertebrates and macro-invertebrates." *Pedobiologia* **1993**, *37*(5), 303-320.

36. Bengtsson, J.; Setälä, H.; Zheng, D. W. "Food webs and nutritive cycling in soils: interactions and positive feedbacks." In *Food webs. Integration of patterns & dynamics.* G. A. Polis and K. O. Winemiller (Eds). New York, Chapman & Hall **1996**, 30-38.

37. Yeates, G. W.; Wardle, D. A.; Watson, R. N. "Responses of soil nematode populations, community structure, diversity and temporal variability to agricultural intensification over a seven-year period." *Soil Biol. Biochem.* **1999**, *31*(12), 1721-1733.

38. Salminen, J.; Setälä, H.; Haimi, J. "Regulation of decomposer community structure and decomposition processes in herbicide stressed humus soil." *Appl. Soil Ecol.* **1997**, *6,* 265-274.

39. Ingham, E.; Coleman, D. C.; Crossley, D. A. "Use of sulfamethoxazole-penicillin, oxytetracycline, carbofuran, carbaryl, naphthalene and temik to remove key organism groups in soil in a corn agroecosystem." *J. Sustain. Agricul.* **1994**, *4* (3), 7-30.

40. Bardgett, R. D.; Cook, R. "Functional aspects of soil animal diversity in agricultural grasslands." *Appl. Soil Ecol.* **1998**, *10* (3), 263-276.

41. Bogomolov, D. M.; Chen, S. K.; Parmelee, R. W.; Subler, S.; Edwards, C. A. "An ecosystem approach to soil toxicity testing: A study of copper contamination in laboratory soil microcosms." *Appl. Soil Ecol.* **1996**, *4* (2), 95-105.

42. Parmelee, R. W.; Phillips, C. T.; Checkai, R. T.; Bohlen, P. J. "Determining the effects of pollutants on soil faunal communities and trophic structure using a refined microcosm system." *Environ. Toxicol. Chem.* **1997**, *16* (6), 1212-1217.

43. van den Berg, M. M. H. E.; Tamis, W. L.M.; van Straalen, N. M. "The food web approach in ecotoxicological risk assessment." *Human Ecol. Risk Assess.* **1998**, *4* (1), 49-55.

44. Wall, R.; Strong, L. "Environmental consequences of treating cattle with the antiparasitic drug ivermectin." *Nature* **1987**, *327,* 418-421.

45. Sommer, C.; Steffansen, B.; Nielsen, B. O.; Jensen, V.; Jespersen, K. M. V.; Springborg, J. B.; Nansen, P. "Ivermectin excreted in cattle dung after subcutaneous injection or pour-on treatment - concentrations and impact on dung fauna." *Bull. Entomol. Res.* **1992**, *82* (2), 257-264.

46. Holter, P.; Sommer, C.; Gronvold, J.; Madsen, M. "Effects of ivermectin treatment on the attraction of dung beetles (Coleoptera, Scarabaeidae and Hydrophilidae) to cow pats." *Bull. Entomol. Res.* **1993**, *83* (1), 53-58.

47. Gunn, A.; Sadd, J. W. "The effect of ivermectin on the survival, behaviour and cocoon production of the earthworm Eisenia fetida." *Pedobiologia* **1994**, *38* (4), 327-333.

48. Madsen, M.; Gronvold, J.; Nansen, P.; Holter, P. "Effects of treatment of cattle with some anthelmintics on the subsequent degradation of their dung." *Acta Veterin. Scandin.***1988**, *29* (3-4), 515-517.

49. Wislocki, P. G.; Grosso, L. S.; Dybas, R. A. "Environmental aspects of abamectin use in crop protection." In *Ivermectin and Abamectin*, W.C. Cambell (Ed.), Springer Verlag: New York, **1989**, pp 182-200.

50. Halley B. A.; Van den Heuvel, W. J. A.; Wislocki, P. G. "Environmental effects of the use of avermectins in livestock." *Veterin. Parasitol.* **1993**, 48: 109-125.

51. Webb, J. D.; Burg, J. G.; Knapp, F. W. "Moxidectin evaluation against *Solenoptes capillatus* (Anoplura, Linognathidae), *Bovicola bovis* (Mallophaga, Trichodectidae), and *Musca autumnalis* (Diptera, Muscidae) on Cattle." *J. Econom. Entomol.* **1991**, *84* (4), 1266-1269.

52. Doherety. W. M.; Stewart, N. P.; Cobb, R. M.; Keiran, P. J. "*In vitro* comparison of the larvicidal activity of moxidectin and abamectin against *Onthophagus gazella (F.) (Coleoptera: Scarrabaeidae)* and *Haematobia irritans exigua* De Meijere (Diptera: Muscidae)." *J. Austral. Entomol. Soc.* **1994**, *33*, 71-74.

53. Strong, L.; Wall, R. "Effects of ivermectin and moxidectin on the insects of cattle dung." *Bull. Entomol. Res.* **1994**, *84* (3), 403-409.

54. Wardhaugh, K. G.; Holter, P.; Whitby, W. A.; Shelley, K. "Effects of drug residues in the faeces of cattle treated with injectable formulations of iveremectin and moxidectin on larvae of the bush fly, *Musca vetustissima*, and the house fly, *Musca domestica*." *Austral. Vet. J.* **1996**, *75*, 370-374.

55. Fincher, G. T.; Wong, G. "Injectable moxidectin for cattle: Effects on 2 species of some dung burying beetles." *Southwest Entomol.* **1992**, *21*, 871-876.

56. Sonnenschein, C.; Soto, A. M. "An updated review of environmental estrogen and androgen mimics and antagonists." *J. Steroid Biochem. Molec. Biol.* **1998**, *65* (1-6), 143-150.

57. Shore, L.; Kapulnik, Y.; Ben-Dor, B.; Fridman, Y.; Wininger, S.; Shemesh, M. "Effects of estrone and 17-beta-estradiol on vegetative growth of *Medicago sativa*." *Physiol. Plantar.* **1992**, *84* (2), 217-222.

58. Shore, L. G.; Gurevitz, M.; Shemesh, M. "Estrogen as an environmental pollutant." *Bull. Environ. Contam. Toxicol.* **1993**, *51* (3), 361-366.

59. Doran, J. W.; Parkin, T. B. "Defining and assessing soil quality." In *Defining soil quality for a sustainable environment*. Doran J. W., Coleman, D. C., Bezdicek, D. F., Stewart, B. A.; Eds; Soil Science Society of America. 1994, Special Publication Number 35.

60. Spaepen, K. R. I., van Leemput, L. J. J., Wislock, P. G., Verschueren, C. "A uniform procedure to estimate the predicted environmental concentration of the residues of veterinary medicines in soil." *Environ. Toxicol. Chem.* **1997**, *16* (9), 1977-1982.

61. Montfors, M. H. M. M., Kalf, D. F., van Vlaardingen, P. L., Linders, J. B. "The exposure assessment for veterinary medicinal products." *Sci. Tot. Environ.* **1999**, *225*, 199-133.

62. "Drugs in the environment." Special Issue of *Chemosphere* **2000**, *40*(7), 691-793. Jørgensen, S.E., Halling-Sørensen, B (Guest Editors).

63. Spratt, D. M. "Endoparasite control strategies: Implications for biodiversity of native fauna." *Internat. J. Parasitol.* **1997**, *27* (2), 173-180.

64. McCracken, D.I. "The potential for avermectins to affect wildlife." *Veterin. Parasitol.* **1993**, *48*, 273-280.

65. Forbes, A. B. "A review of regional and temporal use of avermectins in cattle and horses worldwide." *Veterin. Parasitol.* **1993**, *48*, 19-28.

66. Herd, R. "Endectocidal drugs - Ecological risks and countermeasures." *Internat. J. Parasitol.* **1995**, *25* (8), 875-885.

67. Sherratt, T. N.; Macdougall, A. D.; Wratten, S. D.; Forbes, A. B. "Models to assist the evaluation of the impact of avermectins on dung insect populations." *Ecolog. Model.* **1998**, *110* (2), 165-173.

Risk Assessment

Chapter 17

Environmental Risk Assessment of Pharmaceuticals: A Proposal with Special Emphasis on European Aspects

Jörg Römbke[1], Thomas Knacker[1], and Hanka Teichmann[2]

[1]ECT Oekotoxikologie GmbH, Böttgerstr. 2-14, D-65439 Flörsheim, Germany
[2]Federal Environmental Agency, Bismarckplatz 1, D-14191 Berlin, Germany

From an ecotoxicological point of view, pharmaceuticals were not considered to be a problem until the early 1990s (*1*). The need for data on the fate and effects of veterinary pharmaceuticals in the environment was identified for the first time in The Netherlands (*2*). Since then the environmental risk caused by pharmaceuticals, in general, was re-evaluated (*3*), mainly because an ever increasing number of drugs were detected in the environment (*4*). Therefore governmental authorities began to discuss how medicinal products should be dealt with from an ecotoxicological point of view. Today, prudence and good practice dictate that an environmental risk assessment (ERA) for new pharmaceuticals should be performed. Current practice follows the same rules that have been established for other chemicals (nothing is required for existing drugs). However, there are some large discrepancies in how such an ERA should look in detail. In this paper a short overview on the legal status in the European Union (EU) concerning environmental aspects of the registration procedure for new medicinal products is given. In addition, a modified ERA for (human) medicinal products is proposed and compared with ERA procedures currently used in the EU. Since the situation in the USA is described by Velagaleti and Gill (this volume) it will not be discussed in any detail here.

Legal situation in the European Union

Since 1 January 1995, innovative drugs as well as certain medicinal products manufactured by biotechnological methods are registered by the European Medicines Evaluation Agency (EMEA) located in London, U.K. For all other types of drugs, the rule has been established that the registration of medicinal products by one national competent authority should be approved by all other national competent authorities within the EU. In the case of an objection by one member country, EMEA is asked to settle the case by arbitration. EMEA consists of two committees, of which one (CPMP: Committee for Pharmaceutical Medicinal Products) is responsible for human and the other one (CVMP: Committee for Veterinary Medicinal Products) for veterinary medicinal products. Additionally, these committees develop guidelines for the registration of medicinal products, and these are regularly published (e.g. "Notices for Applicants"; 5). All regulations for drugs within the EU are based on the Directive 65/65/EEC, extended and amended by Directive 75/318/EEC and 93/39/EEC, respectively. In the latter, § 4.6 states that each application for the registration of a medicinal products must provide information on whether the product may cause a risk to the environment.

In the years 1994/95 several EU working groups discussed various test requirements as a pre-requisite for the ecotoxicological risk assessment of human medicinal products. At first it was thought to use the same procedure as described for industrial chemicals in Directive 67/548/EEC including its 7th amendment (Directive 92/32/EEC) and the related Technical Guidance Documents (6) as a starting point. In these documents and guidance papers the required environmental data as well as the environmental risk assessment procedures for existing and new chemicals are laid down. Some of the early draft documents on environmental data requirements for medicinal products were very far reaching and beyond practicability. Later the discussion focused on the value of the Predicted Environmental Concentration (PEC), which should be used as a trigger value. Estimated PECs higher than the trigger value would then initiate further investigations, in particular, on environmental effects of the drug. Estimated PECs lower than the trigger value would lead to the conclusion that the drug is of no concern for the environment. In an earlier version of the draft documents, 0.001 μg/L (later 0.01 μg/L) was chosen as the trigger value for the aquatic medium, while 10 μg/kg was determined to be the trigger value for drugs in soil. However, the various draft documents on human medicinal products have not been finalized.

The requirements for the registration of veterinary medicinal products within the EU were harmonized by Directive 81/852/EEC, amended by Directive 92/18/EEC. In contrast to the Directive 93/39/EEC for human medicinal products, this document for the very first time contained data requirements for "ecotoxicity" (§ 3.5). Based on these Directives, the EMEA/CVMP published a detailed "Note for Guidance" (7) for the registration of all veterinary medicinal products that do not contain genetically modified organisms. Since January 1998, an environmental assessment based on the EMEA Note is necessary for the registration of veterinary drugs in Europe.

International Harmonization

Currently the registration procedure for human and veterinary medicinal products are about to be harmonized within the European Union, Japan, and the United States. The Veterinary International Co-operation on Harmonisation (VICH) is responsible for the harmonization process of veterinary drugs. In 1996 the VICH Steering Committee authorized the formation of a working group to develop harmonized guidelines for conducting the environmental risk assessment (here called environmental impact assessment: EIA) for veterinary medicinal products. The mandate of this VICH group is to elaborate tripartite guidelines on the design of studies and the evaluation of the environmental impact of veterinary drugs. It was suggested to follow a tiered approach based on the general principles of risk analysis. Categories of products to be covered by different tiers of these guidelines should be specified. Existing or draft guidelines in the USA, the EU, and Japan should be taken into account. The working group consists of six members — one from each tripartite member, one expert from industry, and one expert from competent authorities.

Environmental Risk Assessment

Since it was not intended to reinvent the wheel for medicinal products, general principles of Environmental Risk Assessment (ERA), which are widely accepted in the industrialized countries for various chemical groups, are used as a starting point. An ERA consists of the following steps (Fig. 1):
1. Hazard identification
2. Exposure (PEC) and effect (PNEC: Predicted No Effect Concentration) analysis
3. Risk characterization (PEC/PNEC ratio)
4. Risk management.

According to Barnthouse (8) an ERA is "simply a systematic means of developing a scientific basis for regulatory decision making". It has been developed in the USA for the notification of chemicals. The EU started off to use ERAs for the registration of pesticides. The whole ERA is an iterative process, i.e., in a tiered approach, in particular, exposure and effect analysis can be repeated twice with increasing complexity. Probably the example of an ERA best suited for medicinal products is the process currently used in the EU for existing and new chemicals (6).

Figure 1: General principles of an Environmental Risk Assessment (ERA) process (modified according to various authors).

HAZARD IDENTIFICATION
Evaluation, which environmental compartments (e.g. surface water, soil) are likely to be affected, based on the properties of the compound and the use pattern

EXPOSURE ASSESSMENT   EFFECT ASSESSMENT
Calculation of the PEC   Calculation of the PNEC
(incl.safety factors)

RISK ASSESSMENT
Comparison of (measured or estimated) exposure data with (on different investigation levels, e.g. laboratory, microcosms, field) measured or estimated effect data (separately for the main environmental compartments)

RISK CHARACTERIZATION
Assessment of the probability that an environmental risk is likely to occur by calculating the PEC/PNEC ratio (< or > 1)

RISK MANAGEMENT
Measures in order to avoid or to minimize an environmental risk as part of the registration or re-registration decision

ERAs for selected medicinal products

In the following, ERAs for several human and veterinary medicinal products based on literature data are presented. The first step (hazard identification) is not given since all chemicals listed here have already been measured in the environment or are likely to occur in the environment due to their high production volume.

Risk Assessment for Acetylsalicylic acid

Exposure analysis: Measured concentration in surface water: 0.34 µg/L
    ==> PEC 0.34 µg/L
Effect analysis: Acute laboratory test (*Daphnia magna*): $LC_{50} \approx 167.5$ mg/L
    Assessment factor according to EU (*6*): 1000
    ==> PNEC = 167.5 µg/L
Risk characterization: PEC / PNEC - ratio: 0.002 (EU Class 2)

Comment:
There is no indication of a risk for the environment. The same result would be found when using an EC_{50} (reproduction) of 61 mg/L and an assessment factor of 100. However, due to the high production volume, monitoring samples in surface waters should be taken from time to time. Stuer-Lauridsen et al. (*9*) compared the PNEC with a calculated PEC based on the average use of this drug in Denmark and revealed a PEC / PNEC – ratio of 1.3. In addition, data should be provided for the main metabolite salicylic acid, taking into consideration that anthropogenic as well as natural sources for this substance are known.

Low PEC/PNEC ratios (EU-Class 2) were found for the psychiatric drug diazepam, the bronchodilator theophylline, and the antibiotic erythromycin (*3, 9*).

Risk Assessment for Clofibric acid/Clofibrate

Exposure analysis: Measured concentrations in surface water: 0.001 – 1.75 µg/L
    ==> PEC = 1.75 µg/L
Effect analysis: Acute laboratory tests with two species: EC_{50} 12 – 28.2 mg/L
    Assessment factor according to EU (*6*): 1000
    ==> PNEC = 12 µg/L
    Sublethal laboratory tests with 2 species: NOEC 0.01 – 5.4 mg/L
    Assessment factor according to EU (*6*): 50
    ==> PNEC = 0.2 µg/L
Risk characterization: PEC / PNEC - ratio: 8.75 (EU-Class 3)

Comment:
Compared with the fairly large data basis for the exposure analysis, the effect analysis should be improved. Kalbfus & Kopf (*10*) calculated a PEC/PNEC ratio of 5, using own exposure as well as acute and chronic effect data derived from five species and an assessment factor of 50. It should be kept in mind that this ERA was performed with effects data obtained from tests in which clofibrate was

applied while the exposure data were measured for clofibric acid. This might be acceptable because the transformation of the ester to the acid occurs in water as well as in organisms.

Risk Assessment for Ethinylestradiol

Exposure Analysis: Measured concentration in surface water: 0.062 mg/L
         ==> PEC = 0.062 mg/L
Effect Analysis: NOEC from two chronic tests (Algae, Daphnia): 10 - 54 mg/L
           Assessment factor according to EU (6): 100
           ==> PNEC = 0.1 mg/L
Risk-Characterization: PEC / PNEC - Ratio: 0.62 (EU-Class 2)

Comment:
No indication of environmental risk potential. However, endocrine effects on fish were found in non-standardized tests at about 1 ng/L. If these results are assessed as being valid by the authorities, EU-class 1 has to be taken.

Risk Assessment for Ivermectin (Antiparasitic)

Probably the best investigated veterinary drug is the antiparasitic Ivermectin, due to its very high acute toxicity to many invertebrates (11). An ERA had been performed as early as 1986 (U.S. Federal Register, Vol. 51, No. 145). Since then, many studies (but few according to standardized guidelines) have confirmed that exposure and effects are likely to happen in the environment (3). Labels are already in use to avoid damage for the aquatic compartment.

Exposure Analysis: Measured concentration in cow dung: 0.4 - 9 mg/kg
         ==> PEC = 9.0 mg/kg
Effect Analysis: LC_{50} from an acute tests (earthworm): 15.8 - 315 mg/kg
           Assessment factor according to EU (6): 1000
           ==> PNEC = 0.016 mg/kg
Risk-Characterization: PEC / PNEC - Ratio: 562 (EU-Class 1)

Comment:
When reassessing the environmental concentration of Ivermectin (maximum 1.70 mg/kg; Montforts, pers. comm.) and assuming that the low effect concentration for earthworms (15.8 mg/kg) might not be valid, i.e. (Barth, pers. comm.), the PEC/PNEC ratio would still be higher than 1. Nevertheless, refinement of effects data seems to be necessary. For the fauna of the dung sub-compartment, a risk has been identified, whereas the soil compartment, in general, seems not to be affected (12). The results of some field studies confirm

a risk, whereas in other investigations no effects on the meadow ecosystem were found (*13*, *14*). These seemingly contradictory results are probably caused by the application of different, usually not standardized methods.

For 64 selected human drugs Webb (*15*) assessed the environmental risk for the aquatic compartment in the U.K. Based on short-term effects data and worst-case assumptions for assessing the environmental concentrations (no metabolism, no removal in wastewater plants, no dilution in surface water), only seven compounds showed a PEC/PNEC ratio higher than 1 (see also *9*). When taking into account a dilution factor of 10, the number of drugs with a PEC/PNEC ratio higher than 1 was reduced to one (paracetamol). Paracetamol, however, can be metabolized and has not been found in field monitoring programs so far (*16*). However, the author identified data gaps (e.g., concerning fate and effects of metabolites or chronic ecotoxicity tests in general) which should be filled for a final assessment.

When treating medicinal products like "normal" industrial chemicals, i.e., when ignoring that drugs are specially designed to achieve defined biological effects, the following conclusions can be drawn with regard to ERAs:
- few valid data were found in the literature;
- based on established short-term toxicity data, most medicinal products seem to cause no risk for the environment;
- there is an urgent requirement to adjust the ERA to the specific environmental compartments of concern and to investigate organisms and end-points that are related to the biological activity of the medicinal product.

Proposal for an ERA Procedure for Medicinal Products

Human and veterinary medicinal products should be treated separately as well as the main environmental compartments of concern. In the case of many veterinary medicinal products (e.g., antibiotics in aquaculture or antiparasitics in agriculture), the ERA proposed is similar to the ERA of pesticides due to their direct input into the environment, whereas drugs for pets and human medicinal products are more handled like industrial chemicals. In the following, the four steps of an ERA modified for medicinal products are described.

Hazard Identification

In this first step it is investigated whether and, if yes, how the following steps of an ERA should be performed. The most important issues are:
- Is a contamination of the environment likely to occur at all? If not, an ERA should not be performed. An example for such a case are radioactive tracers

for cancer treatment in hospitals. This group of substances is controlled by other regulations to avoid any release into the environment. Another example are naturally occurring minerals, enzymes, and organic compounds which might be used in small amounts for human health purposes.

- Do the inherent properties of an active ingredient indicate environmental risks? For example, high log P_{ow} values (> 3) indicate that a chemical is liable to accumulate in biota.

- Which exposure pathways should be considered? In general, drugs or their metabolites can reach the environment directly or indirectly via the target organism. The most important exposure pathway for human medicinal products is the entry in surface water after passage through a sewage treatment plant. Veterinary medicinal products applied in agriculture might be either released directly or indirectly as a diffuse source via dung or liquid manure on pasture, arable land, and in surface water. Those veterinary drugs used in aquaculture are released directly into limnic or marine compartments. In industrialized countries, releases during production are unlikely to occur, whereas in other regions such releases might happen more often due to the lack of appropriate regulations (e.g., *17*). Further, it should be examined whether the medicinal product of concern is released to the environment due to other uses (e.g., acetylsalicylic acid can occur in surface waters as a metabolite of naphthalene oil (*18*).

- Which environmental compartments (surface water, marine water, sediment, soil) are likely to be at risk? In the case of most human and those veterinary drugs used in aquaculture, most likely are surface waters and sediments in which the drugs are emitted directly or indirectly after passing through a wastewater treatment plant. Other veterinary drugs are released via excretion in dung or via spreading of manure to soil. A risk to the air/atmosphere compartment seems be very unlikely.

Exposure Analysis

Ecotoxicological fate and effect studies should be performed according to standardized and accepted international guidelines (e.g., OECD). If these are not available, test proposals described in the scientific literature should be used; e.g., for the determination of the degradation of veterinary drugs in manure (*19*). In any case, expert knowledge is necessary not only for the assessment of test results, but also for the selection of the best suited test methods (*20*). Additionally, a suitable specific analytical method must be available. Since the active ingredient and the main metabolites of a new drug have to be analyzed in various organic substrates like blood anyway, it should normally be no problem to adapt these methods to environmental compartments like water or soil.

In the following, a tiered scheme for an exposure assessment of human medicinal products is presented for the limnic compartment. Exposure assessments for other environmental compartments could follow the same principal rules but, so far, are less elaborated. At the beginning and on a regional scale, the PEC of a human medicinal product in surface water should be related to the number of inhabitants in a given area, the amount of water used by the population of this area, and the average number of prescriptions. Based on the estimated concentration determined in this step, a tiered scheme can be applied:

Level 1: Evaluation of the PEC after passage through a sewage treatment plant.
In existing formulae, the degradation in the wastewater treatment plant is considered to be zero to simulate the "worst case" situation. A more realistic approach might be to incorporate the results of appropriate biodegradation studies into the assessment of the PEC.

Level 2: Improvement of the exposure assessment.
The PEC assessment can be improved by performing water/sediment simulation tests with environmentally relevant concentrations.

Level 3: Residue analyses in sewage treatment outflow and in surface water.
Based on the experiences and results gained at the first two levels, samples should be analyzed from the environmental compartments of concern. However, level 3 data can only be requested for a product that is expected to have been released into the environment, i.e., if the product is re-registered or if further data are requested to be provided after a preliminary registration has been granted.

Effect Analysis

Due to the usually low concentrations and long exposure periods of medicinal products, long-term studies and the determination of bioaccumulation should be preferred instead of short-term studies normally performed at the first level ("base set") of effect analysis for other chemicals (e.g., *21*). Results obtained from effects studies are transferred into PNEC values by applying assessment factors. Such factors have been introduced to take into account methodological uncertainties of the tests as well as the uncertainties when extrapolating laboratory data to field situations. The assessment factors become smaller the better the environmentally relevant data. According to established guidance (*6*), the assessment factor is 1000 if only acute data are available; if three NOEC tests with species from different trophic groups are performed, the factor is 10 (*6*). Other guidelines (e.g., for pesticides) use different assessment factors, but usually the final result of an ERA would not be different.

It is important to keep in mind that medicinal products are designed to exhibit rather specific biological effects. For most of these specific effects, for example endocrine or neurotoxic effects, test methods are not yet developed or are still in the standardization/validation process. In particular it seems not to be justified to measure effects in cell cultures or by using other *in-vitro* techniques when effects can be expected to occur at higher biological organization levels (e.g., populations). Here, specific trigger values for the performance of tests at level 1 are presented, using the aquatic compartment as an example (Table I). These values are not valid for medicinal products with known mutagenic potential.

Table I: Trigger values for the performance of standardized effects tests (level I)

Evaluated Exposure concentration (μg/L)	Ready Bio-degradation	Low Potential for Bio-accumul. ($\log P_{OW} < 3$)	Effects tests:
PEC < 0.001		Not relevant for decision	No
$0.001 \leq$ PEC < 0.01	Yes	Yes	No
$0.001 \leq$ PEC < 0.01	No	No	Yes
PEC > 0.01		Not relevant for decision	Yes

The determination of the PNEC in surface water follows also a tiered scheme. Due to the reasons given above, e.g., the specific effects mechanisms of medicinal products, new tests need to be developed and incorporated into updated versions of the following approach for the effects analysis:

Level 1:
At least two long-term studies, e.g., Algae Growth Inhibition (*22*) and *Daphnia* Reproduction (*22*), should be performed. An assessment factor of 100 is recommended for level 1. In the case of a persistent drug that might adsorb to the sediment, a sediment dwelling organism like *Chironomus riparius* should also be tested (*23*).

Level 2:
Additional long-term tests, e.g., Fish Early-Life-Stage (*22*) as well as studies related to specific effects mechanisms (e.g. endocrine effects) should be performed. The assessment factor could then be 10. The bioconcentration of the drug in fish (BCF_{Fish}) should be evaluated if $\log P_{ow} \geq 3$ and PEC < 0.01 μg/L. The BCF_{Fish} should be measured if $\log P_{ow} \geq 3$ and PEC $\geq$ 0.01 μg/L. The potential of secondary poisoning should be considered.

Level 3:
Taking into consideration that an ERA is not aimed to evaluate the risk to a

single species, but rather to ecosystems, studies in micro- or meso-cosms might be appropriate at this level. Further, at level 3, biomonitoring studies might be required. As stated earlier, these data can either be requested for a product that is expected to be present in the environment and which is about to be re-registered or for a product that has been registered preliminarily under the condition of providing highly relevant ecological effect data. Biomonitoring of effects of a single medicinal product is extremely difficult taking into account the vast amount of interactions with other chemicals and/or stress factors, the thousands of potentially affected species and the varying environmental conditions.

Risk Characterization

This step follows the rules established by the EU for the ERA of existing and new chemicals (*6*): the PEC / PNEC ratio is calculated. Depending on the result, the medicinal product is classified in one out of three classes.

Initial PEC / PNEC ratio:

PEC/PNEC < 1: No indication of environmental risk potential (EU-Class 2)
PEC/PNEC > 1: Refinement of exposure and effect analysis (EU-Class 1)

PEC / PNEC ratio after data refinement:

PEC/PNEC < 1: No indication of environmental risk potential (EU-Class 2)
PEC/PNEC > 1: Risk management necessary (EU-Class 3)

There is general consensus that in cases where a refinement of data is necessary, preferably a PEC-refinement should be performed.

Risk Management

Recommendations for risk management measures are difficult, but nevertheless some thoughts are indicated, which might need further considerations:
- In a class of drugs that exhibit the same medicinal properties, the use of those with a minimum ecotoxicological risk potential should be preferred.
- Releases into the environment might be reduced by modifying the formulation or by recommending special treatments of non-used amounts.
- In aquaculture the application of veterinary drugs directly into surface waters could either be adjusted to the minimum of the required amount or banned for a limited period of time or be eliminated from the market.

In addition, it seems to be doubtful whether an appropriate labelling (for whom, how?) should be accepted as a genuine risk management tool.

315

Comparison and Discussion of Various ERA Proposals

Currently, various ERAs for medicinal products have been recommended (e.g., *6, 24*). All of these concepts are based on the same principles (e.g., a tiered test system approach). In the following, the EU/CVMP concept is discussed (Fig. 2), but several of its critical points are also true for other concepts.

Figure 2: Simplified schematic flow diagram for the evaluation of an Environmental Risk Assessment (ERA) for veterinary drugs (7).

PHASE I:
> Significant exposure of the product in the environment (> 10 µg/kg soil or dung; 0.1 µg/L groundwater) after PEC calculation and/or fate tests ?
> Fish medicine: Go directly to Phase II – Tier A
>   IF NO: No specific precautions necessary
>   IF YES: Go to Phase II – Tier A

PHASE II-TIER A:
> Harmful effects on the environment likely (based on – mainly acute - standard fate and effect tests) ?
>   IF NO: No specific precautions necessary (taking risk management strategies into consideration)
>   IF YES: Go to Phase II – Tier B

PHASE II-TIER B:
> Harmful effects on the environment identified (based on specific (including, if necessary, field) tests) ?
>   IF YES: Taking risk reduction measures and risk/benefit analysis into consideration ➔ acceptable
>   IF YES: Potential for harmful effects are not outweighed by therapeutic efficiency ➔ unacceptable.

In Europe, the ERA for veterinary medicinal products published in the Note for Guidance (*7*) is required since 1998. It is divided into two phases: In Phase I the potential of exposure of the product, its ingredients, or relevant metabolites is assessed by applying trigger values for concentrations in manure (100 µg/kg), dung (10 µg/kg), soil (10 µg/kg) and groundwater (0.1 µg/L). Products exceeding any trigger value have to be examined in the rather complex Phase II (many different studies can be required). Fish medicinal products to be applied directly to the environment have directly to be tested in Phase II. This Phase consists of Part A in which the fate of the substance in relevant environmental compartments (including bioaccumulation) and, for the first time, its effects on organisms (e.g., earthworms) using short-term tests are studied. However, if high persistence (DT_{90} > 1 year) and low adsorption (Koc < 500) have been determined, even microcosm or field studies can be requested (20). If in Part A of Phase II no harmful effects have been found or an appropriate risk management strategy is proposed by the applicant, no further testing is necessary. Otherwise, in Part B, refined tests for the environmental compartment likely to be affected have to be performed, which to a large extent are subject to expert judgment. In contrast to other ERA proposals, the EMEA/CVMP Note mentions the need of risk management measures – but the guidance on measures is rather general

The EMEA/CVMP Note has been criticized by scientists as well as governmental authorities. In addition, industry has proposed some concern intending to reduce the numbers of medicinal products to be assessed since according to Montforts (pers. comm.), "all veterinary antibiotics intended for herd treatment will enter a Phase II assessment for the soil compartment (because the PEC_{soil} are higher than the trigger)". An ERA encompassing Phase I and II certainly demands considerable efforts with regard to time and costs. In his comments on the EMEA/CVMP Note, Lepper (25) has expressed that especially Phase II "offers potential for an improvement and optimisation with respect to the environmental relevance of the recommended tests and the testing efforts". The same author has proposed to include effects data, possibly based on QSAR estimations, in Phase I which might allow to avoid assessments for drugs in Phase II. In addition, he recommends a shift from various single-species tests to simple microcosm tests since results from such studies are usually more reliable, show a better reflection of environmental conditions, and are more cost-effective, especially if several metabolites have to be considered simultaneously. These comments are in agreement with the opinion of the authors of this paper and should be incorporated into the ERA procedures.

Further aspects of the EMEA/CVMP Note should also be discussed:
- The assessment of the aquatic compartment for veterinary drugs released to the terrestrial compartment should not be explicitly excluded from an ERA since these drugs can reach surface waters via wastewater or run-off.
- It is not feasible that in Phase I of the Note no-effect tests are required at all. Even in Phase II Part A, (mainly acute) effects tests have to be performed only if the trigger values have been exaggerated. However, effects can occur at lower concentrations than those defined for the trigger values. For example, the antiparasitic Ivermectin induces aberrations in wings of a dung fly at 0.5 µg/kg (26).
- Also the exclusion of veterinary drugs for pets from ERAs is not justified since adverse effects can be caused by any product released into the

environment. An example issue is the antiparasitic veterinary drug Bromocyclen which is closely related to some well-known plant protection products like dieldrin. This plant protection product has not been re-registered due to its high persistence and bioaccumulation potential. Bromocyclen, however, for which an ERA was not performed, was found at high concentrations in fish caught for human consumption. After proving that these findings were caused by the use of Bromocyclen as a pet drug, the drug was voluntarily withdrawn from the market (27).

- Montforts et al. (28) summarize their analysis of this approach as follows: "At the moment the triggers that lead to further extensive testing are either useless, or at least inconsistent with each other. The current proposals for effect assessment are too diverse."

Due to this criticism, an improvement of the existing EMEA note for veterinary medicinal products is necessary. The lessons learned from this process should be implemented in any notes for the assessment of human medicinal products.

Summary and Recommendations

There are few data available on the exposure and effects of medicinal products in the environment. Nevertheless these data indicate that some of these biologically highly active chemicals pose a potential risk to the environment (3, 9). There is a common understanding that a harmonized ERA should be applied to human and veterinary medicinal products, based on the following considerations:

- The ERA should be based on existing guidelines including assessment factors (e.g. as developed by the EU for the notification of New and Existing Chemicals, 6) and the Precautionary Principle.
- A tiered test strategy should be used in order to use resources efficiently, including the possibility of using multi-species test systems.
- The performance of an ERA is not necessary, if exposure is unlikely.
- Effect and exposure (incl. biodegradation) tests should start on level 1 of the tiered test strategy. NOECs or EC_x should be used for effects assessment. Persistence, bioaccumulation, and toxicity are the main endpoints.
- The scenarios and models used to calculate the PEC values have to be harmonized since even small changes can have considerable implications on the outcome of the calculation and therefore, for the risk assessment (19).
- New test systems have to be developed for the detection of specific (e.g., endocrine) or indirect (e.g., resistance of bacteria) effects.
- Parent compounds as well as "relevant" metabolites have to be considered.
- Risk management measures in addition to simple labelling should be developed (29).

318

- Finally, the international harmonization of the use of ERAs for medicinal products is supported.

There is also an overall agreement that more and better (i.e., valid) data are necessary; not only for new substances, but also for those existing drugs which have never been assessed up to now (*30*). One possibility to improve the ecotoxicological risk assessment of these existing drugs would be the definition of "priority substances" which are typical for individual classes of medicinal products like cytostatics, antibiotics, or sedatives. Such "priority substances" could be tested intensively in order to identify those groups or classes of drugs which could cause an environmental risk. However, better than all these activities would be to avoid the introduction of persistent chemicals into the environment (*30*).

References

1. Jørgensen, S.E.; Halling-Sørensen, B. *Chemosphere* **2000**, *40*, 691-699.
2. Hekstra, G.P. In *Terrestrial and Aquatic Ecosystems. Perturbation and Recovery;* Ravera, O., Ed.; Ellis Horwood Ltd.: Chichester, U.K. 1991; pp 501-516.
3. Römbke, J.; Knacker, T.; Stahlschmidt-Allner, P. *UBA-Texte* **1996**, *60/96*, 1-361.
4. Daughton, C.G.; Ternes, T. *Environ. Health Perspect.* **1999**, *107* (suppl. 6), 907-938.
5. Irwin, V. In *Environmental Risk Assessment for Pharmaceuticals and Veterinary Medicines;* Wolf, P.U., Ed.; RCC Group: Basle, Switzerland, 1995; pp 48-57.
6. *Technical Guidance Documents in Support of The Commission Directive 93/67/EEC on Risk Assessment for New Notified Substances and The Commission Regulation (EC) 1488/94 on Risk Assessment for Existing Substances;* European Union: Brussels, Belgium, 1996; 734 pp.
7. *Note for Guidance: Environmental Risk Assessment for Veterinary Medicinal Products other than GMO-Containing and Immunological Products;* EMEA/CVMP/055/96. European Union, EMEA: London, 1997.
8. Barnthouse, L.W. *Environ. Toxicol. Chem.* **1992**, *11*, 1751-1760.
9. Stuer-Lauridsen, F.; Birkved, M.; Hanse, L.P.; Holten Lützhøft, H-C.; Halling-Sørensen, B. *Chemosphere* **2000**, *40*, 783-793
10. Kalbfus, W.; Kopf, W. *Münchener Beitr. Abwasser-, Fischerei- Flussbiol.* **1998**, *51*, 628-652.
11. Campbell, W.C. *Ivermectin and Abamectin*; Springer Verlag: New York, 1989.
12. Bloom, R.A.; Matheson, J.C. *Veterin. Parasitol.* **1993**, *48*, 281-294.

13. Barth, D.; Heinze-Mutz, E.M.; Langerholff, W.; Roncalli, R.A.; Schlüter, D. *Appl. Parasitol.* **1994**, *35*, 277-293.

14. Strong, L. *Veterin. Parasitol.* **1993**, *48*, 3-17.

15. Webb, S. Presentation at the SETAC 19th Annual Meeting, Charlotte, NC, 1998.

16. Ternes, T.A. *Water Res.* **1998**, *32*, 3245-3260.

17. Bisaya, S.C.; Patil, D.M. *Res. Industry* **1993**, *38*, 170-172.

18. Richardson, M.L.; Bowron, J.M. *J. Pharm. Pharmacol.* **1985**, *3*, 1-12.

19. Spaepen, K.R.I.; Van Leemput, L.J.J.; Wislocki, P.G.; Verschueren, C. *Environ. Toxicol. Chem.* **1997**, *16*, 1977-1982.

20. Aldridge, C.A. In *Environmental Risk Assessment for Pharmaceuticals and Veterinary Medicines;* Wolf, P.U., Ed.; RCC Group: Basle, Switzerland, 1995; pp 67-79.

21. Wenzel, A.; Schäfers, C. In *Environmental Risk Assessment for Veterinary Medicinal Products*, Lepper, P. Ed., Fraunhofer Institute for Environmental Chemistry and Ecotoxicology: Schmallenberg, Germany, 1998; pp 153-160.

22. *Guidelines for Testing of Chemicals.* OECD (Organisation for Economic Development), Paris, 1993.

23. Streloke, M.; Köpp, H. *Mittl. Biol. Bundesanst. Land- Forstwirtsch.* **1995**, *15*, 1-96.

24. *Guidance for Industry: Environmental Assessment of Human Drug Applications.* FDA (Food and Drug Administration), CDER/CBER CMC 6, rev. 1: Washington, D.C., 1998; 39 pp.

25. Lepper, P. *Environmental Risk Assessment for Veterinary Medicinal Products.* Fraunhofer Institute for Environmental Chemistry and Ecotoxicology: Schmallenberg, 1998; 194 pp.

26. Strong, L.; James, S. *Veterin. Parasitol.* **1993**, *48*, 181-191.

27. Seel, P. In *Arzneimittel in Gewässern?* Touissant, B. Ed., Hesssiche Landesanstalt für Umwelt: Wiesbaden, Germany, 1998; pp 1-9.

28. Montforts, M.H.M.M.; Kalf, D.F.; Van Vlaardingen, P.L.A.; Linders, J.B.H.J. Abstract No. 4H/P006, SETAC Europe 8th Annual Meeting, Bordeaux, France, 1998; p. 300.

29. Gärtner, S. In *Arzneimittel in Gewässern?* Touissant, B., Ed., Hesssiche Landesanstalt für Umwelt: Wiesbaden, Germany, 1998; pp 59-64.

30. Kümmerer, K. In *Arzneimittel in Gewässern?* Touissant, B., Ed., Hesssiche Landesanstalt für Umwelt: Wiesbaden, Germany, 1998; pp 97-104.

Chapter 18

Regulatory Oversight for the Environmental Assessment of Human and Animal Health Drugs: Environmental Assessment Regulations for Drugs

Ranga Velagaleti and Michael Gill

BASF Corporation, 8800 Line Avenue, Shreveport, LA 71106

Environmental Risk Assessment (ERA) of human and animal health drugs is required by regulatory agencies around the world as a part of the drug approval process. The ERA of the manufacture, use, and distribution of human and animal health drugs is required in the United States under the National Environmental Policy Act (NEPA). The Center for Drug Evaluation and Research (CDER) and the Center for Biologics Evaluation and Research (CBER) of the United States Food and Drug Administration (FDA) published a recent guidance for ERA of Human Drugs and Biologics to meet the NEPA requirements. The European Agency for the Evaluation of Medicinal Products (EMEA) has under the auspices of European Commission published a guidance document for ERA of human health drugs in a draft form. The Committee for the Veterinary Medicinal Products (CVMP) of the EMEA also published guidance for the ERA of the veterinary medicinal products. We will discuss the basic elements of two guidances that have been issued as final documents, one representing human drugs (FDA) and the other animal drugs (EMEA/CVMP). Future changes to guidance for conducting animal drug ERAs for the European Union (EU), Japan, and U.S. developed under the auspices of Veterinary International Conference on Harmonization (VICH) will also be presented.

Regulatory Oversight by FDA for ERA of Human Drugs

The environmental assessment of human drugs in the United States is required by the National Environmental Policy Act (NEPA) of 1969. NEPA requires all Federal Agencies to assess the environmental impact of their actions (such as approval of drugs) and to ensure that the interested and affected public is informed of environmental analyses. Under this mandate FDA requires Environmental Assessments (EAs) to be submitted with each drug application, and applicants submit EAs as a part of the Chemistry, Manufacturing, and Controls section of drug applications. However, in 1995, as required under the President's Reinventing the Government (REGO) initiative, FDA reviewed the available knowledge base on environmental behavior of drugs from the previously submitted drug applications. Based on this review, FDA reevaluated the previous requirements for EAs, and revised environmental regulations with the objective of reducing the number of EAs required to be submitted by the industry and redefined the scope and content of EAs (*1*). The guidance document titled "Guidance for Industry – Environmental Assessment of Human Drug and Biologics Application" (*1*) published in July 1998 with the joint efforts of FDA's CDER and CBER, is a result of these deliberations. Strict adherence to and understanding of the expectations of this FDA guidance (*1*) will help determine if the EA component of the application qualifies for categorical exclusion. When a drug application does not qualify for categorical exclusion, submission of an EA is required. The EA should contain all the information required in the FDA guidance (*1*) to enable the agency to determine whether the proposed action (e.g., NDA, ANDA, or other) will significantly affect environmental quality.

Categorical Exclusions

According to FDA guidance (*1*) certain classes of actions are subject to categorical exclusion and, therefore, do not ordinarily require the preparation of an EA because, as a class, these actions individually or cumulatively do not significantly affect the environment. Applications qualify for categorical exclusion under the following conditions: 1) new drug applications (NDAs), abbreviated new drug applications (ANDAs), applications for marketing approval of a biologic product, and supplements to such applications if FDA's approval of the application does not increase the use of the active moiety [FDA (*1*) defines active moiety as "the molecule or ion excluding those appended portions of the molecule that cause the drug to be an ester, salt (including a salt with hydrogen or coordinated bonds), or other noncovalent derivative (such as complex, chelate, or clathrate) of the molecule, responsible for the physiological or pharmacological action of the drug substance. The active molecule is the entire molecule or ion not the active site"]; 2) NDAs, ANDAs, and supplements to such applications, if FDA's approval of the application increases the use of active moiety, but the estimated concentration of the substance at the point of entry into the aquatic environment will be below 1 part per billion (ppb); 3) NDAs, ANDAs, applications for marketing approval of a biologic product, and supplements to such applications for substances that occur naturally in the environment when the approval of the application does not alter significantly the concentration or distribution of the substance, its metabolites, or degradation products in the environment; 4) investigative new drug applications (INDs); and 5) applications for marketing approval of a biologic product for transfusable human blood or blood components and plasma. An applicant filing for approval of a drug is not required to submit an EA if a categorical exclusion is claimed under any one of the five items listed above. The specific item (1 to 5) under which categorical exclusion is claimed should be stated in the application with a statement that to the applicant's knowledge no extraordinary circumstances exist. The extraordinary circumstances are defined below.

Extraordinary Circumstances

FDA will require an EA for any drug application that ordinarily would be categorically excluded if extraordinary circumstances indicate that the specific proposed action could significantly affect the quality of the human environment. The extraordinary circumstances include: 1) actions for which available data establish that there is a potential for serious harm to the environment at the expected level of exposure; 2) actions that adversely affect a species or the critical habitat of a species determined under the Endangered Species Act (ESA) or the Convention on International Trade in Endangered Species (CITES) of Wild Fauna and Flora; or 3) wild fauna or flora that are entitled to special protection under some other Federal law or international treaty to which the United States is a party. FDA guidance (*1*) considers toxicity to environmental organisms and lasting effects on ecological community dynamics at the expected level of exposure harmful to the environment. In addition, the Council on Environmental Quality (CEQ) has defined a list of significant effects to the quality of the human environment [Attachment C of FDA guidance document

(FDA, 1998)] which should be considered by the applicant claiming categorical exclusion.

When the drug or biologic product is derived from plants or animals taken from the wild (in contrast to cultivated plants or laboratory bred or domestic animals), the following would be considered as extraordinary circumstances and will require an EA: 1) NDAs, ANDAs, and applications for marketing approval of a biologic product where the drug or biologic product is derived from plants or animals taken from the wild; 2) supplements to such applications that relate to changes in the source of the wild biomass (e.g., species, geographic regions where biomass is obtained); or 3) supplements to such applications that are considered to increase the use of active moiety or biologic substance that will cause more harvesting than was described in the original EA. CDER and CBER will evaluate INDs (which normally qualify for categorical exclusion) case by case when wild plants or animals are used as a source of drug to determine if the extraordinary circumstance provision should be invoked.

The following information is required with the claim of categorical exclusion when wild flora and fauna are used: 1) biological identification (i.e., common names, synonyms, variety, species, genus, family); 2) statement as to whether wild or cultivated species are used; 3) the geographic region (e.g., country, state or province) where the biomass is obtained; and 4) a statement whether the species or the critical habitat of a species is determined endangered or threatened under ESA or CITES or entitled to special protection under some other Federal law or international treaty to which the United States is a party. CDER and CBER scientists will use this information to evaluate whether the claim of categorical exclusion is appropriate.

Categorical Exclusions for Actions Based on Increased or No-increased Use Scenario

The following actions are not considered by FDA (*1*) to result in an increased use of active moiety and qualify for categorical exclusion: 1) chemistry, manufacturing, and control supplements; 2) abbreviated applications; 3) lower doses (i.e., total daily dose) or shorter duration (e.g., number of days) of use than previously approved for the same indication; 4) exclusion of a patient population in the labeling by age, gender, or medical conditions; 5) a prodrug for which the active metabolite is an approved drug and is intended to substitute directly for an approved product or an active moiety which is the active metabolite of an approved drug; 6) new dosage forms that substitute directly for an approved product; 7) product reformulations in which the labeled amount of active moiety/biologic substance remains constant; 8) packaging changes or dosage form product line extensions that substitute directly for an approved product; and 9) combination drugs in which a single product substitutes directly for two approved products that would be administered separately.

The following types of actions are considered by FDA (*1*) to result in increased use of an active moiety if approved and may not qualify for categorical exclusion unless the estimated concentration at the point of entry into the aquatic environment is below one part per billion (1 ppb): 1) new molecular entities; 2) a new indication for a drug that was previously approved including those actions requesting approval of off-label uses and switches from a first line to second line indication; 3) prescription to OTC switches; 4) higher total daily dose and longer duration (days) of use than previously approved; 5) inclusion of a patient population that had previously been specifically excluded; and 6) new dosage forms/route of administration that increase the amount of active ingredient/biologic substance used.

Estimating Expected Introduction Concentration (EIC) of an Active Moiety in the Aquatic Environment

The EIC at the point of entry into the aquatic environment is estimated in parts per billion (ppb) using the formula provided in the FDA guidance (*1*): EIC-aquatic (ppb) = A x B x C x D, where A = kg/year produced for direct use as active moiety; B = 1/liters per day entering Publicly Owned Treatment Works (POTW), estimated at 1.214×10^{11} in the 1996 Needs Survey by the EPA; C = year/365 days; and D = 10^9 µg/kg (conversion factor). This calculation assumes: 1) that all drug product produced in a year enters the POTW; 2) drug product usage occurs throughout the United States in proportion to the population and the amount of waste generated; and 3) and that there is no metabolism of the active moiety. The kg/year of active moiety (rather than the salt or the complex) should include: 1) the highest quantity of active moiety expected to be produced for direct use by humans (quantities for inventory build up are excluded) in any of the next five years, normally based on marketing projections; 2) the quantity used in all dosage forms and strengths included in the application; and 3) quantities in the applicant's related applications. For example, if the maximum production of an active moiety/year in a five-year marketing estimate is 41,500 kg ("A" in the above equation), the EIC-aquatic for the drug is 0.937 ppb, which qualifies for a categorical exclusion. A majority of the human pharmaceuticals are produced in far lower quantities, and most have an EIC of less than 1 ppb (= 44,300 kg) and, therefore, are generally granted categorical exclusion by FDA. A provision is made in the guidance to consider human metabolism data in estimating EIC. Under a scenario (provided by the authors, not FDA), where human metabolism data suggests that 10% of the drug has been metabolized in the human body to expired gases such as CO_2 or other (known to be possible with some drugs), such depletion from metabolism can be taken into consideration in estimating the EIC. For example, under this scenario, if the maximum production in a five-year production cycle is estimated at 45,650 kg/year because the drug is depleted in the human body to the extent of 10% (41,085 kg), the EIC estimate will be below 1 ppb. The FDA guidance (*1*) also provides an alternative calculation if the drug product is intended for use in a specific geographic location. By substituting the exact quantity (liters per day) of water entering the POTW for that particular geographic location for "B" in the EIC equation given above, the EIC estimate for that particular location can be determined.

Content and Format of an EA Document for Submission to CBER or CDER

When an EA is required, the format described in the guidance document must be followed for timely approval. The outline of the format for an EA is provided in Attachment D of the guidance document (1) and has the following elements: 1) Date; 2) Name of Applicant/Petitioner; 3) Address; 4) Description of the Proposed Action (Requested Approval, Need for Action, Location of Use and Disposal Sites); 5) Identification of Substances that are the subject of Proposed Action [Nomenclature including the Established Name (U.S. adopted name; USAN), Brand/Proprietary/Trade Name; Chemical Names (Chemical Abstract Index Name, Systematic Chemical Name) or Genus/Species in the case of a Biologic Product; Chemical Abstracts Service (CAS) Registry Number; Molecular Formula and Weight; Structural Formula or Amino Acid Sequence in the case of a biologic product]; 6) Environmental Issues; 7) Mitigation Measures; 8) Alternatives to the Proposed Action; 9) List of Preparers; 10) References; and 11) Appendices. A brief description of items 4, and 6 to 8 is provided below. For detailed understanding of FDA expectations for each of the items, readers are referred to the FDA guidance (*1*).

Description of the Proposed Action

Items to be included in the "Requested Approval" subsection are drug or biologic application number, product name, dosage form and strength, and a brief description of product packaging. A description of the drug's or biologic's intended uses in the diagnosis, cure, mitigation, treatment or prevention of disease should be provided in the "Need for Action" subsection. The "Locations of Use and Disposal Sites" subsection should state that the locations of use are typically hospitals, clinics, and/or patients in their homes. If use is expected to be concentrated in a particular region, it should be stated. A statement indicating that at U.S. hospitals, pharmacies, or clinics, empty or partially empty packages will be disposed of according to hospital, pharmacy, or clinic procedures, respectively, should be included. The statement should also include that for uses at home, empty or partially empty containers will form part of domestic waste, which will be disposed of in community or commercial landfills and incinerators. Any exceptions to these scenarios should be described.

Environmental Issues

When the preparation of a full EA is required, FDA recommends the use of a logical, tiered approach to fate and effects testing so that adequate information is available to assess the potential environmental fate and effects of pharmaceuticals [active moiety or structurally related substances (SRS) but not excipients] (1). The EA should list the drug or biologic substance and the predominant SRSs (SRSs of >10% of dose) expected in the environment, describing where possible the chemical and physical property information of these entities. The rationale for selecting an active moiety (preferred over SRS) and/or a SRS for fate and effects testing should be provided. For SRSs, the potential fate and effects can be discussed based on the structural similarities or computerized structure-activity relationships without necessarily generating the data. Relevant available toxicological and pharmacological information should be provided for the SRSs.

Environmental Fate of Released Substances

Water solubility, dissociation constant, octanol/water partition coefficient (Kow), vapor pressure or Henry's Law Constant information generally gives an indication whether a compound is likely to partition predominantly into aquatic, terrestrial and/or atmospheric environments. For example, if Kow is greater than 3, an adsorption/desorption study may be required to establish the partitioning of the drug to sludge solids or soils. The extent of degradation and depletion of the drug should be established through laboratory testing. As stated in the FDA guidance (1), if a rapid and complete depletion mechanism is identified through laboratory tests and the degradates formed are relatively simple polar by-products, no laboratory testing to determine the environmental effects should be performed; only a microbial inhibition or other appropriate test to assess the potential for the compound to disrupt the waste treatment processes at the POTW is required. FDA guidance (1) considers the following half-life ($t_{1/2}$) estimates (equal to or less than) as rapid depletion mechanisms that will warrant no further fate or effects tests: hydrolysis - 24 hours; aerobic biodegradation in water – 8 hours; soil biodegradation – 5 days. When such rapid depletion mechanisms are not identified, adjustments to the EIC may be made based on spatial or temporal concentration or depletion factors to provide an expected environmental concentration (EEC), where EEC = EIC – depletion and/or dilution. The EEC for the aquatic environment would be expected to be significantly less than the EIC for the aquatic environment based on degradation and depletion processes in POTW and the dilution factors of 10 provided by EPA, when the waste water effluents are released from POTW to surface waters.

A summary discussion of the environmental fate of the substances of interest should be provided for each environmental compartment based on the information and data provided in the EA. The environmental compartment(s) into which the substance is expected to partition predominantly should be identified. If the substance of interest rapidly degrades and meets criteria described above, or adsorbs completely and irreversibly to biosolids, then the fate and effects in the aquatic environment should not be considered (*1*). On the other hand, if these processes do not occur, then the aquatic environment should be considered for effects testing. Fate and effects testing in the terrestrial environment should be considered if testing indicates that the substance(s) of interest will significantly adsorb to bio-solids with Koc of equal to or greater than 1000. The atmospheric compartment may be of interest for medical gasses. Any potential for a substance to volatilize and recycle into the aquatic or terrestrial environment should be discussed based on the information and data available for the substance.

Environmental Effects of Released Substances

If rapid and complete depletion mechanisms are identified, FDA guidance (*1*) recommends conducting a microbial inhibition test or other appropriate test (e.g., respiration inhibition test) to assess the ability of the drug or biologic substance(s) of interest to inhibit microorganisms and subsequently disrupt waste treatment processes. Under this scenario no further effects testing is needed. If no rapid and complete depletion mechanisms are identified, a microbial inhibition test is still recommended followed by a tiered approach to environmental effects testing. If log Kow is less than or equal to 3.5, a tier 1 acute toxicity test with one species is recommended. If the EC_{50} or LC_{50} from the tier 1 acute test divided by maximum expected environmental concentration (MEEC = EIC or EEC whichever is greater) is greater than or equal to 1000, no further testing is required. If there are observed effects at concentrations less than or equal to MEEC in the tier 1 acute toxicity test performed, tier 3 testing may be warranted. If the EC_{50} or $LC_{50} \div$ MEEC is less than the assessment factor of 1000, tier 2 testing should be performed, which normally includes an acute toxicity base set (aquatic and/or terrestrial, depending on the partitioning behavior of the drug or biologic substance into water and sludge matrices). The aquatic toxicity base set normally includes: 1) fish acute toxicity test; 2) aquatic invertebrate toxicity test (such as *Daphnia* acute test); 3) algal species bioassay test. The terrestrial toxicity base set includes: 1) plant early growth test; 2) earthworm toxicity test; and 3) soil microbial toxicity test. If the EC_{50} or LC_{50} for the most sensitive test organism in the base set divided by the MEEC is equal to or greater than tier 2 assessment factor of 100 then no further testing is necessary, unless sublethal effects are observed at test concentrations at or below MEEC, in which case tier 3 chronic toxicity testing is required. Tier 3 testing will also be required if the EC_{50} or LC_{50} for the most sensitive test organism in the base set divided by MEEC is less than the tier 2 assessment factor of 100. If the chronic LC_{50} or $EC_{50} \div$ MEEC is equal to or greater than the assessment factor of 10, no further testing is necessary, unless sublethal effects are observed at test concentrations at or below the MEEC. The sponsor of the application is asked to consult with FDA [CDER (drugs) or CBER (biologics)] if the chronic EC_{50} or $LC_{50} \div$ MEEC is less than 10 or sublethal effects are observed at test concentrations at or below MEEC. Results of tests conducted to determine toxicity of the drugs or biologics to terrestrial or aquatic species should be reported in terms of quantity and/or concentration of the active moiety. In the tiered approach to effects testing (1), FDA recommends that the toxicity effects tests be designed appropriately so that a no observed effect concentration (NOEC) can be determined. The toxicity test results (LC_{50}, EC_{50}, or NOEC) should be compared with estimated environmental concentrations (EIC, EEC, or MEEC) and assessment factors provided wherever applicable together with conclusions on risk or no risk to the environment.

Test Methods, Report Formats, and Data Summary Tables

The procedures/methodology for conducting the tests and reporting are provided in the FDA Technical Assistance Handbook (*2*). Other guidance documents for conducting these tests from the U.S. Environmental Protection Agency (U.S. EPA) and the Organisation for Economic Cooperation and Development (OECD) are also acceptable to FDA (*3*). Performance of the tests and reporting of results should meet the FDA Good Laboratory Practice (GLP) standards requirement (*4*). Test reports should be submitted only under the "Confidential Information" section of the EA; no raw data or copies of the notebooks supporting these reports should be submitted. A data summary table including names of all the tests performed for physical and chemical characterization of active moiety, fate (including supporting data for depletion mechanisms) and effects tests, and the salient findings of these should be provided using the format in "Attachment E" of the EA guidance document (*1*).

Use of Resources

In this section any effects on natural resources should be described. For example, when the manufacture of drug or biologic requires use of wild flora or fauna, information relating to the source of the plant or animal, such as biological identification, government oversight of harvesting, geographic region where biomass is obtained, and harvesting methods and techniques should be provided.

Mitigation Measures

Measures taken to avoid or mitigate any potential environmental effects associated with the proposed action should be discussed. If no adverse environmental effects have been identified and that as a consequence no mitigation measures are required, a statement to that effect should be made. When using wild flora and fauna for the active moiety or biologic substance, measures taken before harvesting (e.g., developing a process that uses renewable part of a plant), during harvesting (e.g., limiting/selecting specimens to be harvested), and after harvesting (e.g., reforesting) should be discussed under this section.

Alternatives to the Proposed Action

If there are no potential adverse effects to the environment, no alternatives to the proposed action are required. If adverse effects are noticed, the EA should provide a discussion on an alternative course of action that offers less environmental risk or an action that is environmentally preferable to the proposed action.

Confidential and Non-Confidential Information

The EA document submitted to the FDA should contain three distinct parts: 1) the EA Summary Document, which is non-confidential; 2) Non-confidential appendices; and 3) appendices with confidential information to support the EA, such as production data, proprietary product information, test reports etc. "Attachment F" of the FDA guidance (*1*) document provides a table distinguishing confidential information from non-confidential information.

Environmental Assessment of Animal Health Drugs - Regulatory Oversight in the European Union by EMEA/CVMP

The EMEA/CVMP published a guidance document titled "Environmental Risk Assessment for Veterinary Medicinal Products Other than GMO-Containing and Immunological Products," which came into operation January 1, 1998 (5). The ERA is included in an application for marketing authorization for veterinary medicinal products other than those submitted with abridged applications.

Phased Approach to Environmental Risk Assessment

The ERA is normally conducted in two phases, Phase I and Phase II. In Phase I, the potential exposure of the environment to the product, its ingredients, or relevant metabolites is considered. The Phase I review should lead to identification of those products that are unlikely to result in significant exposure of the environment and consequently be of low environmental risk and, therefore, qualify for exemption from further testing. If potential exposure of the environment is identified in Phase I, the Phase II investigations will identify the effect of the product on a particular ecosystem.

Phase I Assessment

During Phase I, risk assessment decisions are based on a straightforward decision tree concept. Phase I investigations and/or review of existing data will help determine whether the use and disposal of the veterinary drug will lead to significant exposure of any environmental compartment. Phase I evaluations could exempt a drug from further testing under the following conditions: 1) the active substances consist of vitamins, electrolytes, natural amino acids or herbs; 2) the target animals for the active substance(s) are companion animals or if only a small number of animals are treated; 3) the active substance(s) is used externally (pour-ons, dips, fumigation etc.,) and the substance(s) are not expected to enter the environment; 4) the active substance(s) are present in manure or slurry, for spreading on land, in concentrations lower than 100 μg/kg; 5) the substance(s) used for pasture animals are excreted in concentrations lower than 10 μg/kg; 6) the DT_{50} (degradation half life of active moiety) in manure is less than 30 days; 7) the predicted environmental concentration (PEC) in soil demonstrated by a worst case calculation is below 10 μg/kg; and 8) the predicted environmental concentration (PEC) in ground water is below 0.1 μg/L. If adverse environmental effects are anticipated under these exemptions, further assessment may be required under Phase I or Phase II scenarios.

Phase II Assessment

Phase II assessment is required under the following conditions: 1) the active substance is used to treat a large number of animals; 2) there is a potential for direct entry into the aquatic environment; 3) the drug substance is used for internal application and the drug or its major metabolite is excreted on pasture in concentrations of >10 µg/kg (10 ppb); 4) the manure is spread on to the land and the drug or its major metabolite exceeds 100 µg/kg (100 ppb); and 5) the PEC in soil and ground water exceeds 10 µg/kg (10 ppb) or 0.1 µg/L, respectively.

In Phase II, testing is conducted in a tiered approach. Phase II, tier A would require studying the degradation rate of active substance and adsorption coefficients (Koc) in three soils. The earthworm LC_{50} (median lethal concentration), and the phytotoxicity tests are required under conditions 3, 4, and 5 (soil PEC) stated above, and acute toxicity test in *Daphnia* is required under conditions 2 and 5 as a part of tier A testing. If DT_{50} is greater than 60 days, a microbial growth inhibition test is required. If DT_{50} is less than 60 days and DT_{90} is less than a year, tier B testing is not required. If the PEC/EC_{50} (median effective concentration) for phytotoxicity study or PEC/LC_{50} for earthworm is less than 0.1, no tier B testing is needed. If the PEC and the Predicted No Effect Concentration (PNEC) ratio (PEC/PNEC) for ground water is less than 1, no tier B testing is required. If Koc is less than 500, aquatic tests such as *Daphnia*, alga, and fish toxicity tests are required. If the PEC/PNEC ratio from these tests is less than 1, no tier B testing is required.

The following end points may lead to appropriate risk management strategies or Tier B testing: 1) the PEC/EC_{50} is >0.1 for phytotoxicity; 2) the PEC/LC_{50} is >0.1 for earthworm toxicity; 3) the PEC/LC_{50} is >0.01 for earthworm toxicity and PEC/MIC for soil microorganisms is >0.1; 3) the PEC/PNEC is >1 for ground water; 4) PEC (for surface water)/PNEC (for Daphnid, alga, fish) is >1; and 5) DT_{50} and DT_{90} in soils is >60 days and 1 year, respectively.

All applicants are required to submit a complete report with a conclusion on ERA based on the characteristics of the product, its potential environmental exposure, environmental fate and effects, and risk management strategies that support ERA conclusions. The report should also take into account the pattern of use, administration of the product, excretion of the active substance and the major metabolites and the disposal of the product.

References

1. *Guidance for Industry – Environmental Assessment of Human Drugs and Biologics Applications;* U.S. Food and Drug Administration, 1998.

2. *Environmental Technical Assistance Handbook;* U.S. Food and Drug Administration, NTIS Pub. No. P87-175345, Springfield, VA, 1987.

3. Stamm, J; Velagaleti, R. *J. Haz. Mater.* **1993**, *35*, 313-329.

4. *Good Laboratory Practice for Nonclinical Laboratory Studies. 21 CFR Part 58;* U.S. Food and Drug Administration, Revised 09/20/99.

5. *Environmental Risk Assessment for Veterinary Medicinal Products Other Than GMO-Containing and Immunological Products, Committee for Veterinary Medicinal Products (EMEA/CVMP/O55/96-Final, which came into operation on January 1, 1998);* The European Agency for Evaluation of Medicinal Products, London, U.K. 1998.

Thanks are due to Dr. Joseph Robinson, Pharmacia Corporation, for valuable discussions on this topic, and Mr. Phil Burns of BASF Corporation, for review and comment on this manuscript

Chapter 19

Degradation and Depletion of Pharmaceuticals in the Environment

Ranga Velagaleti and Michael Gill

BASF Corporation, 8800 Line Avenue, Shreveport, LA 71106

Manufacture of pharmaceutical compounds is conducted under Good Manufacturing Practice Regulations (GMPs), which require contained manufacture, and process and product accountability for both the bulk drug substance and the drug product. Cleaning of equipment may contribute drug or related substances to the manufacturing process waste effluents. These effluents are processed on-site and discharged into domestic sewage and then into Publicly Owned Treatment Works (POTW) in compliance with prevailing regulations. The on-site processing facilities and the POTW are designed to facilitate extensive degradation and depletion of drug residues. According to the United States Food and Drug Administration's (FDA's) guidance on Environmental Assessments, regulated articles produced and disposed of in compliance with all applicable regulations are considered to have no effect on the environment. During use, human and animal health drugs are metabolized in the target species, sometimes extensively resulting in partial depletion of drugs. For human drugs, the remaining drug residues containing parent compound and metabolites are released through the excreta into the domestic sewage, which is processed in POTW, where further degradation and depletion of the residues are facilitated. When animal (as manure applied to soil) or human health drug residues (from POTW) are released into the environment, dilution in soil or water plays a major role in their decline, followed by degradation and depletion through chemical and biological processes. The degradation and depletion processes and a rationale for zero risk to the environment from drug manufacture and use are discussed in light of current regulations.

Introduction

The manufacture, use, and disposal of pharmaceutical active ingredients and drug products are contained and their exposure to the environmental matrices is limited. This is in contrast to crop protection and other external use chemicals which by their very nature of intended use and application may be directly released into the environment (air, soil, and water) from the target or non-target crops and soil, or other domestic and industrial uses. This manuscript discusses how the release of drug active ingredients into the environment could occur under their contained manufacture, use and disposal scenarios. It subsequently describes the general degradation and depletion mechanisms in various environmental matrices that could lead to zero risk exposures for the majority of drugs. Finally, a discussion of the current regulatory perspective driving the Environmental Risk Assessments (ERA) is provided. Manufacture and disposal issues are well regulated and are largely common to human and animal health drugs, and, therefore, are discussed under one topic. The disposition of drug residues resulting from human and animal health use are unique to each use and, therefore, discussed separately.

Drug Manufacture

The manufacture of pharmaceuticals in general is contained, with minimal release of raw materials, intermediates or bulk drug substance during the manufacture of active pharmaceutical ingredient (API) and finished drug product (tablets, capsules, etc.). API and drug product manufacture is governed by Current Good Manufacturing Practice (CGMP) regulations (1,2). CGMPs require accountability of all raw materials that go into the production of drug substance or drug product. For example, the CGMPs for API manufacture state: "Raw materials used for manufacturing APIs and intermediates should be weighed and measured ... Weighing and measuring devices should be of suitable accuracy for the intended use." Further, in terms of the accountability of final yields from raw materials the CGMPs state: "Actual yields and percentages of expected yields should be determined at the conclusion of each appropriate phase of manufacturing or processing of an API or intermediate" (*1*). Similar controls for the accountability of raw materials (API and excipients) are required under CGMPs for drug product manufacture, which state "Actual yields and percentages of expected yields should be determined at the conclusion of each appropriate phase of manufacturing, processing, packaging, or holding of the drug product" (*2*). These strict controls and accountability for API and drug product manufacture under CGMPs as described above preclude any significant release of pharmaceutical compounds into the environment.

API Manufacture

Intermediates used in API manufacture and/or APIs can become a part of the process waste effluent stream during the cleaning of the equipment used in manufacture. Most commercial API manufacturing plants have holding tanks for process waste effluents, where the effluents are processed to encourage biodegradation. Specifications are set and approved by regulatory authorities for the limits on carbon loading of these effluents. These waste streams must meet regulatory specifications on discharge limits upon release into municipal sewer collection systems. Extensive treatment and stringent discharge limits are applied to effluents when directly discharged from API manufacturing facilities to surface waters. Depending on the chemical nature of the API or intermediate, and the physical and chemical condition of the collection systems and treatment facilities, biological and chemical degradation processes could reduce or even deplete the concentrations of these components in the waste effluents. For example, aerobic biodegradation could result in reduction in concentration of these components through biotransformation (metabolism to transformed products mediated by microorganisms present in the holding tanks) or complete depletion of these components through mineralization (conversion to CO_2 and H_2O). Biotransformation of an organic molecule can occur through one or more of the following mechanisms: oxidation, oxidative dealkylation, decarboxylation, epoxidation, aromatic hydroxylation, aromatic non-heterocyclic ring cleavage, aromatic heterocyclic ring cleavage, hydrolysis, dehalogenation, and nitroreduction (*3*). Also, chemical degradation processes such as hydrolysis and photolysis could reduce the concentration of these components in effluents. Hydrolysis is a key reaction of organic compounds with water. The chemical reaction is mediated by a direct displacement of a chemical group by the hydroxyl group (*4*). The components in the waste effluents can also be degraded if they are exposed to natural sunlight, either by direct photodegradation or indirect photodegradation. Direct photodegradation by natural sunlight occurs in pharmaceutical compounds or their intermediates that have absorbance in the range of 290 to 800 nm. Pharmaceutical compounds that do not have absorbance in 290 to 800 nm range can also be photodegraded through indirect photodegradation if one or more of the chemical components (sensitizer) present in the waste effluent can absorb light in this region and transfer energy to and facilitate the degradation of those non-absorbing pharmaceutical residue components. Photodegradation of API or intermediates can occur through any of the following mechanisms: fragmentation into free radicals or neutral molecules, rearrangement and isomerization reactions, photoreduction, dimerization and other addition reactions, photoionization and electron transfer (4).

Through the use of appropriately designed process waste treatment facilities that facilitate biological and/or chemical degradation, the concentration of the components present in the process waste effluents can be reduced to acceptable levels. Degradation processes such as hydrolysis and biodegradation could reduce components discharged into municipal domestic sewage before they reach the Publicly Owned Treatment Works (POTW). The POTWs are designed to facilitate degradation of organic molecules present in the domestic sewage. Extensive biodegradation is possible in the activated sludge aeration tanks where the microbial load is very high due to aerobic multiplication of microbes. Also, the open aeration tanks at the POTW facilitate extensive hydrolysis and photodegradation of organic molecules. The wastewater and the sludge solids (containing predominantly microbial biomass) are separated during processing at POTW. Organic molecules that have a high octanol/water partition coefficient (Kow) tend to partition into sludge solids. The sludge solids at many POTWs are subjected to anaerobic digestion where anaerobic biodegradation of most organic molecules typically results in the production of methane, CO_2, and biotransformed products. These degradation processes at the POTW will account for decline and depletion of any drug residues present. A combination of these processes should eliminate most drug residues before the effluents enter the aquatic or terrestrial environmental compartments. Wastewater effluent from POTW is discharged to rivers or streams where dilution will reduce residues further. Similarly, residues in sludge solids are diluted when applied to soil.

Drug Product Manufacture

During drug product manufacture, the only source of release of drug residue will be through the cleaning of equipment used for coating, blending, tablet compressing and packing operations. These residues are minimal and they may be discharged into domestic municipal sewage systems, especially when water is used for the cleaning process. When organic solvents are used, the washes are collected and disposed as hazardous wastes or the solvent is recycled. Most residues released through the cleaning operation will enter the domestic sewage where they are extensively diluted (possibly depleted) before entering the POTW. Residues entering the POTW undergo degradation and depletion through chemical (hydrolysis and photolysis) and biological (aerobic and anaerobic biodegradation) degradation processes. The minimal releases during drug product manufacture and their degradation, depletion, and dilution may lead to minimal or non-detectable residue components related to drug product manufacture in the wastewater effluents or sludge solids released from the POTW.

Drug Disposal

The material accountability under CGMPs applies to the unused, expired, or returned APIs or drug products. CGMPs for APIs state: "Records of returned APIs and intermediates should be maintained and should include the name, batch or lot number, reason for the return, quantity returned, date of disposition, and ultimate disposition" (*1*). CGMPs for finished drug products state: "Records of returned drug products shall be maintained and shall include the name and label potency of the drug product dosage form, lot number (or control or batch number), reason for the return, quantity returned, date of disposition, and ultimate disposition of the drug product" (*2*). When the ultimate disposition status is to destroy, a majority of the pharmaceutical compounds are disposed of through incineration or landfilling in a certified incinerator or landfill, respectively. Both these methods of disposal are designed specifically to eliminate the exposure of subject articles to the environment. The end users for the drug products are hospitals, pharmacies, clinics, and domestic users (*5*). Empty or partially empty packages are disposed of according to hospital, pharmacy, or clinic procedures. Typically, such procedures will include collection in appropriate containers and ultimate disposition through certified landfill or incinerator. Domestic users will typically dispose of the empty or partially empty containers through a community solid waste management system, which may include certified landfills or incineration (*5*). No significant environmental releases are likely to occur from disposal by end users of drug products.

Regulatory Perspective on Environmental Issues Related to Manufacture and Disposal

FDA guidance (*5*) states that the regulated articles produced and disposed of in compliance with all applicable emission requirements do not significantly affect the environment. Therefore, FDA's Center for Drug Evaluation and Research (CDER) and Center for Biologics Evaluation and Research (CBER) will not routinely request submission of manufacturing and disposal information as a part of the Environmental Assessment (EA) requirement of drug applications (5). If there are unique emissions resulting from the manufacture or disposal of a particular drug that may harm the environment, FDA may request manufacturing and disposal information as well as the evaluation of the environmental impact of such emissions in an EA. The degradation and depletion mechanisms for such emissions should be provided to support a "no impact" scenario.

Use of Human Health Drugs

Human health drugs or biologics are likely to be metabolized by the patient and excreted through urine and feces. The excreted compounds may include the substance and/or its structurally related substances (SRSs) such as the dissociated parent compound, metabolites, conjugates, or degradates (drug residues), which will form part of the domestic sewage. The domestic sewage is carried through the municipal sewer system, which then enters the POTW. Sand and grit are removed from the domestic sewage at the POTW by screen filtration after which it is subjected to primary and secondary treatments. After processing at the POTW, the domestic sewage is partitioned into wastewater effluent and sludge solids. The wastewater from the POTW is released into nearby streams or rivers, thus exposing the surface water and underlying sediment. The sludge solids (biosolids) are landfilled, incinerated, or applied to agricultural soils as an organic supplement, the later practice being more common.

Degradation, Depletion and Dilution Processes

The API or biologic substance and their SRSs may undergo degradation and depletion at every step beginning with metabolism by the patient. Almost all drugs are metabolized to some extent prior to excretion. Extensive metabolism (and some depletion) in the human body is reported for many drugs. After they are excreted, the drug residue components could be diluted in the domestic sewage and undergo aerobic and/or anaerobic degradation and hydrolysis en route to the POTW. After the domestic sewage reaches the POTW, aerobic biodegradation, hydrolysis, and photolysis could result in degradation and depletion of these components in the activated sludge aeration tanks and in the settling tanks. The sludge solids are typically digested anaerobically after they are separated from the wastewater, where additional depletion of the drug residues may occur via anaerobic biodegradation. Upon discharge of POTW effluents to receiving waters, a typical dilution of an order of magnitude occurs resulting in additional decline in the concentration of drug residues (5). Further decline may occur through aerobic and anaerobic degradation in surface water and anaerobic degradation in sediment. Approximately 6.8 million tons of biosolids are generated per year with 54% of that quantity applied to the land as an organic supplement (5). Application of sludge solid to agricultural lands results in decline of drug residues due to dilution with soil. Photodegradation and hydrolysis could result in further degradation and depletion of drug residue components in the upper layers of soil to which the sludge is applied. The microorganisms present in the soil (i.e., bacteria, such as actinomycetes, and fungi) facilitate extensive degradation of organic molecules. In all these matrices, biodegradation may result in biotransformation to small molecules or complete mineralization of an organic molecule to CO_2 and H_2O. Products formed as a result of hydrolysis and photodegradation could also be mineralized by a secondary degradation process of biodegradation. The biotransformed degradation products could also undergo mineralization by further microbial attack. When drug residue components are completely mineralized to CO_2 and H_2O, they are considered completely depleted. The potential matrices and the chemical and biological degradation and depletion processes that could facilitate elimination of drug residues are presented in Table I.

Regulatory Limits on Estimated Concentrations of Human Drugs in the Environment and Their Relation to Degradation and Depletion

Regulatory Limits for Estimated Introduction Concentrations

The vast majority of new human drug applications qualify for any one of the five categorical exclusion criteria listed in FDA guidance (5). The most commonly applied categorical exclusion is based on the maximum annual use, which is derived from a five-year marketing projection. If FDA's approval of the application increases the use of active moiety but the estimated introduction concentration (EIC) of the substance at the point of entry into the aquatic environment is below 1 part per billion (ppb), the drug application will qualify for a categorical exclusion. If the EIC is >1ppb, submission of a full EA is required. Estimation of the EIC is, therefore, very important prior to addressing the EA component of drug applications. The EIC at the point of entry into the aquatic environment is estimated using the formula provided in FDA guidance (5): EIC-Aquatic (ppb) = A x B x C x D, where A = kg/year produced for direct use as active moiety; B = 1/liters per day entering POTW, an estimate of 1.214 x 10^{11} from the 1996 Needs Survey by the EPA; C = year/365 days; and

Table I. Degradation and Depletion of Human Drugs

Human Body
Metabolism – Transformation and Depletion (CO_2 or other expired gases)
Adsorption and Assimilation – Depletion
Excretion – Parent Compound and its Metabolites/Degradates (contribution to domestic sewage)

Domestic Sewage
Hydrolysis – Chemical Transformation/Degradation
Biodegradation – Microbial Mediated Biotransformation/Degradation
Mineralization – Depletion
Dilution – Residue Decline/Depletion

POTW
Dilution – Residue Decline/Depletion
Hydrolysis – Chemical Transformation/Degradation
Photolysis – Degradation
Aerobic Biodegradation – Degradation/Mineralization (Depletion)
Anaerobic Biodegradation – Degradation/Mineralization/Methane Production (Depletion)

Surface Water
Dilution – Residue Decline/Depletion
Hydrolysis – Chemical Transformation/Degradation
Photolysis – Degradation
Aerobic Biodegradation - Degradation/Mineralization (Depletion)

Sediment
Dilution – Residue Decline/Depletion
Hydrolysis – Chemical Transformation/Degradation
Anaerobic Biodegradation – Degradation/Mineralization/Methane Production (Depletion)

Soil
Dilution – Residue Decline/Depletion
Hydrolysis – Chemical Transformation/Degradation
Photolysis – Degradation
Aerobic Biodegradation – Degradation/Mineralization (Depletion)
Anaerobic Biodegradation – Degradation/Mineralization/Methane Production (Depletion)

$D = 10^9$ µg/kg (conversion factor). Using this equation, an EIC of one part per billion evaluates to 44,300 kg/year. Use of the active moiety below this EIC limit qualifies the application for a categorical exclusion. This calculation assumes: 1) that all drug product produced in a year enters the POTW; 2) drug product usage occurs throughout the United States in proportion to the population and the amount of waste generated; and 3) that there is no metabolism of the active moiety. The kg/year of active moiety (rather than the salt or the complex) should include: 1) the highest quantity of active moiety expected to be produced for direct use by humans (quantities for inventory build up are excluded) in any of the next five years, normally based on marketing projections; 2) the quantity used in all dosage forms and strengths included in the application; and 3) quantities in applicant's related applications. Very few human pharmaceutical compounds with annual production of >44,300 kg are produced in the United States. A vast majority of the drugs are, therefore, granted categorical exclusions by FDA.

EIC Based on Drug Metabolism, Degradation, Dilution and Depletion

In a hypothetical situation, where human metabolism data suggests that 10% of the ingested drug has been metabolized to CO_2 or other expired gases (known to occur with some drugs), then this 10%, representing a known depletion mechanism, can be used to adjust the EIC estimate downward by 10%. The extent, to which an administered dosage is converted to pharmacologically inactive metabolites, is, therefore, an important mechanism to consider when addressing the EA component of the drug application. The FDA guidance (5) considers the following half-life ($t_{1/2}$) estimates (equal to or less than) as rapid depletion mechanisms that will warrant no further fate or effects tests: hydrolysis - 24 hours; aerobic biodegradation in water – 8 hours; soil biodegradation – 5 days. The guidance (5) also advises photodegradation to be considered as a depletion process. If rapid and complete depletion is not achieved, further evaluation is necessary including the conduct of toxicity tests. When the wastewater effluents from the POTW are released into surface waters, an order of magnitude of dilution factor is applied [dilution factor of 10 is suggested in the FDA guidance (5), but may vary with each POTW], and as a result of which, the concentration in surface water could be further reduced. Further degradation and dilution in surface waters and sediment (Table I) could possibly deplete the drug completely or reduce it to undetectable levels (using modern analytical methodology). Similarly dilution of residues in sludge solids when applied to soil and subsequent degradation and depletion could result in undetectable levels. As stated in the FDA guidance (5), an estimated environmental concentration (EEC) should be evaluated taking into account the degradation, depletion, and dilution processes (EEC = EIC – degradation, depletion, dilution). The maximum expected environmental concentration (MEEC = EIC or EEC, whichever is greater) is compared against the no observed effects (NOEC) from the toxicity tests, to evaluate the risk to environmental organisms. For example, if MEEC is lower than NOEC, then no risk is anticipated.

Use of Animal Health Drugs

Like human drugs, animal health drugs or biologics are likely to be metabolized, partially depleted and/or excreted through urine and feces. The excreted compounds may include the parent substance and/or its SRSs, such as the dissociated parent compound, metabolites, conjugates, or degradates (drug residues). The urine and feces along with the bedding will constitute the manure, which is either stored in manure pits for subsequent application or is immediately applied to the agricultural field as an organic matter supplement and fertilizer. A significant number of animal health drugs are used to control various diseases of domestic animals such as dogs and cats. The urine, feces and bedding are disposed of as a part of domestic municipal solid waste, which is sent to either licensed incinerators or landfills. Exposure from these drug residues is, therefore, not an environmental concern. The use of animal health drugs in farm animals (cattle, poultry, swine, and sheep) may result in the unchanged parent molecule (if any) and the SRSs in the manure being exposed to soil when manure is applied to agricultural land.

Degradation, Depletion and Dilution Processes

When manure is applied to soil, there is significant dilution of any drug residues present. Aerobic soil biodegradation is the major pathway of degradation and depletion of animal health drugs in soil. Biotransformation (degradation to polar or non-polar degradates) and mineralization (degradation to CO_2 and H_2O) are the main mechanisms of biodegradation. Complete mineralization leads to complete depletion of a chemical from soil. Other degradation and depletion mechanisms include soil photolysis, especially in the upper few centimeters of soil where the manure may be dispersed. The photodegradation products may also be subjected to aerobic biodegradation in upper soil layers. Soil hydrolysis could occur in moist soils at any given depth. The degradation products of hydrolysis could undergo aerobic biodegradation in upper soil layers or anaerobic degradation in deeper soil layers with sufficient moisture that facilitates both hydrolysis and anaerobic microbial growth and consequent biodegradation. The run-off from soil may carry the drug residues to surface water. Photolysis and aerobic biodegradation may be the major degradation and depletion mechanisms in surface water while anaerobic biodegradation is the major mechanism of degradation and depletion in sediment. The major degradation and depletion mechanisms for animal health drugs are summarized in Table II.

Regulatory Limits on Estimated Concentrations of Animal Drugs in the Environment and Their Relation to Degradation and Depletion

The Committee for Veterinary Medicinal Products (CVMP) under the direction of the European Agency for the Evaluation of Medicinal Products (EMEA) published a guidance document (EMEA/CVMP/055/96-FINAL, which came into operation on January 1, 1998), titled "Environmental Risk Assessment for Veterinary Medicinal Products Other than GMO-Containing and Immunological Products", which describes a phased (Phase I and Phase II) approach to ERA of animal drugs (6). This is the primary guidance document currently under review by the Veterinary International Conference on Harmonization (VICH) in conjunction with other guidances [U.S. FDA and Animal Health Institute (AHI)] and published literature, to arrive at a global harmonized guideline for performing ERA for animal health drugs. The following discussion pertains to the CVMP/EMEA guidance. Phase I ERA evaluations could exempt a drug from further testing under the following conditions: 1) the active substances consist of vitamins, electrolytes, natural amino acids or herbs; 2) the target animals for the active substance(s) are companion animals or if only a small number of animals are treated; 3) the active substance(s) is applied externally (pour-ons, dips, fumigation, etc.,) and the substances are not expected to enter the environment; 4) the active substance(s) are present in manure or slurry, for spreading on land, in concentrations lower than 100 μg/kg; 5) the substance/s used for pasture animals are excreted in concentrations lower than 10 μg/kg; 6) the DT_{50} (degradation half life of active moiety) in manure is less than 30 days; 7) the predicted environmental concentration (PEC) in soil demonstrated by a worst case calculation is below 10 μg/kg; and 8) the predicted environmental concentration (PEC) in ground water is below 0.1 μg/kg. No degradation and depletion mechanisms need to be established under these criteria. Phase II assessment is required under the following conditions: 1) the active substance is used to treat a large number of animals; 2) there is a potential for direct entry into aquatic environment; 3) the drug substance is used for internal application, and the drug or its major metabolite is present on pasture in concentrations of >10 μg/kg (10 ppb); 4) the manure is spread on to the land and the drug or its major metabolite concentration exceeds 100 μg/kg (100 ppb); and 5) the PEC in soil and ground water exceeds 10 μg/kg (10 ppb) and 0.1 μg/L, respectively. The Phase II

Table II. Degradation and Depletion of Animal Health Drugs

Animal Body

    Metabolism – Transformation and Depletion (CO_2 or other expired gases)
    Adsorption and Assimilation – Depletion
    Excretion – Parent Compound and its Metabolites/Degradates
            (Contribution to Manure and Soil)

Manure

    Dilution – Residue Decline/Depletion
    Hydrolysis – Chemical Transformation/Degradation
    Photolysis – Degradation
    Aerobic Biodegradation – Degradation/Mineralization (Depletion)
    Anaerobic Biodegradation – Degradation/Mineralization/Methane
            Production (Depletion)

Soil

    Dilution – Residue Decline/Depletion
    Hydrolysis – Chemical Transformation/Degradation
    Photolysis – Degradation
    Aerobic Biodegradation – Degradation/Mineralization (Depletion)
    Anaerobic Biodegradation – Degradation/Mineralization/Methane
            Production (Depletion)

Surface Water

    Dilution – Residue Decline/Depletion
    Hydrolysis – Chemical Transformation/Degradation
    Photolysis – Degradation
    Aerobic Biodegradation – Degradation/Mineralization (Depletion)

Sediment

    Dilution – Residue Decline/Depletion
    Hydrolysis – Chemical Transformation/Degradation
    Anaerobic Biodegradation – Degradation/Mineralization/Methane
            Production (Depletion)

---

evaluations should establish dilution, degradation, and depletion mechanisms in the manure, soil, and water to estimate the final PEC to compare with the predicted no effect concentrations (PNEC) for environmental species to provide an ERA.

References

1. *Current Good Manufacturing Practice for Active Pharmaceutical Ingredients: Guidance for Industry – Manufacturing, Processing, or Holding Active Pharmaceutical Ingredients;* U.S. Food and Drug Administration, 03/98.

2. *Current Good Manufacturing Practice for Finished Pharmaceuticals. 21 CFR Part 211;* U.S. Food and Drug Administration, Revised 02/95.

3. *Expert Systems Questionnaire. Survey Concerning Biodegradation.* Prepared by Gregg, B.; Gabel, N.W.; Campbell, S.E. (Versar, Inc). For Office of Toxic Substances, Exposure Evaluation Division, U.S. Environmental Protection Agency, Wahington, DC, 1987.

4. *Handbook of Chemical Property Estimation Methods.* Lymann J.W.; Rosenblatt, D.H., Eds.; American Chemical Society: Washington, DC, 1990.

5. *Guidance for Industry – Environmental Assessment of Human Drugs and Biologics Applications;* U.S. Food and Drug Administration, 1998.

6. *Environmental Risk Assessment for Veterinary Medicinal Products Other Than GMO-Containing and Immunological Products* (EMEA/CVMP/055/96-FINAL, Operational Effective January 1, 1998), *Committee for Veterinary Medicinal Products;* The European Agency for Evaluation of Medicinal Products, London, U.K. 1998.

Thanks are due to Dr. Joseph Robinson of Pharmacia Corporation for discussions on this topic, and Mr. Phil Burns of BASF Corporation for review and comment on the manuscript.

Epilog: Linkage between Social Sciences and Environmental Monitoring

Chapter 20

Illicit Drugs in Municipal Sewage

Proposed New Nonintrusive Tool to Heighten Public Awareness of Societal Use of Illicit–Abused Drugs and Their Potential for Ecological Consequences

Christian G. Daughton

Chief, Environmental Chemistry Branch, ESD/NERL, Office of Research and Development, Environmental Protection Agency, 944 East Harmon Avenue, Las Vegas, NV 89119 (Telephone: 702-798-2207; Fax: 702-798-2142; email: daughton.christian@epa.gov)

Even though a body of data on the environmental occurrence of medicinal, government-approved ("ethical") pharmaceuticals has been growing over the last two decades (the focus of this book), nearly nothing is known about the disposition of illicit (illegal) drugs in the environment. Whether illicit drugs are similarly discharged to and survive in the environment (as discussed for medicinal drugs in the previous chapters of this book), and if so, whether they have adverse effects on native biota, is completely unknown. Regardless, with the newly acquired ability of environmental chemists to monitor for medicinal drugs in environmental samples, science is now afforded the rare opportunity to simultaneously advance the understanding of a pollution process (i.e., the inadvertent discharge of illicit drugs to the environment via their purposeful use) and to also have the ability to impact public discourse and social policy on a highly controversial subject — namely, the pervasive manufacture, trade, and use of illegal drugs and abused controlled substances.

The idea proposed in this chapter provides a rare bridge between the environmental and social sciences. The central aspect to the proposal centers on the use of non-intrusive drug monitoring at sewage treatment facilities and the use of the resulting non-incriminating data to determine *collective* drug usage parameters at the community level as well as to provide exposure data for the aquatic realm. This is the first feasible approach to obtaining real-time data that truly reflects community-wide usage of drugs — while concurrently assuring the inviolable confidentiality of every individual. At the same time, this approach yields environmental data for a class of potential pollutants never before considered as such.

This proposal, which merely capitalizes on science's existing technical capabilities in analytical chemistry, is ground-breaking in its parallel objectives of (i) advancing our understanding of the intimate, immediate, and inseparable connection between humans and their environment (through personal use of chemicals), and (ii) furthering a national debate regarding the use/abuse of illicit and recreational drugs; any national discussion on the purported adverse effects of illicit drugs on a wide spectrum of societal concerns and issues should be helped with the availability of hard data regarding usage. In this sense, the capability outlined here is unique in its purpose of linking social discourse with the furthering of our understanding of environmental processes (such as fate and effects of pollutants). Implementation of this proposal on either limited local levels, nationwide, or internationally could provide a radically innovative approach and totally new dimension to the decades-old quest of understanding the overall issue of illicit drug use as well as informing the public on the perceived widespread use and the many purported consequences of illicit/recreational drugs.

Introduction

Human actions and activities can impact the environment in many ways — deforestation and the potential for global warming being but two obvious examples. Through the interconnectedness of humans and their environment, these impacts can in turn affect our daily lives. Another possible aspect of our society that might harbor the potential for impacting both the environment and the lives of ourselves and others is the purported widespread and continually escalating use of illicit drugs and the misuse/abuse of certain medicinal pharmaceuticals. Drugs comprise a myriad of chemical and therapeutic classes, all of which are specifically designed to elicit potent pharmacological effects (mediated through hundreds or possibly thousands of unique biochemical receptors or targets), as well as numerous toxicological side effects (e.g., adverse effects). A greatly underappreciated aspect of environmental pollution is that the personal, individual use of these substances can lead to both their direct (purposeful) and indirect (inadvertent) discharge (as well as that of any associated bioactive metabolites) to the environment via excreta (through untreated sewage and sewage treatment systems) and by illegal disposal.

Illicit drugs are considered a world-wide concern, harboring the potential for profoundly affecting societies in a myriad of ways. They can have profound economic and political ramifications in locales where they are manufactured or transported — by weakening of government authorities and fostering corruption, eventually leading to political instability. They also elicit intense debate over moral issues. These problems, coupled with the problems attendant to smuggling into consumer countries, can pose concerns with respect to national security. Potential ecological aspects of manufacturing, disposal, and usage also cannot be ignored.

In spite of an awareness of the general problem, the actual types and quantities of illicit drugs used across the U.S. are ill-defined, and numerous new illicit drugs appear annually. The quantities discharged to sewage systems are totally unknown. The relative contributions from individual ingestion versus disposal by clandestine drug labs are also unknown. Clandestine manufacturing and use of illicit drugs, and abuse of controlled substances, constitute a potentially large and highly dispersed source of a chemically diverse (both natural and synthetic), bioactive group of pollutants (both the economic drug and bioactive metabolites, as well as synthesis by-products), differing in many respects from the conventional (priority) pollutants.

Overviews of the types of illicit drugs used in the U.S. are available from the White House Office of National Drug Control Policy (ONDCP) (at: http://www.whitehousedrugpolicy.gov/), from the National Institute of Drug Abuse (NIDA) (at: http://www.nida.nih.gov/NIDAHome1.html), from the Drug Enforcement Administration (at: http://www.usdoj.gov/dea/concern/concern.htm), and from "Streetdrug. org" (at: http://www.mninter.net/%7epublish/index2.htm). The major common classes of abused drugs include stimulants, hallucinogens, opioids, depressants, inhalants, and steroids; these are augmented by several other classes of lesser importance. Terminology of psychoactive recreational and prescription drugs by generic, street, and trade names is available at: Office of National Drug Control Policy (Street Terms: Drugs and the Drug Trade), http://www.whitehousedrugpolicy.gov/drugfact/terms/index.html; and http://www.behavenet.com/capsules/treatments/drugs/drug.htm. Certain illicit drugs are used not just recreationally, but also for self-prescribed medical purposes.

Sensitive chemical analysis methodologies to identify and quantify medically prescribed and over-the-counter drugs in community sewage treatment plant influents and effluents have been successfully employed (see review: Daughton and Ternes [1]; also other chapters in this book). Expanding this approach to illicit drugs could accomplish a variety of additional objectives not relevant to medicinal drugs. Objective data that reflect overall illicit drug manufacturing, disposal, and consumption on a **community-wide** basis could be gained without the risk of implicating or incriminating individuals.

The community-scale surveillance tool proposed in this chapter has multiple objectives. The idea centers on the technological ability to sample community sewage treatment systems (preferably influents, where concentrations are highest) and to analyze for trace chemicals – in this case illicit drugs. This idea has never before been considered and is an outgrowth of the recent review and overviews of Daughton and Ternes (*1*) and Daughton (*2*). It more fully develops the question briefly first posed in those articles as to the significance of illicit (recreational, street) drugs in the environment. Never before has the opportunity presented itself for gaining a window into the collective personal chemical-use activities of a community – without implicating or incriminating the individual. The proposed "surveillance" approach could serve as an unobtrusive, non-invasive means of assessing collective community-wide manufacturer, use/abuse, and disposal of illicit drugs (both types and quantities) and controlled substances while assuring the anonymity and privacy of the individual. As to which of these three major sources is the major contributor to overall loading of a particular drug in sewage would be a function of each individual locale.

Drugs are known to enter the environment via several routes, the most significant one with respect to the work proposed here being domestic sewage. Simply stated, any chemical ingested by humans has the potential to be excreted via the feces and urine — usually passing directly into sewage treatment systems and septic systems and sometimes simply being directly "straight-piped" (without treatment) into surface waters. Direct disposal of unused/unwanted drugs and synthesis by-products to domestic sewage is another source. The amounts that are discharged to the environment via excreta are a function of pharmacokinetics (absorption, distribution, metabolism, and excretion — the last two of which are also influenced by transformation by gut microbiota), biodegradation or biotransformation (including conversion of conjugates back to the parent drug) during conveyance of sewage to treatment plants, and amounts removed during physicochemical/biological treatment (these additional removals do not occur with direct discharge of untreated sewage).

Quantification data acquired for drugs identified in STW influents would be used to compute an average consumption rate normalized across the population serviced by the sewage treatment facility. Using pharmacokinetic data and environmental transformation rate data, coupled with the occurrence concentrations in sewage influent, the population-normalized total consumption/use/production for a given drug within the population of a sewage treatment service area could be back calculated; alternatively, minimum population-based consumption could be calculated simply by using uncorrected influent concentrations. By determining the concentrations of drugs or key metabolites (e.g., benzoylecgonine or ecgonine methyl ester as a surrogate for cocaine, *d*-amphetamine for *d*-methamphetamine, 11-nor-9-carboxydelta-9-tetrahydrocannabinol for delta-9 tetrahydrocannabinol, morphine for several opiates) representing the important classes of illicit drugs and drugs of abuse that enter municipal sewage treatment systems, it should be possible to conservatively estimate (i.e., provide lower estimates) the daily influx of each drug (or chemical class, for example, opiates). Dividing the daily loads of each drug (or class) by the serviced population of the sewage treatment plant would yield the lower estimate of normalized-population usage of each drug. By applying knowledge of biodegradation within sewage conveyance systems and biodegradation from gut microbiota and human metabolism (pharmacokinetics), the original concentrations of each drug could be back-estimated, yielding upper-estimates of community-wide usage. Furthermore, by monitoring drug metabolites instead of the parent drug, amounts resulting from human usage could be largely distinguished from synthesis/disposal. Provision to the public of daily, population-normalized equivalents of drug usage would offer an entirely new perspective on the overall magnitude and extent of illicit drug usage. Actual usage by the user-individual would obviously be higher than normalized total-population usage since only an unknown fraction of the populace uses illicit drugs; the approach proposed here is not capable of addressing this factor.

This proposed strategy could be implemented on the local level by whatever organizations so choose to monitor and analyze sewage influents. If successful, it could be expanded nationwide by government, public, or private organizations. Logical collaborations could be established among local/state/federal environment agencies, wastewater districts, and law enforcement forensic laboratories. As discussed later, such sewage-treatment-plant monitoring networks could be expanded to include schools and other organizations, who could participate in environmental monitoring for educational enrichment.

Objectives

The proposal described here addresses one of the 10 Goals that compose the U.S. EPA's *2000 Strategic Plan* (http://www.epa.gov/ocfopage/plan/plan.htm). Among these goals, the objective of Goal 7 *(Quality Environmental Information: Expand the American Public's Right to Know About Their Environment)* is to facilitate easy public access to local/regional/national environmental information and thereby expand citizen involvement by providing the data required to make decisions in protecting their families, their communities, and their environment – as they see fit. The goal is to increase information exchange between scientists, public health officials, businesses, citizens, and all levels of government, with the hope of fostering greater knowledge and awareness of the environment and thereby enhance the involvement of a better-informed, empowered citizenry in formulating lasting solutions to environmental problems – all within the complex interrelationships of risks and the associated tradeoffs involved in making decisions. One specific approach to Goal 7 that has been implemented by the EPA is the President's "Environmental Monitoring for Public Access and Community Tracking" (EMPACT) Initiative (see: http://www.epa.gov/empact/). This initiative focuses on improving environmental data collection and on deploying new analytical technologies for real time / automated monitoring and promulgation of data. The monitoring approach proposed in this chapter is illustrative of these goals, and could even serve (if desired) as a possible program under the EMPACT initiative.

The two major objectives of this proposed monitoring approach are to (1) heighten the public's awareness and understanding of the many issues involving pollution by synthetic chemicals — but especially, how the collective personal actions/activities of individuals can have a direct impact on the environment, and (2) for the first time ever, make available to the public, community-scale data on "real-time" actual consumption/usage and ultimate environmental disposal/disposition of illicit drugs. Lacking to date has been the ability to know (in a timely manner) what specific drugs are being used and in what communities. Current consumption data are hard to verify and are often qualitative at best. In the absence of such data, public discourse over the last decades on the various issues involving illicit drugs has been less than fully informed. Sewage monitoring would provide the ability to directly measure real drug use. Monitoring of sewage influents could be done on any time line desired (weekly, daily, or more frequent), and the resulting data could be conveyed to the public in any way desired (e.g., via the various mass media) with the potential for daily updates, in a manner analogous to the reporting of air or water quality parameters or the weather.

Public Awareness of Chemical Pollution Caused by Individual Actions:
Periodic reporting of community-scale drug consumption would foster a better understanding and appreciation by the public that actions of individuals impact the environment — especially with regard to the use of chemicals — in this instance drugs. Historically, it has proved difficult for the individual to perceive their small-scale activities or actions as having any measurable impact on the larger environment – personal actions are often deemed minuscule, or discounted as inconsequential in the larger scheme. Yet it is the combined actions and activities of individuals that indeed can significantly impact the environment in a myriad of ways. A factor making it difficult for individuals to perceive the interconnectedness of their lives with the health of the environment is the lack of perceived immediacy of any untoward (or beneficial) effects — as most actions have delayed consequences that make cause-effect relationships difficult or impossible to perceive. This temporal disconnection obscures the ultimate causes of effects, thereby negating the possibility of modifying/reinforcing behavior via feedback (positive or negative).

In addition to inadvertent discharge of bioactive drugs to the environment simply via their ingestion/administration, the disposal/dumping of precursors, catalysts, and by-products used in illicit drug manufacturing and trade could have adverse environmental consequences not yet realized. While documentation of direct environmental impacts of these illicit-drug secondary materials is sparse, obvious consequences include damaging farm practices (e.g., deforestation and over use of fertilizers/pesticides for drug crops) and resultant water pollution, all of which can be caused by the drug trade itself or by anti-drug "eradication" programs. Dumping of large quantities of toxic substances used in, or generated by, the production process (e.g., diethyl ether, acetone, and sulfuric and hydrochloric acids used in the production of cocaine hydrochloride; kerosene, diesel oil, and calcium oxide used in the production of coca paste) or dumping of unwanted product are other major sources of potential ecological damage of unknown magnitude.

While illicit drugs (DEA Schedule I) have yet to be reported in environmental samples (perhaps largely because they have not been monitored for), certain Schedule II drugs of abuse are now reported in chapters of this book — by Snyder et al. in lake water at the tens-to-hundreds of ng/L range, including phenobarbital, primidone, hydrocodone, and codeine, and by Möhle and Metzger in sewage influents/effluents in the µg/L range (hydrocodone and dihydrocodeine) (3,4).

Limitations and Problems of Current Drug Tracking Methodologies:
One can probably safely assume that the known instances of illicit drug use (by both chronic and recreational users) and manufacturing that are derived from criminal arrest data represent only a small but unknown portion of overall incidence of use. Also assumed is the difficulty in obtaining an accurate picture of societal drug use via the only three currently available means: (1) self-administered surveys of personal use, (2) by drug-testing programs, or (3) hospital admission data. Furthermore, despite the growing popularity of drug-testing programs, the legality and ethics of workplace/school drug testing continues to be debated and challenged in court. With the means proposed here to finally by-pass drug-testing programs centered on the individual, we could also avoid the potential of impairing worker morale and inadvertently providing the impetus to drug users to switch to yet more dangerous drugs that are not subject to, or amenable to, screening. Drug-testing programs are also fraught with many technical and legal difficulties with respect to interpreting the meaning of positive results – namely, whether any particular concentration of a drug, other than ethanol, actually constitutes physical impairment at a particular time.

Guidelines for drug testing also result in gross under-reporting of drug usage by most drug-testing procedures because all drugs present at lower-than-established (regulatory) "cut-off" concentrations (e.g., those specified by DOT — 49 CFR Part 40: "Procedures for Transportation Workplace Drug and Alcohol Testing Programs, Proposed Rules": http://www.datia. org/legislative/49cfr40text.htm) are reported as "not present"; only those procedures using "zero tolerance" (i.e., results reported as a function of method detection limit) avoid this substantial built-in negative bias. Finally, drug-testing programs targeted on the individual are severely limited to acquiring random discrete snap-shots of excreted drug/metabolites. Very recent drug usage, for example, can escape undetected in a urine drug test because the drug/metabolite has yet to appear in the urine, while older previous intake can escape because of prior metabolic clearance. In contrast, sewage monitoring would yield a more comprehensive, integrated reflection of continual/cumulative usage at the scale of an entire community. Replacement of drug-testing programs that are solely focused on the individual with one based on collective, anonymous community-scale screening would yield a more accurate reflection of total drug usage in part because pressure to circumvent drug screening tests would disappear, eliminating an historic negative bias in estimating usage.

Public Discourse of Drug Use/Abuse: Monitoring of illicit drugs in community-wide sewage could advance the stalemated, decades-old national discourse that revolves around the use of illicit drugs. By bringing to the fore the real extent and magnitude of nationwide manufacturing/trade and usage of both street (illicit) and abused prescription drugs, the debate could be furthered and better informed. Also, the levels of street drugs in sewage influents and effluents must be documented before their potential to enter natural waters and drinking waters (via treated and untreated sewage) can be evaluated; the environmental fate and transport of prescription drugs, leading to their ultimate disposition to surface, ground, and sometimes even drinking waters would not be expected to differ dramatically from that of illicit drugs. Only then can their risk to aquatic biota and to consumers of drinking water be assessed. Ultimately, should illicit drug residues (regardless of concentration) appear in domestic drinking water (as with prescription drugs), they would serve as a definitive, continual reminder of not just the magnitude of illicit drug use, but also the fact that society's combined use of drugs by multitudes of individuals has the potential to impact aquatic biota, as well as anyone who drinks water that has not been sufficiently upgraded. This monitoring approach will foster the recognition of collective and individual responsibility with respect to environmental pollution.

Implementation and Public Participation in Monitoring: A nationwide semi-quantitative screening program at the community-scale could be inexpensively established with the participation of primary, secondary, and upper school systems. Such a program would focus solely on monitoring community-wide sewage and receiving waters for illegal/abused (and perhaps certain legal) drugs. To facilitate low-cost analysis, such a network could adapt the use of readily available "home test kits" (largely based on immunoassay), which are otherwise designed for screening of individuals (see: "SAMHSA Evaluation of Non-Instrumented Drug Test Devices," http://www.health.org/workplce/summary.htm). With cooperation of local municipalities, such an approach could inexpensively and quickly supply nationwide monitoring data for sewage-treatment facilities and surface waters (to augment the quantitative data supplied by laboratories).

Such a volunteer monitoring program could also provide a unique opportunity for students of all grade levels to delve more deeply into environmental science — promoting further participative, hands-on study of a wide spectrum of issues involving environmental pollution, including chemical analysis, identification of source/occurrence, environmental fate, exposure, effects, risk assessment, mitigation, pollution prevention, and regulation. Such a program would foster a continuing, and more objective, dialog on drug use throughout the country and would gain added meaning once historical "baselines" had been established and upon which status and trends could be amassed and inter-community comparisons made. Should home test kits prove capable of sufficiently sensitive drug detection in sewage (without sample preconcentration), then they could prove of direct use by schools in monitoring treated sewage effluent (non-sterilized influent is unsafe for school work because of safety issues involving infectious pathogens; it would be suited, however, if screened for, or sterilized of, pathogens by the treatment plant operators). If sample preconcentration is required, the drug-test kit manufacturers could redesign their assays with increased sensitivity (they are currently set for the relatively high regulatory levels). Otherwise, simple solid-phase extraction techniques could be developed (for amenable drugs) in order to lower the method detection limits.

An important adjunct tool in making use of the resulting monitoring data would be geographic information system (GIS) technology, whose role has been expanding in the public health arena. With GIS, drug monitoring data could be overlaid with health, demographic, environmental, crime, and any other conventional data tied to geographic location and sewage-service areas. As one of many examples, public health professionals would have access to another measurement to correlate with disease outbreaks (e.g., because of drug use involving shared-needle use). Another example is that data on community-wide usage could be used to answer major, long-unresolved questions such as whether illicit drug use affects economic productivity and overall health (for example, on absenteeism, overuse of health benefits and cost for rehabilitation, accidents/fatalities, disciplinary actions, turnover, and crime/delinquency – all purportedly caused by both the need to buy drugs and by the use of drugs). Questions such as these have simply not been addressable in a satisfactory manner using the only existing but highly controversial analytical tool available – drug-testing programs aimed at the individual. While these issues have long been, and continue to be, actively debated, it is widely recognized that drug testing programs are extremely costly and enormously controversial.

New Societal Perspective and Insight: With public access to actual, community-scale drug usage on a continual, "real-time" basis, this new monitoring approach could have far-ranging (and perhaps difficult-to-foresee) ramifications for society. By providing data never before available, this program would foster the recognition of the importance of collective and individual responsibility with respect to environmental pollution. More effective drug prevention could result from a greatly heightened public awareness of the actual nature and extent of drug usage at local levels and nationwide — an awareness based on non-obtrusive scientific measurements rather than on invasive testing (which can inadvertently lead to biased data) and subjective surveys. More objectivity could be infused into what has historically proved to be a highly charged, politicized debate (5).

A continuous but non-intrusive, anonymous indicator of drug use would serve to keep the issue in the fore of public discourse and awareness. Society could gain a new tool for gaging the use of illicit drugs and mapping their use down to the community level (more specifically, the service population for a given sewage district). These collective data could serve as the input for a numeric indicator of community "health" (much as other socioeconomic and pollutant index measures are used), as a tool for educators to better develop the content (and better target the audience) for their drug-abuse programs, as well as a tool for law enforcement to better understand the types and quantities of drugs it needs to be aware of in individual communities. Finally, with an objective of promoting a national/local discourse, sewage-influent data could be obtained by (or used by) schools, health care/social workers, law enforcement, courts, public health professionals, and public officials, all of whom would then have the ability to compare usage (types, quantities, consumption patterns, trends) of drugs between and among communities. Usage patterns may vary dramatically not just from country to country, but also region to region, subculture to subculture.

Benefits to the public and law enforcement include (1) dramatically improving public awareness of (i) how the individual can have a direct impact on the environment and (ii) the magnitude of drug usage, and (2) promoting the use of "home" drug testing kits by schools to perform nationwide environmental sewage monitoring – serving as an educational exercise in both environmental and social sciences. If such data acquired from sewage treatment plant effluent or raw sewage effluent were uploaded to a nationwide web-based database, it could greatly assist environmental scientists in determining the frequency of occurrence, extent, and magnitude of drugs in the aquatic environment. The analogous data from sewage treatment plant influent could possibly offer drug and health enforcement organizations a new tool in the early detection of new drug usage activities. A nationwide program would provide perspective not just for the national prevalence of illicit drug use/manufacturing, but also on how each metropolitan area of the U.S. compares and differs with respect to types and amounts of drugs used, and would serve as a sentinel system for new trends in usage of "emerging" drugs as well as current illicit drugs. Numerous other uses can be envisioned, including elucidation of status, fluctuations, and trends, the latter of which could be used in part to help verify the effectiveness of drug prevention, interdiction, or enforcement programs.

Context

National Drug-Use Surveys: There are numerous periodic drug-use surveys conducted at local, state, and national levels. The primary source of nationwide information in the U.S. for use/abuse of drugs has been the surveys conducted by the Substance Abuse and Mental Health Services Administration (SAMHSA) (e.g., see their most recent report: "1999 National Household Survey on Drug Abuse," released 31 August 2000, at: http://www.samhsa.gov/oas/household99.htm). Since 1971, these surveys have resulted in estimates of the prevalence and incidence of illicit drug use across the U.S. Like all surveys, the resulting statistically derived estimates have numerous caveats and shortcomings. Relying largely on population self-administration, these surveys can be costly, and they are subject to a plethora of biases, resulting from sampling error and non-response/reporting errors. The latter derive from the respondents' truthfulness and accuracy of memory, fear of self-reporting, and under-reporting or inability to report. The former includes an inability to capture heavy drug users (a small sub-population that may be responsible for a disproportionately large portion of environmental discharge of illicit drugs, especially via terrestrial run-off), exclusion of youths under 12 years old, and economic status (e.g., exclusion of the homeless from surveys).

The second SAMHSA data collection program is administered by the Office of Applied Studies (OAS). Since 1988, the Drug Abuse Warning Network (DAWN) data have been collected from a representative sample of eligible hospitals located throughout the U.S. This survey captures data from emergency room admissions where the subject's condition has been induced by or is related to the "use of an illegal drug or the nonmedical use of a legal drug." These DAWN data therefore "do not measure prevalence of drug use in the population." DAWN reports can be found at: http://www.samhsa.gov/oas/p0000018.htm. Much of SAMHSA's materials can also be found via the National Clearinghouse for Alcohol and Drug Information web site "Prevention Online" (PREVLINE), available at: http://www.health.org/.

The proposed new sewage monitoring program could be used to augment these surveys and possibly serve to "calibrate" or quality-assure the traditional survey results. Monitoring would provide real-time data, which would fulfill the need for identifying new trends in emerging drug use (as well as detecting trends in existing use). Existing surveys are merely retrospective, and therefore their timeliness lags current use by one or more years (e.g., see http://www.samhsa.gov/oas/household99.htm). An extensive discussion and survey of drug trend detection can be found in Griffiths et al. (5). Additional web resources for illicit drugs are available via the links provided at: http://www.epa.gov/nerlesd1/chemistry/pharma/index.htm.

Environmental Fate of Anabolic Androgenic Steroids – Special Significance: The disposition and fate of sex steroids (both natural and synthetic) in the aquatic environment has captured a large portion of the attention cast on the overall issue of medicinal drugs in the environment as pertaining to "endocrine disruptors" (see: "Environmental Estrogens and other Hormones" at: http://www.tmc.tulane.edu/ecme/eehome; NAS report "Hormonally Active Agents in the Environment" at: http://www.nap.edu/books/0309064198/html). This debate has focused, however, on the estrogenic sex steroids at the exclusion of the hundreds of commercially available and illicit anabolic androgenic steroids (related to androgens, most being derivatives of testosterone, and which promote skeletal muscle growth coupled with male sexual characteristics). These steroids of abuse have experienced escalating use, especially among athletes and, more widely and recently, youths (see NIDA's "SteroidAbuse.org" at: http://165.112.79.55/, and "Anabolic Steroids Community Drug Alert Bulletin" at: http://165.112.78.61/SteroidAlert/Steroidalert.html; DEA's "Steroids" at: http://www.usdoj.gov/dea/concern/steroids.htm). Even though some of the testosterone derivatives (e.g., esters) have medicinal uses as prescription drugs (primarily for male androgen-replacement therapy), the non-prescription abusive consumption of others can be at a significantly higher (and sustained) frequency and duration and at several orders of magnitude higher doses (and in multiple combinations – referred to as "stacking"). These high doses, coupled with their acting directly at the endocrine level, could pose particular concern with respect to the health of aquatic organisms that may suffer exposures; the documented side effects in both human sexes are numerous and profound.

While it has been established that sub-ppb levels of estrogenic steroids in sewage effluent can affect aquatic and marine biota, practically no work has been performed on low-level exposure to androgens. Androgenic steroid-spiked feeds have been long-used, however, for growth enhancement and for forcing desirable sex-ratios (via sex-inversion) in aquacultured fin fish. With the issue of potential androgenic steroid release to the environment via aquaculture aside, the published aquaculture literature clearly documents that androgenic steroids (such as could be expected from abuse of illicit anabolic steroids) have the potential at low levels (ppb) to elicit aquatic effects (e.g., skewed sex-ratios), especially in light of the fact that short-term, brief immersion of fish fry in androgenic steroid-spiked water (17α-methyltestosterone and 17α-methyldihydrotestosterone) can induce sex inversion (6); aromatizable androgens, such as 17α-methyltestosterone, can also lead to "paradoxical feminization". Although the short-term concentrations required for total sex inversion in fish are orders of magnitude higher than what could be expected in sewage, it begs the question as to what other possible effects (e.g., tumorigenic/teratogenic) could occur from continual exposures to androgens from sewage effluent, especially during critical developmental milestones. Comprehensive listings of steroids can be found in catalogs from commercial suppliers of analytical standard reference materials, such as "Steraloids Online" (http://www.steraloids.com). In addition to the completely uninvestigated burden of illicit androgens in the environment (primarily as a result of use by athletes and bodybuilders), new routes of delivery for testosterone (primarily dermal) are leading to exponentially increasing use of testosterone in replacement therapy.

Caveats: The proposed approach for back-calculating drug use is subject to the following limitations and biases, among others. Septic leach fields would not be incorporated in a collective measure (leading to underestimation of community-wide capita use). Disposal to domestic sewage of unusable clandestine manufactured product(s) or product hurriedly disposed of during drug raids – but not to landfills – would lead to further bias (but could be ascribed as such if detected as a transient spike); disposal from clandestine labs can occur by numerous routes — many terrestrial (7). As mentioned previously, unless surrogate metabolites are monitored, the amounts contributed by use may not be distinguishable from those contributed by manufacturing/disposal. Further limitations to the proposed approach include the diminished ability (except as available through GIS or demographic means) to assign calculated usage according to age, gender, education, race/ethnicity, or employment.

Chemical Analysis Considerations

The two major limitations to successfully analyzing sewage influent for drugs are (1) the need to acquire time-integrated samples as opposed to discrete "grab" samples to avoid bias caused by episodic fluctuations in concentrations (e.g., highest drug concentrations in urine are during the day's first void) or flow rates and (2) the necessity of suitable pre-concentration methods, especially for polar analytes, being that drug concentrations in the influent will be reduced orders of magnitude from when they are excreted (by way of microbial degradation, dilution, and sorption to particulates). Both of these needs have been addressed for polar analytes in a newly proposed sampling/enrichment approach (Polar Organic Chemical Integrative Sampler, POCIS), a methodology that deserves further attention (8). Known sewage marker chemicals of human chemical consumption (such as caffeine or nicotine) or activity (e.g., coprostanol, a major fecal sterol, or the major urinary product, creatinine) should also be monitored partly as an internal form of quality assurance but also as a means of providing normalized data — especially useful for comparison across sewage treatment plants (and therefore across communities).

The levels of drugs that can easily be measured in urine are reflected by the confirmatory cut-off concentrations for those drugs subject to DOT screening/confirmation. These range from 15 ng/mL (15 ppb) for 11-nor-delta-9-tetrahydrocannabinol-9-carboxylic acid to 500 ng/mL for amphetamine (and 2,000 ng/mL [2 ppm] for morphine). While these levels are orders of magnitude higher than would occur in sewage influent, they are also considerably higher than method detection limits for conventional GC/MS and immunoassay. Once the target analytes (either the parent drug or a surrogate metabolite) can be sufficiently preconcentrated to meet analytical detection limits, analysis can proceed by any of a variety of approaches, ranging from mass spectrometric to immunochemical. Analytical standard solutions (including deuterated standards) of most DEA Schedule I-IV Controlled Substances are readily available commercially without a DEA licence (e.g., see: "The United States Pharmacopeia" at: http://www.usp.org/; "Radian Services Analytical Reference Materials Group" at: http://www.radian.com/html/services/arm/genforen.htm; "The Promochem Group" at: http://www.promochem.com/pro_frames.html). Whatever approaches are eventually developed for definitive analysis, a nationwide "volunteer" monitoring network would incur lower cost.

Finally, it must be noted that the metabolism of certain illicit drugs can yield other drugs (some of which may even be legal). For example, both isomers of amphetamine and methamphetamine can result from the metabolism of other drugs (e.g., see *9*). If amphetamine were monitored in sewage, its origin would be nebulous – its occurrence could not be ascribed solely, for example, to the use of methamphetamine being that there are numerous amphetamine-derivative drugs; see Meatherall (*10*) for an example approach to GC/MS chiral determination of methamphetamine. Many drugs yield a plethora of metabolites, a fact that holds the potential to confound any monitoring plan if not properly recognized and accounted for.

Dissemination of Data

Much like readily accessible weather data, it is anticipated that any real-time illicit drug data generated from local sewage treatment plants would prove of keen interest to the public. These data would perhaps prove more popular than any other measure of environmental pollution. It is also anticipated that the mass media would be anxious to broadcast the data through print, websites, television, and radio, or any other means already proposed or used under the EPA EMPACT program (http://www.epa.gov/empact/). A nationwide web-based database of collected data could prove extremely useful.

Possible Future Work

The removal efficiencies of drugs from both waste treatment and water-upgrade facilities are known to vary according to the technologies employed as well as from plant to plant. Depending on the occurrence concentrations and the effectiveness of waste treatment and drinking water treatment technologies (if any) for a particular locale, drugs could be searched for in domestic drinking waters. These concentrations would prove to be very low (ppt — ng/L — and below). While the presence of even minute quantities of any type of parent drug or bioactive metabolite in drinking water (which provides a source of continual exposure for people) has totally unknown consequences for human health (especially fetuses, infants, children, metabolically impaired/abnormal individuals, and other at-risk sub-populations), such occurrences would serve as the ultimate reminder that the linkage between wastewater and human reuse can be very direct.

Further Perspective

In Bruce Alberts' *2000 President's Address* to the National Academy of Sciences ("Science and Human Needs"), he addresses in part "the responsibilities of scientists" with the following words, which lend further, and final, perspective on the idea presented in this chapter (*11*).

"... (B)ecause political will is often short term, and misinformation about science abounds, we scientists ourselves must become much more engaged in the everyday life of our governments and our communities." As 'civic scientists', "...in the 21st century, science and scientists will be judged on how well they help solve local and world problems, not only on how well they generate new knowledge. The impact of our research is everywhere, and we must step out and make sure that our work is understood and appropriately used by the world. ...We also need to be explicit about what is not known, and be clear about the questions that science cannot answer."

REFERENCES

1.   Daughton, C.G.; Ternes, T.A. "Pharmaceuticals and personal care products in the environment: agents of subtle change?" *Environ. Health Perspect.* **1999**, *107*(suppl 6), 907-938.
2.   Daughton, C.G. "Pharmaceuticals and Personal Care Products in the Environment: Overarching Issues and Overview," in Pharmaceuticals and Personal Care Products in the Environment: Scientific and Regulatory Issues; Daughton, C.G. and Jones-Lepp, T., Eds.; American Chemical Society: Washington, DC, 2001 (see introductory Chapter in this book).
3.   Snyder, S. Kelly, K.L.; Grange, A.; Sovocool, G.W.; Synder, E.; Giesy. J.P. "Pharmaceuticals and Personal Care Products in the Waters of Lake Mead, Nevada," in Pharmaceuticals and Personal Care Products in the Environment: Scientific and Regulatory Issues; Daughton, C.G. and Jones-Lepp, T., Eds.;

American Chemical Society: Washington, DC, 2001 (see chapter in this book).
4. Möhle, E.; Metzger, J. "Drugs in Municipal Sewage Effluents — Screening and Biodegradation Studies," in Pharmaceuticals and Personal Care Products in the Environment: Scientific and Regulatory Issues; Daughton, C.G. and Jones-Lepp, T., Eds.; American Chemical Society: Washington, DC, 2001 (see chapter in this book).
5. Griffiths P.; Vingoe L.; Hunt N.; Mounteney J.; Hartnoll R. "Drug information systems, early warning, and new drug trends: Can drug monitoring systems become more sensitive to emerging trends in drug consumption?" *Substance Use Misuse* **2000**, *35*(6-8), 811-844.
6. Gale, W.L; Fitzpatrick, M.S.; Lucero, M.; Contreras-Sánchez, W.M.; Schreck, C.B. "Masculinization of Nile tilapia (*Oreochromis niloticus*) by immersion in androgens," *Aquaculture* **1999**, *178*(3-4), 349-357.
7. The Joint Federal Task Force of the Drug Enforcement Administration, the U.S. Environmental Protection Agency, the U.S. Coast Guard "Guidelines for the Cleanup of Clandestine Drug Laboratories," EPA540P90005, March 1990 (43 pp.).
8. Alvarez, D.A.; Petty, J.D.; Huckins, J.N. "Development of an Integrative Sampler for Polar Organic Chemicals in Water," paper #4, presented at the 219th National Meeting of the American Chemical Society, San Francisco, CA, 27 March 2000 (published in "Issues in the Analysis of Environmental Endocrine Disruptors", Preprints of Extended Abstracts, vol. *40*(1), pp. 71-74, 2000).
9. Kraemer, T.; Theis, G.S.; Weber, A.A.; Maurer, H.H. "Studies on the metabolism and toxicological detection of the amphetamine-like anorectic fenproporex in human urine by gas chromatography-mass spectrometry and fluorescence polarization immunoassay," *J. Chromatogr. B: Biomed. Sci. Applicat.* **2000**, *738*(1), 107-118.
10. Meatherall, R. "Rapid GC-MS confirmation of urinary amphetamine and methamphetamine as their propylchloroformate derivatives," *J. Anal. Toxicol.*, **1995**, *19*(5), 316-322.
11. Alberts, B. "Science and Human Needs," 2000 President's Address to the National Academy of Sciences, 137th Annual Meeting, Washington, D.C., 1 May 2000 (available at: http://www.nas.edu).

ACKNOWLEDGMENTS: The author thanks the following people for taking their valuable time to provide helpful review of both technical and policy aspects of this manuscript: James Quackenboss, Andrew Grange, Tammy Jones-Lepp, John Lyon, and Gareth Pearson (U.S. EPA's National Exposure Research Laboratory), Alan Gallaspy (Las Vegas Metropolitan Police Department Criminalistics/

Forensics Laboratory), and the American Chemical Society's anonymous reviewers, all of whom contributed to improving the quality of the manuscript.

NOTICE: The U.S. Environmental Protection Agency (EPA), through its Office of Research and Development (ORD), funded and performed the research described. This manuscript has been subjected to the EPA's peer and administrative review and has been approved for publication. Mention of trade names or commercial products does not constitute endorsement or recommendation by EPA for use.

More Best Sellers from ACS Books

Microwave-Enhanced Chemistry: Fundamentals, Sample Preparation, and Applications
Edited by H. M. (Skip) Kingston and Stephen J. Haswell
800 pp; clothbound ISBN 0–8412–3375–6

Designing Bioactive Molecules: Three-Dimensional Techniques and Applications
Edited by Yvonne Connolly Martin and Peter Willett
352 pp; clothbound ISBN 0–8412–3490–6

Principles of Environmental Toxicology, Second Edition
By Sigmund F. Zakrzewski
352 pp; clothbound ISBN 0–8412–3380–2

Controlled Radical Polymerization
Edited by Krzysztof Matyjaszewski
484 pp; clothbound ISBN 0–8412–3545–7

The Chemistry of Mind-Altering Drugs: History, Pharmacology, and Cultural Context
By Daniel M. Perrine
500 pp; casebound ISBN 0–8412–3253–9

Computational Thermochemistry: Prediction and Estimation of Molecular Thermodynamics
Edited by Karl K. Irikura and David J. Frurip
480 pp; clothbound ISBN 0–8412–3533–3

Organic Coatings for Corrosion Control
Edited by Gordon P. Bierwagen
468 pp; clothbound ISBN 0–8412–3549–X

Polymers in Sensors: Theory and Practice
Edited by Naim Akmal and Arthur M. Usmani
320 pp; clothbound ISBN 0–8412–3550–3

Phytomedicines of Europe: Chemistry and Biological Activity
Edited by Larry D. Lawson and Rudolph Bauer
336 pp; clothbound ISBN 0–8412–3559–7

For further information contact:
Order Department
Oxford University Press
2001 Evans Road
Cary, NC 27513
Phone: 1-800-445-9714 or 919-677-0977

Bestsellers from ACS Books

The ACS Style Guide: A Manual for Authors and Editors (2nd Edition)
Edited by Janet S. Dodd
470 pp; clothbound ISBN 0–8412–3461–2; paperback ISBN 0–8412–3462–0

Writing the Laboratory Notebook
By Howard M. Kanare
145 pp; clothbound ISBN 0–8412–0906–5; paperback ISBN 0–8412–0933–2

Career Transitions for Chemists
By Dorothy P. Rodmann, Donald D. Bly, Frederick H. Owens, and Anne-Claire Anderson
240 pp; clothbound ISBN 0–8412–3052–8; paperback ISBN 0–8412–3038–2

Chemical Activities (student and teacher editions)
By Christie L. Borgford and Lee R. Summerlin
330 pp; spiralbound ISBN 0–8412–1417–4; teacher edition, ISBN 0–8412–1416–6

Chemical Demonstrations: A Sourcebook for Teachers, Volumes 1 and 2, Second Edition
Volume 1 by Lee R. Summerlin and James L. Ealy, Jr.
198 pp; spiralbound ISBN 0–8412–1481–6
Volume 2 by Lee R. Summerlin, Christie L. Borgford, and Julie B. Ealy
234 pp; spiralbound ISBN 0–8412–1535–9

The Internet: A Guide for Chemists
Edited by Steven M. Bachrach
360 pp; clothbound ISBN 0–8412–3223–7; paperback ISBN 0–8412–3224–5

Laboratory Waste Management: A Guidebook
ACS Task Force on Laboratory Waste Management
250 pp; clothbound ISBN 0–8412–2735–7; paperback ISBN 0–8412–2849–3

Good Laboratory Practice Standards: Applications for Field and Laboratory Studies
Edited by Willa Y. Garner, Maureen S. Barge, and James P. Ussary
571 pp; clothbound ISBN 0–8412–2192–8

For further information contact:
Order Department
Oxford University Press
2001 Evans Road
Cary, NC 27513
Phone: 1-800-445-9714 or 919-677-0977

Highlights from ACS Books